LUFTWAFFE SPECIAL WEAPONS 1942–45

OSPREY
PUBLISHING

LUFTWAFFE
SPECIAL
WEAPONS
1942–45

ROBERT FORSYTH

OSPREY PUBLISHING
Bloomsbury Publishing Plc
PO Box 883, Oxford, OX1 9PL, UK
29 Earlsfort Terrace, Dublin 2, Ireland
1385 Broadway, 5th Floor, New York, NY 10018, USA
E-mail: info@ospreypublishing.com
www.ospreypublishing.com

OSPREY is a trademark of Osprey Publishing Ltd

First published in Great Britain in 2021

ISBN: HB 9781472839824; eBook 9781472839831;
ePDF 9781472839800; XML 9781472839817

21 22 23 24 25   10 9 8 7 6 5 4 3 2 1

Edited by Tony Holmes
Artwork by Jim Laurier and Steve Zaloga
Line Drawings by A. L. Bentley

Cover images (from left to right):
Me 262A-1a/U4 Wk-Nr 111899. (EN Archive)
Armourers heave a 21 cm mortar shell into the outer launch tube
of a pair fitted to the underside port wing of a Bf 110. (EN Archive)
'Fritz X' glide bomb. (EN Archive)
Ju 88A-4 BF+YT. (EN Archive)

Index by Zoe Ross
Printed and bound in India by Replika Press Private Ltd.

Osprey Publishing supports the Woodland Trust, the UK's leading woodland conservation charity.

To find out more about our authors and books visit **www.ospreypublishing.com**. Here you will find extracts, author interviews, details of forthcoming events and the option to sign up for our newsletter.

FSC
www.fsc.org

MIX
Paper from
responsible sources
FSC® C016779

CONTENTS

INTRODUCTION AND ACKNOWLEDGEMENTS

On 30 December 1943, the crews of USAAF B-17 Flying Fortresses executing a bombing raid to Ludwigshafen in Germany described being attacked by an enemy fighter trailing a length of steel cable through their formation. In its post-mission report, the Eighth Air Force's 3rd Bomb Division stated impassively that the 4th Combat Wing 'reported one single-engined enemy aircraft carrying a bomb attached to a 100-ft cable flying 800 ft over the lead group'.

To the crews, this represented an alarming development; the reinforced steel cable was designed to cut into wings or coil itself around engines. And this was in addition to the massed conventional fighter attacks, made from all 'clock' directions, which were experienced in the target area, as well as air-to-air mortars being fired into the bomber 'boxes' – or the dreaded *Pulks* if you were a German fighter pilot.

Indeed, only two weeks before, B-24 Liberator and B-17 crews had reported incidents of enemy aircraft dropping conventional bombs into their formations. However, such accounts were met with caution and even some degree of scepticism by USAAF Technical Intelligence, but the intelligence officers should have paid more heed to the reports of their airmen, for their observations were entirely accurate. These were just further examples of operational trials using a range of inventive, albeit radical and unconventional weapons deployed by the Luftwaffe in its attempt to inflict carnage on the increasing numbers of Allied *Viermots* (for *Vier Motor* – four-engined – as the heavy bombers were referred to colloquially by the *Jagdflieger*) targeting the Reich around the clock.

By late 1943 Germany was confronted by draining, multi-dimensional warfare fought across the massive Eastern and Mediterranean Theatres, as well as by the increasingly intensive battles of attrition being waged in the *Reichsverteidigung* – the aerial defence of the homeland itself.

Such was the rapid growth of Allied military strength, combined with technological development, that Germany was being forced, with its limited and narrowing resources, to demand from its ballistics engineers and weapons designers ever more inventive, radical and effective armament with which to combat the enemy in, and from, the air.

At around the time the crews of the 4th Combat Wing witnessed cable being dragged into their formation, on the Eastern Front German aircraft were being fitted with a range of heavy calibre 30 mm, 37 mm and even 75 mm anti-tank cannon intended to punch back at Russian tanks, while from mid-1943 at sea, Luftwaffe bombers deployed state-of-the-art Hs 293 and *Fritz X* guided glide bombs against Allied shipping off the coast of Italy.

And it would not stop there. As the stakes grew higher, so Germany's armaments designers produced ever more sophisticated weapons such as the innocuously named *Sondergeräte* (special devices) with photoelectric cell triggering systems, and the 55 mm R4M rocket and X-4 guided missiles for use against enemy bombers. At the fringes of weaponry development, there were no bounds to what was put forward for serious consideration, from chemical sprays to artificially generated squalls of air intended to destabilise enemy formations.

I have long been intrigued by the inventiveness of the German weapon manufacturers, and the lengths to which their designers and engineers would go to bring down enemy bombers from mid-1943 onwards, as well as their attempts to make some impact against masses of Soviet armour on the battlefield. But this is not to say that the production of weapons was trouble-free. As one example, I cite the following post-war report written by Oberst Ingenieur Johannes Mix, who, in January 1943, was in charge of *Abteilung* (*Abt.*) E6, the weapons section of the Development Group in the *Technisches Amt* (Technical Office) within the *Reichsluftfahrtministerium* (RLM – German Ministry of Aviation), and who held similar positions until war's end. The report highlights the Machiavellian connivance which affected the manufacture of heavy calibre cannon such as the 30 mm MK 108 and MK 103:

> The tendency of military agencies to demand that new weapons developed by industry be put into full-scale series production as soon as the first sample weapons had been test-fired with satisfactory results was confirmed repeatedly in the case of the newly developed equipment and – just as repeatedly – this habit gave rise to setbacks.

Mix was of the opinion that:

> . . . the intervention of the Ministry for Armaments and War Production towards the conclusion of development work was not favourable in the case of the MK 103 and MK 108. [Mix's] reproaches in this connection are directed not against the *Waffenkommission* (Ordnance Commission) and its special sub-committees, but rather against the office of the *Sonderbeauftragter Für Waffen* (Special Coordinator in Charge of Weapons) in the Ministry for Armaments and War Production. This office, run on the lines of Party politics and staffed by non-experts, supervised developmental work with a peculiarly unhealthy kind of coordination. In this connection it may be worthwhile to mention that this agency took steps to have the development work on the MK 103 and MK 108 transferred to itself shortly before its scheduled conclusion, requisitioning one of the chief experts to take over on behalf of the office. Subsequently, development work was officially brought to a close under the auspices of the Specialist's

office, without, however, absolving the *Entwicklungsabteilung* (Development Department) of ultimate responsibility for the equipment.

This book offers an insight into the Luftwaffe's 'special' weapons. I must stress 'insight' – space is finite and, regrettably, it is impossible to include all 'special' weapons, so I have elected to focus on the most radical or the most advanced or the most used, all depending on the availability of source material. I have opted to structure the book by intended role and deployment, though inevitably there is some crossover. With regard to cannon, I have decided to cover only weapons of 30 mm and greater calibre. The 20 mm MG 151/20 was a fine automatic cannon which equipped large numbers of Bf 109s and Fw 190s in both the fighter and the ground-attack roles but, unlike the much rarer 75 mm gun or steel cable, it cannot be regarded, in the context of this work, as a 'special' weapon.

I began assembling material for this book more than 30 years ago, and the process of returning to, and rediscovering, documents which I sourced during the late 1980s and early 1990s has been an enriching, enjoyable and educational undertaking.

It seems somehow indecently tardy to thank people after so long but, as a priority, I must offer my thanks to Phil Reed. Back in 1988, he introduced me to the wonders of the German Document Collection (GDC) and USAAF T-2 reports, as well as original reports prepared by Rheinmetall-Borsig at Unterlüss and the *Luftfahrtforschungsanstalt* (LFA – 'Institute of Aeronautical Research') *Hermann Göring* at Braunschweig, and the collection of translations produced by the Halstead Exploiting Centre, the British Intelligence Objectives Sub-Committee (BIOS) and the Combined Intelligence Objectives Sub-Committee (CIOS) held at the Imperial War Museum (IWM) in London. I spent many happy hours left to my own devices with a microfilm-reading machine in a room off the main reading room at Lambeth searching for nuggets. On occasion Phil was even good enough to assist with translation.

A few years later, Stephen Walton kindly allowed Eddie Creek and myself to explore the then still 'off-limits' document archive at IWM Duxford on more than one visit. The expression 'little boys in a toy shop' comes to mind because that is what we felt like as we pulled down box after box of uncatalogued, but fascinating, files from the shelves. None of this would have been possible without the support and patience of Phil Reed and Stephen Walton.

Also overdue for thanks are Hanfried Schliephake, whose book *Flugzeugbewaffnung* first aroused my interest in the aircraft armament of the Luftwaffe many years ago. Inspired, I corresponded with *Herr* Schliephake for some time, and eventually visited him at his home in Königsbrunn in 1990, where he arranged for me to view copies of his collection of weekly reports from the *Erprobungsstelle* (*E-Stelle*) Tarnewitz. It was *Herr* Schliephake who also made me aware of the seminal and ground-breaking work on German air armament of World War II, *Deutsche Geheimwaffen 1939–1945* by Fritz Hahn.

Another well-known German aviation historian whom I was fortunate enough to meet at his home in Mainz was the late Manfred Griehl. He kindly gave me copies of the

Arbeitsberichte of *Erprobungskommando* (E.Kdo – Test Command) 25, the Luftwaffe weapons testing and operational evaluation unit, and fragments of those prepared by its successor, *Jagdgruppe* 10.

I was also fortunate to correspond with, and to then meet, the former commanding officer of E.Kdo 25, Horst Geyer, at his home near Hamburg in 1993, where I spent a day talking to him about his interesting wartime career and memories. Other Luftwaffe veterans who kindly responded to my questions about Luftwaffe armament included Adolf Galland, Oscar Boesch, Franz Stigler, Walter Hagenah, Willi Unger, Gen. a.D. Walter Windisch, Fritz Buchholz, Rudi Riedl and Heinz Frommhold.

Closer to home, I would like to thank Eddie J. Creek, J. Richard Smith, Steve Coates, Martin Pegg, Nick Beale, Arthur Bentley, Mike Norton, Christopher Shores, Martin Streetly and Chris Goss for their kind assistance, suggestions and goodwill over the years, while information and support from beyond British shores came from Dr. James H. Kitchens III, Jürgen Rosenstock, Stephen Ransom, Hans-Hermann Cammann, Richard Chapman, John O. Moench, Eric Larger, Eric Mombeek, Hans-Heiri Stapfer, Friedemann Schell, Dr. Brett Gooden and Ted Oliver.

And finally I must thank Osprey Publishing for offering me the opportunity to write this book, and to Tony Holmes and Marcus Cowper whose faith in me has enabled it to materialise.

Any errors are mine alone, and it is my hope that I have done all the foregoing individuals sufficient justice for their kind help over the years.

Robert Forsyth
Sussex
September 2020

CHAPTER ONE

THE SPECIALISTS

On 17 November 1943, Reichsmarschall Hermann Göring arrived at Achmer airfield in northwest Germany to inspect the Luftwaffe units based there. On this occasion, the Reichsmarschall was accompanied by the commanding general of the Luftwaffe Fighter Arm, Generalmajor Adolf Galland, a 96-victory fighter ace and holder of the Diamonds to the Knight's Cross, Nazi Germany's highest military decoration. Despite any outward display of confidence and geniality towards the assembled airmen they had come to visit, both men were under pressure.

The daunting sight of a formation of B-17 Flying Fortresses on their way to bomb a target in occupied Europe. The USAAF's Eighth Air Force continually evolved the structures of its massed formations of 'combat boxes', known to the Luftwaffe as *Pulks*, to maximise self-defence against fighter attack, and these became extremely effective in their composition. Armed with Bendix chin turrets mounting two 0.50cal Browning M2 machine guns to defend against frontal attack, these aircraft are B-17Gs from the 532nd and 535th BSs of the 381st BG, based at Ridgewell, Essex. (Forsyth/Nelson)

By the autumn of 1943 the signs were of escalating enemy air operations – everywhere. The Luftwaffe had been drawn into a multi-front war of attrition against enemies of superior strength, onto which was layered the demand of defending the skies over the Reich. In the daylight air war over Germany, the losses of experienced pilots and formation leaders was rising at an alarming rate as the frequency, strength and depth of USAAF bombing raids began to increase – along with their numbers of escort fighters. To combat a formation of enemy heavy bombers armed with massed defensive firepower in the form of hundreds of 0.50cal Browning M2 machine guns required experience, tenacity and skill, as Franz Stigler, a pilot with Bf 109-equipped *Jagdgeschwader* (JG) 27 and an old 'Africa hand', recollected:

The B-17s took a lot of punishment. It was terrifying. I saw them in some cases with their tail fins torn in half, elevators missing, tail gun sections literally shot to pieces, ripped away, but they still flew. We found them a lot harder to bring down than the B-24 Liberators. The Liberators sometimes went up in flames right in front of you.

Attacking bombers became a very mechanical, impersonal kind of warfare; one machine against another. That's why I always tried to count the parachutes. If you saw eight, nine or ten 'chutes come out safely, then you knew it was okay, you felt better about it. But when you flew through a formation, the B-17s couldn't miss you. If they did something was wrong. I never came back from attacking bombers without a hole somewhere in my aircraft.

Wire frames and wooden models of a Flying Fortress and an Fw 190 are used to demonstrate to Luftwaffe fighter pilots the cones of fire that they would experience from a B-17's defensive chin, nose, top, waist, ball and tail guns. One B-17 fielded 13 0.50cal machine guns in nine positions. (EN Archive)

Indeed, Adolf Galland had experienced the personal impact of this new warfare. On 17 August, his brother, Major Wilhelm-Ferdinand 'Wutz' Galland, the *Kommandeur* of II./JG 26, a respected formation leader who had 55 victories to his credit, including eight *Viermots*, was shot down and killed in his Fw 190 by P-47 Thunderbolts during the USAAF raids on the ball-bearing works around Schweinfurt and the Messerschmitt plant at Regensburg. In that savage air battle, the Luftwaffe shot down 60 B-17s and damaged another 168 for a cost of 17 German pilots killed and 14 wounded, along with the loss of 42 aircraft. A week later the *Generalluftzeugmeister*, Generalfeldmarschall Erhard Milch, told subdued aircraft industry chiefs at a production conference:

It is clear from this that the struggle will not be without cost. Enemy bomber losses in May and June amounted to about 4.4 per cent of the total raiding force. In July there was a slight increase, the figure

being 6.4 per cent. It is clear these losses are not enough to deter an enemy as resolute as ours. You know that the defence of our homeland is now in the forefront of our strategy. A large number of single-engine and twin-engine fighter *Gruppen* have been brought back to Germany. In my opinion, this is absurdly late in the day, but at least it has been done. Reichsmarschall Göring too, is now bringing pressure to bear in this matter. And as a result of the raid on five of our two largest repair centres, we shall be at least 150 fighters down on last month, even with no further raids being made. We are therefore about 220 fighters short of our actual programme. This is very serious.

When the USAAF returned to Schweinfurt on 14 October, once again the Flying Fortresses were mauled – the Eighth Air Force's VIII Bomber Command (BC) lost 60 B-17s and 600 aircrew. Seventeen more bombers were seriously damaged and a further 121 were damaged but repairable. It was a body blow. Despite this catastrophe, however, a buoyant Brig Gen Ira C. Eaker, commander of VIII BC, claimed that the Luftwaffe's response was 'pretty much as the last final struggles of a monster in his death throes'.

The Luftwaffe had lost 31 aircraft destroyed, 12 written off and 34 damaged – between 3.4 and 4 per cent of available fighter strength in the West. These were acceptable figures in themselves, but the point was that the Americans could absorb, sustain, repair and reinforce. The Luftwaffe was not in that position. To this perfect storm could be added the fact that standards of Luftwaffe pilot training at the flying schools was suffering on account of a decrease in the time allowed to train a pilot, fewer instructors and lowering stocks of fuel and aircraft.

In a typical late-war scene at Leipheim airfield in Germany, Bf 109Gs of Major Walther Dahl's III./JG 3 undergo maintenance in March 1944. The fighter in the foreground carries the white *Geschwader* fuselage identification band and the black vertical bar denoting a III. *Gruppe* aircraft. (EN Archive)

Another disturbing fact was that by the early autumn of 1943, Göring had begun to isolate himself from the reality of the Luftwaffe's predicament. Throughout that year he became increasingly dependent on drugs and more absorbed in expanding his collection of art treasures and jewellery. When he turned his mind to the air war, he assumed that the lack of any decisive victory over the Americans, a nation which he considered capable only of manufacturing 'fancy cars and refrigerators', was down to nothing but lack of fighting spirit on the part of his fighter pilots. He had heard Galland talk of *Jägerschreck* – 'Fighter Fear'. Thus, there was an evident strain in the relationship with his fighter commanders.

It was amidst such an environment that necessity became the mother of invention.

STURMSTAFFEL 1

At Achmer, Göring and Galland had arrived to observe the activities of *Sturmstaffel* 1 and E.Kdo 25. The first-mentioned of these units had been the brainchild of Major Hans-Günther von Kornatzki, a long-serving Luftwaffe officer married to one of Göring's secretaries and a member of Galland's staff, who proposed the creation of a dedicated and specially equipped assault unit or *Sturmstaffel*. In the late summer of 1943, Kornatzki spent four weeks attached to E.Kdo 25, with whom he studied tactics and weapons intended for close-range operations against enemy bombers. During subsequent meetings with Göring and Galland in the autumn, Kornatzki advocated adopting radical new tactics involving massed rear attacks against the bomber *Pulks* by tight formations of heavily armed and armoured Fw 190s.

Kornatzki had studied reels of gun-camera film, read combat reports describing attacks on *Viermots* and interviewed fighter pilots. He reasoned that during a rearward attack against an American heavy bomber formation, one German fighter was potentially exposed to the defensive fire of more than 40 0.50cal machine guns, resulting in only the slimmest

Generalmajor Adolf Galland arrives at Achmer in his Fw 190A-6 with its distinctive triple chevron, to be met by Hauptmann Horst Geyer, commander of E.Kdo 25, standing at right. (Forsyth)

chance of escaping damage during attack. Under such circumstances, it was even less likely that a lone fighter could bring down a bomber.

However, if a complete *Gruppe* could position itself for an attack at close range, the bomber gunners would be forced to disperse their fire, and thus weaken it, allowing individual fighters greater opportunity to close in, avoid damage and shoot a bomber down. The loss of speed and manoeuvrability incurred by the extra armament and armour carried by these *Sturm* aircraft would be countered by the presence of two regular fighter *Gruppen* which would keep any enemy escort at bay.

Kornatzki also suggested to Galland that, if necessary and as a last-ditch resort, in instances where pilots were close enough, and if ammunition had been expended, a bomber could be rammed in order to bring it down. He further proposed that, initially, a *Staffel* rather than a *Gruppe* be established to train up volunteer pilots who would test and evaluate the new tactics under operational conditions. Under pressure from Göring, Galland needed little convincing. He quickly authorised the establishment of *Sturmstaffel* 1 and appointed Kornatzki as its commander. Furthermore, Galland advised his formation commanders of the rationale behind this development, judiciously placing the responsibility on Göring's shoulders:

Oberleutnant Franz Frodl of E.Kdo 25 (far right) demonstrates an item of weaponry to Generalmajor Adolf Galland during his visit to Achmer in November 1943. Also seen here, second from left, is Hauptmann Hans-Günther von Kornatzki, the commander of *Sturmstaffel* 1, whose heavily armed and armoured Fw 190s were intended for close-range anti-bomber missions. (Forsyth)

The main reason for this is the failure of formation leaders to lead up whole formations for attack at the closest possible range. Göring has therefore ordered the establishment of a *Sturmstaffel* whose task will be to break up Allied formations by means of an all-out attack with more heavily-armed fighters in close formation and at the closest range. Such attacks that are undertaken are to be pressed home to the very heart of the Allied formation whatever happens, and without regard to losses until the formation is annihilated.

TOP Know your enemy – a scale representation of a B-17 Flying Fortress painted on the doors of a hangar at a Luftwaffe airfield. This would have been used for gunnery and range estimation. Note the line-up of groundcrew to the bottom left of the photograph. (EN Archive)

RIGHT Major Erwin Bacsila, one of Hauptmann Hans-Günther von Kornatzki's senior officers in *Sturmstaffel* 1, blocks his ears as he talks to groundcrew at Dortmund following one of the unit's first missions in January 1944. Behind is Fw 190A-7 'White 7' with its armoured cockpit and 5 mm side panels fitted for close-range operations against USAAF bombers. (EN Archive)

In early November, *Sturmstaffel* 1 received its first aircraft in the form of Fw 190A-6s built with lighter wings capable of accommodating increased armament in the form of four 20 mm MG 151/20 cannon located in the wing-roots and the outer panels, thus phasing out the old, slow-firing MG FF 20 mm cannon. Two fuselage-mounted 7.9 mm machine guns were also retained, and the aircraft featured additional protective armour around the cockpit.

These initial machines constituted what was probably the first batch of operational aircraft to be armour-adapted for close-range anti-bomber work. They featured 30 mm *Panzerscheiben* (armoured glass panels) fitted around the standard glass cockpit side panels and a 50 mm plate of strengthened glass that would protect the pilot from fire from dead ahead. The installation of external 5 mm steel plates to the fuselage panelling around the cockpit area and the nose-cockpit join offered further protection from defensive fire. Additionally, the pilot's seat was fortified by 5 mm steel plates and a 12 mm head protection panel.

Sturmstaffel 1's subsequent combat experiences with the Fw 190 would influence the Luftwaffe's decision to adopt the 30 mm MK 108 electrically ignited, automatic cannon – a weapon that would see widespread use in the subsequently formed *Sturmgruppen* and later in the Me 262.

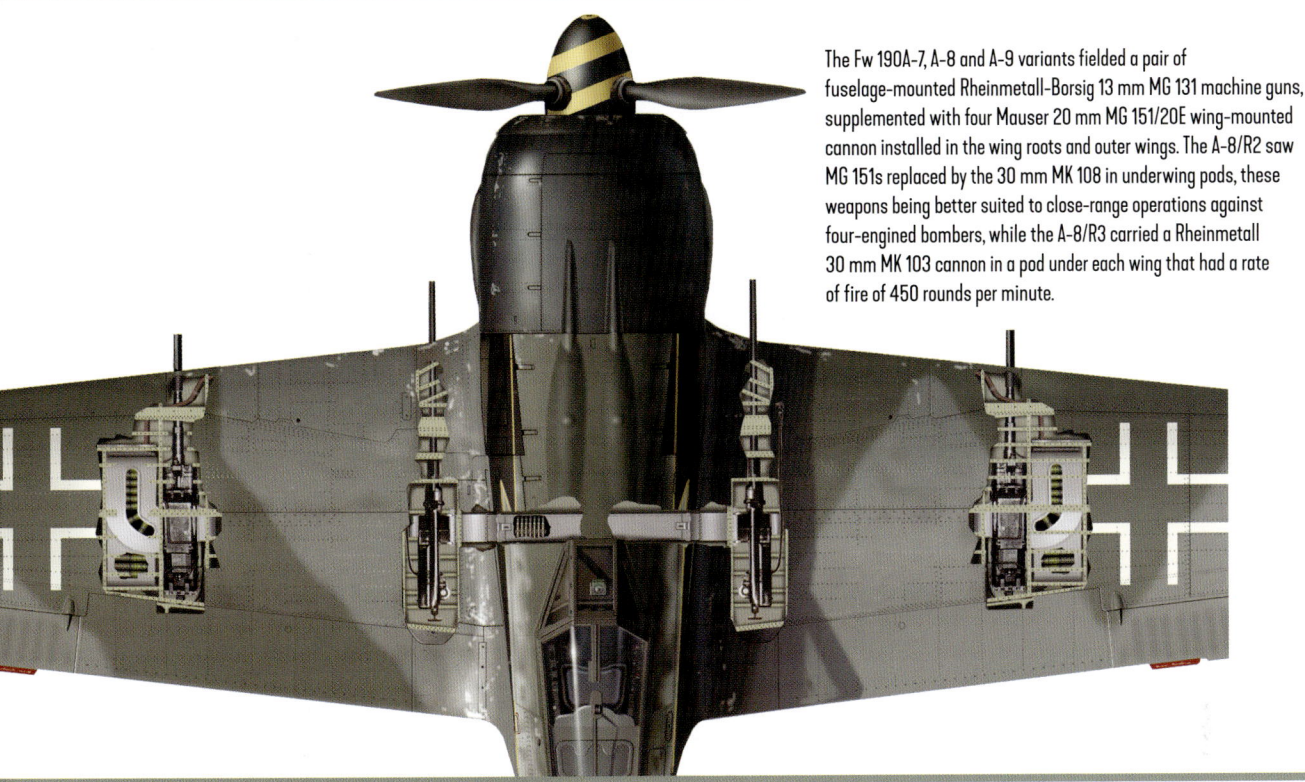

The Fw 190A-7, A-8 and A-9 variants fielded a pair of fuselage-mounted Rheinmetall-Borsig 13 mm MG 131 machine guns, supplemented with four Mauser 20 mm MG 151/20E wing-mounted cannon installed in the wing roots and outer wings. The A-8/R2 saw MG 151s replaced by the 30 mm MK 108 in underwing pods, these weapons being better suited to close-range operations against four-engined bombers, while the A-8/R3 carried a Rheinmetall 30 mm MK 103 cannon in a pod under each wing that had a rate of fire of 450 rounds per minute.

ERPROBUNGSKOMMANDO 25

Meanwhile, sharing Achmer with the new *Sturmstaffel* was E.Kdo 25. This unit had formed just seven months earlier on 17 April at Wittmundhafen under the command of Major Heinz Nacke, a very experienced airman and previously the *Kommandeur* of the nightfighter unit III. *Nachtjagdgeschwader* (NJG) 3. A veteran of the Spanish Civil War, Nacke had been awarded the Knight's Cross in November 1940 for his 12th aerial victory while flying Bf 110s with 6. *Zerstörergeschwader* (ZG) 76. His tenure in command of E.Kdo 25 was brief, however, and he was replaced, on a temporary basis, within a matter of weeks by another equally experienced *Zerstörer* pilot, Hauptmann Eduard Tratt, erstwhile *Staffelkapitän* of 1./ZG 1 in the East. Tratt was also a recipient of the Knight's Cross, having been decorated in April 1942 for his 20th victory.

Once at Wittmundhafen, Tratt went about arranging the establishment of three *Staffeln* for the embryonic *Kommando*. Firstly, a *Jagdstaffel* (fighter squadron) was formed under Leutnant Wilhelm Sbresny and equipped with three Bf 109Gs and seven Fw 190As. Its primary role was to conduct trials with numerous weapons and equipment for combatting heavy bombers, including rearward-firing armament, periscopes, acoustic fuses and wing-mounted RZ 65 rockets that had originally been intended for use by Bf 109s in the ground-attack role against locomotives on the Eastern Front.

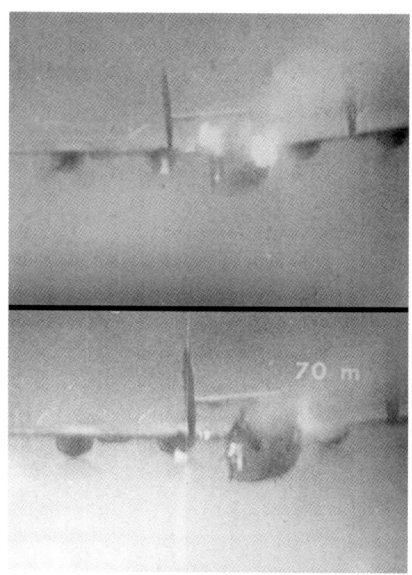

Stills from gun-camera film capture the moment that Feldwebel Otto Erhardt of 10./JG 3 closes his Fw 190A-8 to within 70 m of the tail of a B-24 Liberator on 7 July 1944. In the upper still it is possible that the flames seen are those of the bomber's left waist and tail guns, while in the lower still fire appears to have broken out around the right wing or engine area. This would be Erhardt's eighth victory. (Forsyth)

Secondly, a *Zerstörerstaffel* (twin-engined heavy fighter squadron) was set up under Leutnant Vossel. It was equipped with around ten Bf 110s, a single Me 210 and a pair of Me 410s and intended to trial heavy-calibre armament such as the 3.7 cm *Flak* 18 and *Flak* 43, and the 5 cm *Flak* 41 anti-aircraft cannon. These weapons were tested in a variety of hand-fed, belt-fed and automatic configurations. Generally, however, results were not encouraging, and the eventual operational losses suffered by aircraft fitted with such armament were disproportionately high in the relatively few missions flown, with their envisaged capability nullified by a loss of aircraft speed and the defensive fire of enemy bombers.

Finally, a *Kampfstaffel* (bomber squadron) was formed, equipped with two Do 217s, three Ju 88s, a solitary He 177 and four Bf 109Gs for escort purposes. This latter *Staffel* was intended to assess air-burst bombs, towed bombs, the radio-guided Henschel Hs 293 glide bomb and underwing mortars and rockets, as well as conducting experiments in air-to-air bombing.

Like Nacke, Tratt would remain in command at Wittmundhafen for only a short time, until his permanent replacement arrived in the form of Oberleutnant (soon to be promoted to Hauptmann) Horst Geyer. The latter had begun his wartime flying career as a test pilot at the *Erprobungsstelle der Luftwaffe* at Rechlin. He recalled:

In the summer of 1939, I finished flying school at Altenburg, and I was qualified to fly both fighter and larger, multi-engined aircraft. I was then assigned as a test pilot to Department E6 at Rechlin – the department responsible for the development of air armament, rockets, bombs and bombing systems. When war broke out – and because I was licensed to fly larger aircraft – they posted me to KG.z.b.V. 1, a specialist Ju 52/3m transport unit flying supplies into Poland.

After the Polish campaign, I returned to Rechlin and became involved in developing a new automatic sight and bomb release system for the Ju 88. One day, the Director of Luftwaffe Supply and Procurement, *Generalluftzeugmeister* Ernst Udet, visited the *Erprobungsstelle* and saw me flying and asked to inspect the new bomb system. Udet later asked my commanding officer if I could join him for lunch at a nearby barracks. Over lunch, I explained the workings of the bomb system to Udet, and also told him that I had, in fact, been trained as a fighter pilot since 1936! He promptly offered me a position on his staff, and in December, I found myself at a desk in Udet's office at the RLM in Berlin.

In January 1940, Geyer was appointed to the post of Adjutant to Udet, a man whom he came to admire greatly. However, by the early summer of 1940, with the air campaign commencing against England, Geyer longed for an operational posting. Anxious not to be excluded from the ranks of young *Jagdflieger* already carving themselves reputations on the

Channel Front, he was prepared to seize the first opportunity to escape Berlin so as to 'climb into the cockpit of a Bf 109':

> In the early summer of 1940, I had the opportunity to meet Werner Mölders, the first fighter pilot to have won the *Ritterkreuz* [Knight's Cross], whilst he was discussing certain technical matters in Berlin. On this occasion, Udet took Mölders and I to lunch, and whilst dining, I quietly asked Mölders if I could join his *Jagdgeschwader* based on the Channel Front. He replied that I should ask my boss who was sitting right next to me! Udet agreed to my request, but on the basis that I was to return to Berlin after three months' secondment to JG 51. I was honoured to join the *Geschwader Stab* and served with Mölders, Köpke, Balfanz, Erwin Fleig, Friedrich Beckh and, later, Hartmann Grasser.
>
> I duly returned to Berlin in November to continue my duties as adjutant to Udet, but in September 1941, I was again allowed to join JG 51, which by this time was serving in the East. I was assigned to the II. *Gruppe*, now commanded by Grasser.

Later that month, Geyer was appointed *Staffelkapitän* of 5./JG 51, flying the Bf 109F. At this time, JG 51 was the most successful *Jagdgeschwader* in Russia, and was actively involved in supporting Army Group Centre during Operation *Taifun*, the drive on Moscow:

> On one occasion, while returning to our base at Schatalowka from an operational sortie, my Me 109 was jumped by Russian fighters and I was wounded in the hand and the foot. The cockpit was full of blood. I made it back, but was sent to hospital in Königswinter, and after a period of recovery was ordered back to Udet.

Geyer was credited with 18 victories while flying with JG 51, and was awarded the *Ehrenpokal der Luftwaffe* (Honor Goblet of the Luftwaffe). However, whilst he was serving in Russia, in Germany, Udet's position as *Generalluftzeugmeister* had become increasingly

TOP LEFT The leading *Zerstörer* ace Hauptmann Eduard Tratt (left) briefly commanded E.Kdo 25 in May–June 1943. He is seen here wearing a British Irvin jacket on 22 February 1944 talking with pilots of the Me 410-equipped II./ZG 26 – the *Gruppe* he commanded from September 1943. It would be a fateful photograph, for Tratt was killed in action the same day it was taken. (EN Archive)

ABOVE From June 1943, Hauptmann Horst Geyer was successor to Eduard Tratt as commander of E.Kdo 25. In the same style as Tratt, he is seen here wearing a prized Irvin jacket, with the port engine nacelle of one of the unit's Me 410s behind him. (Forsyth)

Generalmajor Galland looks somewhat bemused by the range of experimental weaponry on display at E.Kdo 25's base at Achmer in November 1943. There appears to be a multi-barrel device on the bench behind the officers, while in the foreground is a large-calibre mortar tube. (Forsyth)

untenable as a result of severe disagreements with Göring and Milch over aircraft production policy. Udet's handling of a demanding post that was wholly inappropriate to his flamboyant and artistic personality had resulted in friction between him and the ambitious Milch, a situation further exacerbated by Göring's refusal to mediate in what had become a bitter power struggle within the most senior levels of Luftwaffe planning and procurement. On 17 November 1941, exhausted and depressed, Ernst Udet, Germany's second-ranking flying ace of World War I, committed suicide.

A state funeral was to be held in Berlin. Göring would deliver a gushing address. A selection of the Reich's finest pilots would form an honour guard. Mölders, now the *General der Jagdflieger*, was returning from Russia especially. Horst Geyer was to carry Udet's baton. 'So I lost my boss, a great man', recalled Geyer. 'Before the funeral, I was advised that I had been appointed to a position on Mölders' staff, but soon afterwards, and tragically, I took a call from Oberst Edgar Petersen, the commander of the *Erprobungsstellen*, who just said, "You can't go to Mölders, there's been an accident."'

On its way to Berlin on 22 November, the He 111 carrying the young *General der Jagdflieger* had crashed in bad weather near Breslau. Mölders and the aircraft's pilot were killed:

So I went to a second funeral, and was subsequently assigned to the staff of the new *General der Jagdflieger*, Oberst Adolf Galland, whom I did not know at that time. Although I was based in Berlin, and was able to be with my wife, it was a very busy time and we worked very hard. My role was that of *Technisches Offizier im Stab* and, amongst other things, I dealt with complaints from frontline fighter and *Zerstörer* units on matters of armament and technical equipment.

Suddenly, one day in May 1943, Galland strode into my office and said, 'Geyer, I want you to go to Wittmundhafen tomorrow and take over *Erprobungskommando* 25. You'd better tell your wife.' That was that! I was to take over from Hauptmann Tratt, an officer I had not previously met and who was returning to I./ZG 1. I think Galland viewed my new appointment as a kind of thank you for flying a desk for so long.

Horst Geyer would rise quickly to the challenge of his new command, and would oversee the *Kommando*'s activities at the most active and urgent period of its existence, assessing and testing new weapons intended for the war against the American *Viermots*. These would include heavy cannon, mortars, rockets, steel cable, cable bombs, chemicals and obliquely mounted, automatic salvo weapons.

ERPROBUNGSSTELLE TARNEWITZ

Understandably, E.Kdo 25 would liaise closely with the main Luftwaffe armaments testing centre at Tarnewitz, on the Baltic coast, 40 km northeast of Lübeck. Development of the *E-Stelle* Tarnewitz had commenced in late 1935, initially as a seaplane experimental station, on reclaimed land between the waters of the Boltenhagen Bucht and the Wohlenberger Wiek and was completed by late 1939. Built with five large hangars, including one for firing tests, the well-equipped but weather-battered site also contained grass runways built on reclaimed land, servicing areas, concrete slipways, fuel and oil tanks, extensive equipment and armament workshops. There was a seaplane harbour with a hangar southwest of the site at the Wohlenberger Wiek, and it boasted a mole, crane, jetties, a tower, barracks, a command centre, administration buildings and a rifle range, all of which were subject to further expansion until 1945.

British Air Intelligence maps from October 1943 of the 'Tarnewitz Aerodrome and Sea Station' showing the *Erprobungsstelle*'s coastal location and well laid-out facilities. (Forsyth)

It was at Tarnewitz that all types of fixed gun armament and cannon were tested and assessed, as were gun sights, periscopes, target indicators, target-finding systems, automatic offensive and defensive weapons systems, rockets and missiles intended for day and night air-to-air and ground-attack deployment. Much of the testing saw weapons aimed at the Baltic Sea, or with flights carried out over it during airborne trials. Other work undertaken included low-temperature research on equipment for aircraft gunners, and the production of training films for various armament and weapons. Such work reached its peak in 1944.

From May 1942, the commander of the *E-Stelle* was Oberstleutnant der Reserve Maximilian Bohlan, with the testing department, *Abt.* E6 (to which Horst Geyer, later commander of E.Kdo 25, had once been assigned), under Oberstleutnant Walter Segitz. The latter unit was comprised of various sections, including IIA (Firearms, Airborne Rockets and Electrics) under Stabsingenieur Walter Denzinger, IID (Turrets, Drives, Remote Steering and Control, and Hydraulics) under Stabsingenieur Rudolf Pätz, IIE (Ballistics, Measurement Systems, Targeting Devices and Optics) under Ludwig Röhm and IIF (Weapon Installation) under Stabsingenieur Ernst Pfister. Flight tests were overseen by Ekkehard von Guenther.

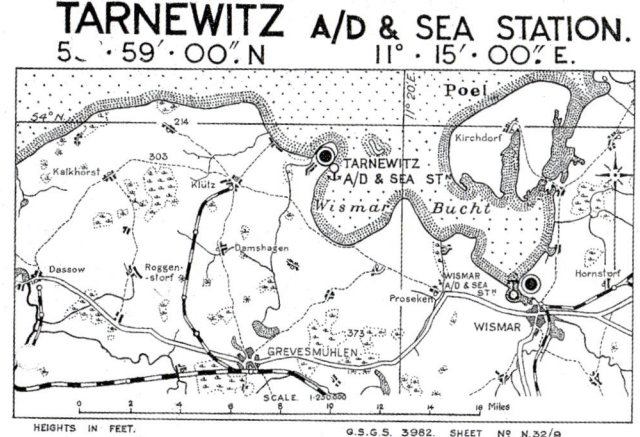

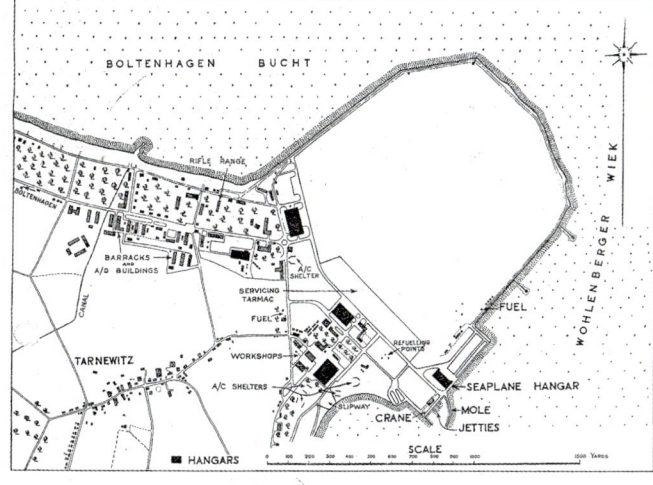

RHEINMETALL-BORSIG AG, UNTERLÜSS AND THE *LUFTFAHRTFORSCHUNGSANSTALT HERMANN GÖRING*

Playing a significant role in the design, development and testing of many aircraft weapons was the *Waffenkonstruktion-Waffe* (WKW) *Gruppe* at the Rheinmetall-Borsig works and proving grounds at Unterlüss, situated some 50 km northwest of Celle in northern Germany.

One of the leading German firms in the research, design and development of modern weapons during World War II, Rheinmetall-Borsig manufactured most of the heavier calibre *Maschinenkanone* (machine cannon) for aircraft including the 30 mm MK 101, MK 103 and MK 108, the 37 mm *Flak* 18, the 50 mm BK 5, the 55 mm MK 112, MK 114 and MK 115, and the 75 mm BK 75. From the end of 1942, to identify a machine cannon manufacturer, the first digit of the three-digit weapon number denoted the manufacturer, while the last two denoted the sequence of development. Rheinmetall-Borsig was assigned 1, Mauser 2, Krieghoff 3 and Krupp 4. Thus, the MK 114 was the 14th cannon specification to be developed (other companies could also have been involved) and built by Rheinmetall.

In addition to manufacturing machine guns and cannon, Rheinmetall-Borsig/WKW also developed a range of electrically operated and automatic *Rohr Batterie* (tube battery) weapons or *Sondergeräte* (special devices), several of which incorporated photoelectric cell technology which were tested at Unterlüss. These horizontally, obliquely or vertically mounted experimental projects were intended as one-time-use salvo weapons for deployment against enemy bombers or tanks.

As the Luftwaffe's main armaments testing facility, the *Erprobungsstelle* at Tarnewitz was progressively expanded and enhanced throughout the war. Here, the He 177 V12 GI+BL is seen in one of Tarnewitz's extensive hangars fitted with an experimental nose installation housing a single 30 mm MK 101 cannon intended as a ground-attack weapon. For tests, the MK 101 was fitted into a L 101/1A mount and fired against two target discs at various ranges with the Heinkel's engines running. (EN Archive)

As testimony to the work carried out at Unterlüss, one post-war Allied report stated:

> In general it may be said that the chief characteristic was the attitude of mind, 'We do not know this will not work, [so] let us try it', rather than the stultifying converse, 'We do not think this will work, [so] we will not try it.'

In charge of all weapons development at Unterlüss was Professor Dr. Ingenieur Carl Wanniger. Reporting to Wanniger and responsible for overseeing the development of automatic aircraft cannon, ammunition, mountings, sights and accessories was Fritz Herlach, an automatic-cannon specialist, who managed a small team of ballistics engineers and armament specialists that worked on a range of air-to-air and ground-attack weapons.

For most of the war Fritz Herlach, a weapon designer and engineer, was Director of the multi-department WKW at the Rheinmetall-Borsig works and proving grounds at Unterlüss. In this capacity, he oversaw all development of automatic weapons and ammunition, as well as negotiation of contracts with foreign Axis countries. (Forsyth)

Herlach's team included ballistics expert Dr. Kokott, who produced several reports of a detailed scientific nature, and Oberingenieur Theodor Rakula, who specialised in automatic cannon of 13–55 mm calibre, plus associated belts, magazines and clips, as well as investigating electrics. Oberingenieur (DPhil) Walter Grasse was a graduate in physics and mathematics from the University of Berlin in 1934, where he had performed experimental work in infrared spectroscopy. He had joined Rheinmetall in 1935 as a physicist, and was appointed to head the development of ammunition for automatic aircraft cannon. Oberingenieur Kuppe and Linke worked on the development of the electrification of various armament projects, as well auxiliary drives and remote controls, while Diplom-Ingenieur Koch handled the general installation of armament into trial aircraft.

Together, this team ensured efficient cooperation between the various design departments, the scientific experimental management and experimental production techniques. Indeed, American author Lt Col George M. Chinn, regarded as an authority on the history and development of the machine gun, has noted:

> Under the guidance of these men, hit-and-miss methods in research and development were replaced by more efficient procedures geared to produce more certain results. The genius of Fritz Herlach and Theodor Rakula is shown in a unique approach to the design of weapons, as well as in their own professional skill in developing actual working principles.

The Unterlüss complex covered an area of some 12 km by 30 km and consisted of extensive firing ranges with seven main firing points and more than 300 buildings, including machine shops, development installations, laboratories, magazines, administrative offices and air raid shelters, while a filling plant was located three kilometres away at Neulüß.

A post-war Allied Intelligence report noted how, with the production methods employed:

> . . . the time required was reduced to a minimum from the day of development to the day of mass production. It was, of course, important that the finished experimental unit should resemble in every respect the future mass production article. It was on this account necessary that the experimental plant should have at its disposal first class machines by which the miscellaneous production could be realised.
>
> Well-equipped physical, chemical, metallographical and X-ray laboratories, durability test benches with the necessary auxiliary devices, shooting benches for cold and hot trial shots etc., are at the disposal of the scientists. For aircraft armament testing, there was available a special aviation section at the Tarnewitz testing station.
>
> The chief sections were assisted by expert engineers in the examination of drawings, measurements, and materials. Special engineers furnished the technical descriptions and servicing instructions.

After ground-firing tests of a particular weapon had been carried out and it was ready for air testing, it was taken to the *E-Stell e* Tarnewitz, where Rheinmetall-Borsig maintained a large hangar, workshop and staff engineers who oversaw the tests. The group was led by chief test pilot and engineer Flugbaumeister Diplom-Ingenieur Reicker. The development team at Unterlüss relied greatly on Reicker's reports, as well as those of Luftwaffe observers at Tarnewitz, as to the performance of new weapons.

Another organisation involved in the development and testing of special aircraft weapons was the *Luftfahrtforschungsanstalt* (LFA – 'Institute of Aeronautical Research') *Hermann Göring* near Völkenrode, 6.5 km west of Braunschweig, under the directorship of Professor Blenk. The LFA had come into being in 1938 when the *Deutsche Forschungsanstalt für Luftfahrt* (DFL) was renamed after Göring, despite the fact that the Reichsmarschall never visited the highly secret LFA. It reported directly to the RLM, and by war's end employed some 1,500 personnel – including 150 scientists and 150 university graduates – at its well-camouflaged 485-hectare site. The LFA comprised five sub-institutes which, respectively, specialised in the fields of aerodynamics, structures and materials, engines and weapons and ballistics, with a sixth that maintained the extensive workshops. A British post-war intelligence report on the LFA noted that 'the scale of the equipment, in quality and quantity, in both laboratories and workshops, was most impressive.'

The weapons research institute, which was headed by Professor Rossmann who joined from Krupps in 1942, was, in turn, sub-divided into three departments and designed automatic triggering devices similar to those developed at Unterlüss. Some of the advanced weapons systems developed by the LFA are covered in this book.

From left to right, Dr. Walter F. Grasse, Dr. Richard H. Braun and Oberingenieur Theodor Rakula were among the team of highly qualified ballistics experts who worked at Rheinmetall-Borsig's development centre and firing ranges at Unterlüss. Grasse was senior engineer for ammunition development for all automatic cannon and weapons. As head of the *Waffenkonstruktion-Waffe Forschungs*, Braun oversaw basic research, while Rakula specialised in automatic cannon from 13 mm to 55 mm calibre, belts, magazines, clips and electrics, and worked on several armaments designs, including the SG 113A installation in the Hs 129. (Forsyth)

TORPEDO SCHUL- UND ERPROBUNGSSTELLE DER LUFTWAFFE/KAMPFSCHULGESCHWADER 2 AND ANTI-SHIP GUIDED MISSILE KOMMANDOS

Hs 129B-0 Wk-Nr 0016, fitted with the SG 113A experimental anti-tank weapon, was photographed at the LFA *Hermann Göring*, Völkenrode in July 1943. (EN Archive)

Further along the Baltic coast from Tarnewitz, north of the *Erprobungsstelle* (*See*), the main maritime testing base at Travemünde, was the Luftwaffe's *Torpedo Schul- und Erprobungsstelle* (Torpedo School and Testing Centre) at Grossenbrode. This facility had been established in the spring of 1941 under Oberstleutnant Karl Stockmann, who had previously been *Kommandeur* of *Küstenfliegergruppe* (Kü.Fl.Gr.) 406, together with his head of training, Hauptmann Werner Klümper. Klümper was a very experienced officer, having previously been *Staffelkapitän* of 3./Kü.Fl.Gr. 906 and the Operations Officer on the *Stab* of *Kampfgeschwader* (KG) 30. He was assisted by three torpedo specialists, Oberleutnant Klaus Toball, the former Operations Officer of 3./Kü.Fl.Gr. 906, Oberleutnante Lorenz and Fritz Müller.

The school was originally based at Travemünde, but as the Luftwaffe's use of torpedoes increased, so it moved to a dedicated station. However, Grossenbrode's position on the Baltic Coast meant that the winter weather, which frequently brought snow and ice, hampered Klümper's efforts, and so a decision was made to open a sub-station at Grosseto, in Tuscany, at the end of 1941. Oberst Martin Harlinghausen, the Luftwaffe's *Bevollmächtigter für die Lufttorpedowaffe* at the RLM, had recognised the advances in Italian torpedo design and wanted to combine and centralise German and Italian efforts in that area. Both Grossenbrode and Grosseto were concerned with training crews in the techniques of aerial minelaying and torpedo dropping.

Thus expanded, the school was then redesignated *Kampfschulgeschwader* (KSG) 2, which comprised a *Stab* from the *Torpedoschule* and two *Gruppen*, one formed from the school's

ABOVE Hauptmann Werner Klümper photographed wearing the Knight's Cross he received on 29 August 1943 for his record in attacking enemy shipping in the Mediterranean and off North Africa. Klümper played an early and pivotal role in training Luftwaffe bomber crews for torpedo operations. (Goss)

ABOVE CENTRE *Legion Condor* veteran Oberst Martin Harlinghausen became the senior officer for torpedo development and deployment at the RLM. A very experienced commander, tactician and anti-shipping pilot who devised the 'turnip' method of broadside-bombing ships, he was awarded the Knight's Cross on 4 May 1940, followed by the Oak Leaves on 30 January 1941 for his night-bombing missions over North Africa. (Forsyth)

ABOVE RIGHT He 111H-5 Wk-Nr 3891 BK+CD drops its starboard torpedo over the waters of the Baltic off the Luftwaffe torpedo training school at Grossenbrode. Strictly speaking, this has been done in contravention to operational and technical requirements which stipulated that the port side torpedo should be released first. (EN Archive)

Lehrgruppe 1 and equipped with He 111s, the other with Ju 88s. Between 1942–43, five to ten crews from the school were tasked with training crews from KGs 26 and 77 and KüFlGr 906 in day and night torpedo attacks. Following heavy fighting in the Mediterranean theatre, KSG 2 was renamed KG 102 on 1 March 1943 under Oberst Horst Beyling and withdrawn to Riga-Spilve, close to the Latvian coast, in July 1943 to train KG 77 and crews intended for the aircraft carrier *Graf Zeppelin*. When the carrier project was abandoned, KG 102 was disbanded in July 1944 and the aircrews dispersed. Its aircraft and groundcrews were sent to IV./KG 26.

The development of sophisticated, guided bombs and missiles designed primarily, but not exclusively, for anti-shipping work, such as the Henschel Hs 293 and the Ruhrstahl PC/SD 1400 *Fritz X*, initiated the formation of specialist *Kommandos* to undertake testing on a specific weapon or weapons prior to operational deployment. The *Erprobungs- und Lehrkommando* 21 was established at the end of 1941 at Schwäbisch Hall, before relocating to Garz on Usedom on 1 August 1942 to trial the *Fritz X*, where it was commanded by Major Ernst Hetzel. The *Versuchsstaffel* Hs 293 and E.Kdo 15, the latter formed in the winter of 1941–42 under experienced anti-shipping pilot, Hauptmann Franz Hollweg, were assigned to test the Hs 293. Later, E.Kdo 36 was set up at Garz to conduct trials with more advanced versions of the Hs 293 (see Chapter Five).

VERSUCHS KOMMANDO FÜR PANZERBEKÄMPFUNG AND ERPROBUNGSKOMMANDO 26

Formed at the *E-Stelle* Rechlin in December 1942 to test heavy calibre, aircraft-mounted, anti-tank weapons, the *Versuchs Kommando für Panzerbekämpfung* (*Vers.Kdo. für Pz.Bekämpfung* – Test Command for Anti-Tank Warfare) was established by the former *Kommodore* of *Schlachtgeschwader* (Sch.G., later abbreviated to SG) 1 and the *Inspekteur für Schlacht- und Zerstörerflieger*, Knight's Cross-holder Oberstleutnant Otto Weiß.

The *Kommando* comprised three *Staffeln* equipped with Ju 87G-1s (1./*Vers.Kdo. für Pz.Bekämpfung*), Ju 87s and Hs 129s (2. *Staffel*) and Ju 88s (3. *Staffel*), respectively. At Rechlin, the Ju 87s, under another Knight's Cross-holder, the former *Kommandeur* of I. *Stukageschwader* (St.G.) 5, Hauptmann Hans-Karl Stepp, carried out test-firings using heavy calibre weapons against captured Soviet tanks. There was an urgency to this since the often fluid and fast-moving conditions on the Eastern Front meant that German forces were vulnerable to being isolated and cut off. The Wehrmacht was finding that its conventional anti-tank guns were insufficient to cope with such situations and were unable to move quickly enough to cover large areas of territory, or to prevent enemy armoured breakthroughs. Thus, there were increasing calls from commanders on the ground for cannon-armed anti-tank aircraft, or *Panzerjäger*, to destroy enemy tanks.

The cannon-fitted Ju 87Gs of the *Vers.Kdo. für Pz.Bekämpfung* were moved to Bryansk in Russia for operational testing, where one pilot to be assigned briefly was a certain Oberleutnant Hans-Ulrich Rudel from St.G. 2, who accompanied the unit when it relocated to Kerch in the Crimea.

Oberstleutnant Otto Weiß, seen at left, was appointed to lead the *Vers.Kdo. für Pz.Bekämpfung* in late 1942. He was a greatly experienced ground-attack pilot, having flown around 520 missions in the Hs 123 and Bf 109 and been awarded the Knight's Cross on 18 May 1940 after 430 missions. He also served as the *Führer der Panzerjägerkommando* – a tactical command intended to control all Hs 129 units on the Eastern Front. He later handed command to another Knight's Cross-holder, Hauptmann Bruno Meyer, seen here to Weiß's left. (Pegg)

RIGHT As an extremely experienced Stuka pilot, Hauptmann Hans-Karl Stepp was assigned briefly to the *Vers.Kdo. für Pz.Bekämpfung* in early 1943. He had previously been *Staffelkapitän* of 7./St.G. 2 and received the Knight's Cross on 4 February 1942 after 418 combat missions. In September 1942 he was appointed as *Kommodore* of that *Geschwader*, ending the war credited with around 900 missions in the Ju 87 and Fw 190 and as a recipient of the Oak Leaves. (EN Archive)

FAR RIGHT The Luftwaffe's most successful anti-tank pilot, Oberst Hans-Ulrich Rudel (left), reads notes on the latest types of Soviet armour while his radio operator/gunner, Oberfeldwebel Erwin Hentschel, holds a large wooden or card identification model of a Russian KV-2 heavy assault tank. Together, Rudel and Hentschel flew more than 1,000 combat missions. (EN Archive)

These aircraft, and the Hs 129 *Staffeln* in southern Russia, would be directed in combat by a specially established tactical command, the *Panzerjagdkommando Weiß*, under the control of Oberstleutnant Weiß.

At some point, cannon-armed Bf 110s were also sent to Bryansk, joining the similarly armed Ju 87Gs, Hs 129s and Ju 88P-1s there. Subsequently, the 3. *Staffel*'s 75 mm PAK 4-fitted Ju 88s went to Poltava.

In January 1944 another anti-tank air warfare testing unit was set up at the *E-Stelle* Udetfeld, an ordnance-testing facility near Gleiwitz in Poland. The E.Kdo 26 was formed from a cadre of the 11.(Pz.)/SG 9 under Major Herbert Eggers, who had previously led the Hs 129B-equipped *Panzerjägerstaffel*/JG 51. The unit was established with four Ju 87s, four Fw 190s, four Hs 129s and four nightfighters, plus two transports. Amongst the unit's

RIGHT A German soldier examines a still smouldering and abandoned Soviet T-34 medium tank. Luftwaffe pilots found the art of destroying tanks with bombs challenging since only a direct hit or a close miss promised any significant effect. Despite the development and use of hollow-charge bomb containers, much investigation was undertaken into 30 mm-plus calibre cannon with armour-piercing ammunition which was found to be more effective against sloping armour when used by well-trained pilots of the *Schlachtgeschwader* from 1943. (EN Archive)

various activities, its aircraft engaged in testing and assessing heavy calibre anti-tank cannon, such as the 75 mm BK 7.5. E.Kdo 26 remained in existence until 14 February 1945, when its personnel were assigned to the staff of the *General der Schlachtflieger* and its outstanding operational tasks handed to the *Ergänzungsstaffel*/SG 151.

The various weapons used by these units and organisations will be examined in the following chapters.

Armourers tighten a new barrel for a 37 mm Flak 18 cannon fitted to a Ju 87G, possibly of 10.(Pz)/SG 2, at an airfield on the Eastern Front in the spring of 1944 while behind, another pair of armourers load high-explosive shells into the breech tray. (EN Archive)

CHAPTER TWO

HEAVY CANNON

By 1943 the Luftwaffe's challenge in all its theatres of operation was to optimise its fighter defences and to increase the number of enemy aircraft shot down, whilst simultaneously reducing its own losses. It was therefore necessary to shorten the time required to shoot down an enemy aircraft so as to minimise the time the enemy pilot/crew had to bring their own weapons into action.

The solutions lay in increasing the range at which the Luftwaffe's weapons could be fired, as well as reducing the firing time necessary to bring about a 'kill' – i.e., increasing the number of strikes possible in the smallest measure of time. This could be achieved by increasing the rates of fire of Luftwaffe airborne weapons, as well as their number, and by deploying heavier calibre

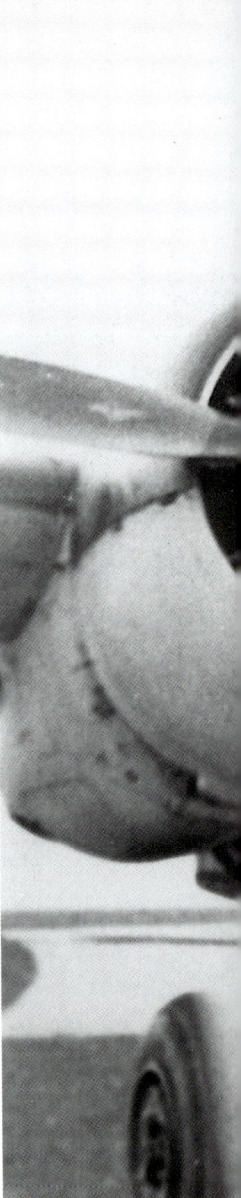

A Rheinmetall-Borsig 30 mm MK 101 machine-cannon is cradled in its centreline fairing beneath an Hs 129B-1. The cannon was fitted on a detachable mount, and by pulling away four pins from lugs on the bottom flange of each wing spar it could be quickly removed, or by pulling away just two pins, the gun would hang in a hinged position. (EN Archive)

weapons that would ensure greater destructive effect, along with improvements in aiming accuracy through more sophisticated sighting devices.

Oberst Stabsingenieur Paul Ossenbühn, head of the *Technische Luftrüstung* (TLR – Air Technical Equipment) *Abt*. Fl.E6 at the *E-Stelle* Tarnewitz in 1944, where he was responsible for the development of aircraft guns and rockets, and his colleague, Oberst Stabsingenieur Franz, had calculated that to destroy a B-17 the following numbers of high-calibre projectiles were required:

20 mm = 16 projectiles
30 mm = 4 projectiles
50 mm or 55 mm = 1 projectile

An armourer loads a clip of six 37 mm high-explosive, self-destroying rounds into the loading tray of a *Flak 18* cannon fitted to a Ju 87G for anti-tank deployment. The white area at the tip of the shell contained the fuse, behind which was a cardboard ring and delay detonator, then the explosive. The rear part of the shell contained the main propellant, powder bag, igniter and a steel C/33 primer. (EN Archive)

RHEINMETALL-BORSIG MK 101 30 mm *MASCHINENKANONE*

Developed from the Waffenfabrik Solothurn (Rheinmetall's Swiss subsidiary) 20 mm S-18-1000 anti-tank rifle in 1936, the 30 mm electro-pneumatically charged, air-cooled MK 101 *Maschinenkanone* was the first weapon to be so-titled with the 'MK' three-digit classification. It was initially turned down by the Luftwaffe because of its size and weight of 185 kg including the mount, and the fact that its method of operation would have required 'the complete rebuilding of a twin-engined aircraft and would have reduced its flying performance considerably'.

Despite this, it was a weapon that had particularly impressed Adolf Hitler and Hermann Göring when they attended an exhibition of the latest German air armament held at the *E-Stelle* Rechlin on 3 July 1939. Indeed, after watching a static firing test with an MK 101 fitted beneath a Bf 110, the *Führer* made special reference to the cannon in his departing speech.

The Rheinmetall MK 101 was 2.9 m long and weighed 180 kg. It featured a long chamber and extremely high-velocity tungsten-core ammunition with a rate of fire of 260 rounds per minute and a muzzle velocity of 69.5 m per second. It could be fed either from a six-round magazine or a 30-round drum.

It had first been envisaged as a weapon to be used against bombers, and the earliest air trials with the MK 101 had also been carried out using a Bf 110 with a semi-rigid installation augmented by a telescopic sight as a means of assessing heavier calibre guns for long-range air combat. In a report prepared by Diplom-Ingenieur Johannes Linke, an ignition specialist at Rheinmetall-Borsig's plant in Sömmerda, it was stated that:

High-Explosive, Incendiary and Mine shells

On their initial introduction for cannon of 30 mm calibre, the Luftwaffe could draw upon two basic types of shells: the high-explosive, self-destroying, tracer shell designed to cause blast effect, and the high-explosive, self-destroying incendiary shell intended to cause both a blast and incendiary effect.

However, in consultation with the Luftwaffe's principal test centre at Rechlin, ballistics specialists at Rheinmetall-Borsig's main testing ground at Unterlüss had calculated that maximum destruction to an enemy aircraft could be created by causing the largest possible explosive effect in its interior. They had also worked out that the latter, in turn, was dictated by the size of the enemy aircraft and by the quantity of explosive that could physically be placed into a projectile. The thicker the shell wall, the more energy was needed for the destruction of the shell itself, and thus less energy remained for the destruction of the target by the ensuing explosion.

This theory led to the development of the *Minen Geschoß* ('Mine Shell' – M-shell), which combined a minimum thickness in shell casing with a maximum load of explosive to produce the greatest possible blast effect. Such a shell was extremely destructive to the stressed skin of a bomber aircraft. Using such ammunition, the entire enemy aircraft could be regarded as the target area, it making no difference where the hit was actually made. As such, with M-shells a fighter pilot had an inherently greater chance of scoring a 'kill'.

Following tests carried out at Rechlin, in terms of deployment against enemy heavy bombers, it was discovered that five hits from a 30 mm M-shell carrying 85 g of explosive were needed for the destruction of either a B-17 or B-24. One post-war report prepared by a team of former German ballistics experts noted, 'Protection against "M" shells is practically impossible.'

Incendiary shells were also considered an extremely potent form of ammunition, but only when targeted at fuel tanks. The vulnerability of an enemy aircraft could therefore be measured by the area and/or size of its tanks. However, a certain degree of penetrative force was still needed in order to break through the airframe or any protective armour carried by the target without breaking up and igniting until actually striking the fuel. To overcome this problem, the 30 mm incendiary shell was fitted with a hydrodynamic fuse which activated only when making contact with a fluid.

When attacked by a fighter directly from behind, the area of a B-17 taken up by its fuel tanks was approximately one-fifth of the total, and it was assumed that by the time an attack was made the tanks would be half empty. Thus, in combat conditions, this area was reduced to one-tenth of the surface area. It was calculated that between five and ten 30 mm incendiary shells were needed to cause inextinguishable burning. However, in the case of the B-24, an effective attack using incendiaries was slightly more difficult since the bomber's main fuel tanks were located in the fuselage, with only reserve tanks located in the wings behind the engines.

A 30 mm MK 108 cannon with, positioned upright, examples of an incendiary shell (at left) and a *Minen Geschoß* (mine shell) marked with an 'M'. Such ammunition was used, respectively, to cause fire in fuel tanks and blast effect within a bomber's fuselage. In the M-shell the upper part of the round contained the fuse, detonator, booster, explosive and closing plug. (Forsyth)

This was a 30 mm MK 101 with a velocity of 900 m per second, a gun which, in external ballistics effect, presented the best expectations for long-range combat, as there was a high velocity and a sufficient flatness in trajectory. The gunner aimed through a paneratic telescopic sight by means of which the range could be determined. The purpose of the trial was to obtain a comparison between the semi-rigid and the rigid methods of firing. The results fully confirmed the superiority of the semi-rigid firing method and its use with large-calibre aircraft guns for long-range combat. Accuracy was more than 60 per cent greater than with rigid firing.

Göring placed an urgent order for 3,000 examples. However, although a limited number of MK 101s were fitted to some Bf 110C-6s of the *Zerstörer* trials unit, 1./*Erprobungsgruppe* 210, in the summer of 1940, the opportunity for deployment against bombers had not arisen. Furthermore, the *Gruppenkommandeur*, Hauptmann Walter Rubensdörffer, reported that the extra weight of the MK 101 had an adverse effect on the Bf 110's performance. Rubensdörffer was of the opinion that the Bf 110's existing 20 mm armament was adequate for deployment against the air and land targets then being engaged, and that other than certain ships and heavily armoured tanks, the MK 101 offered no discernible advantage. Whilst he believed only a small number of Bf 110s should be fitted with the cannon, he did concede that 20 mm armament may not be sufficient or effective against the modern bomber of the future.

Rubensdörffer's views were supported by the *Generalluftzeugmeister*, Generalleutnant Ernst Udet, who also felt that weapons of 20 mm calibre were sufficient for the Luftwaffe's needs – a belief that was shared by members of the Luftwaffe General Staff at the time. Another black mark against the MK 101 was the fact that it was complex to build and was regarded as an experimental, interim design pending availability of the improved 30 mm Rheinmetall MK 103, which meant that no tooling had been readied for series production.

Thus, although a significantly reduced order for 220 cannon was placed, further delivery of the MK 101 to the Luftwaffe was delayed. Of this order, half was assigned for installation into the Henschel Hs 129B-1 and B-2 dedicated anti-tank aircraft for service in Russia (although small numbers were also used in North Africa). However, of the 115 completed cannon, 71 had been used for trials, while the remainder had already been earmarked for the He 115, He 177 and Do 217 types. As the A-1/U2 sub-type, 12 He 177A-1s were modified as intended long-range *Zerstörer* incorporating a pair of MK 101 installed in faired housings mounted just below their noses.

Eventually, in early 1942, a concerted effort was made to gather as many MK 101s as were available at the test centres for the Eastern Front-based Hs 129 *Schlachtgeschwader* (ground-assault units), but even then, while they waited at a depot in Lippstadt for delivery to the East, the VIII. *Fliegerkorps* apparently declined them.

The MK 101 was fixed to a removable mounting beneath the fuselage centreline of an Hs 129, with the breech mechanism enclosed within a gently rounded fairing. The gun and mounting could be quickly detached by withdrawing four pins – one from each corner of the mounting – from lugs on the bottom flange of each wing spar. Similarly, by withdrawing only two of the pins on one side, the remaining two acted as hinges, allowing the cannon and mounting to be swung down for loading and servicing.

In his 1951 study for the US Navy's Bureau of Ordnance, Lt Col George M. Chinn of the US Marine Corps described the firing process of the MK 101:

To fire the MK 101, a loaded drum is attached in its fastening latches on top of the receiver. Air from a cylinder is admitted to the charging mechanism through an electric solenoid valve. This allows compressed air to enter the piston housing against the charging piston, forcing it back with the bolt assembly until it engages a rear-searing device after the contact button of the charging valve has been broken. The release of air depresses the sear and allows the bolt assembly to be driven forward under energy of the compressed driving springs.

The bolt face picks up a round from the mouth of the feed, chambering it at the same time the rotating sleeve locks the piece in the battery. Firing is accomplished by depressing the button that engages the solenoid; this moves a lever that in turn disengages the front sear from the firing pin grooves. At this time a heavy spring drives the firing pin into the primer, to fire the chambered round. The barrel, bolt, and locking sleeve are all firmly joined while the projectile is traveling through the bore, and remain that way until an inch-and-a-half of recoil takes place.

During the first bit of travel the cocking lever is pivoted, at first withdrawing the firing pin within the bolt face and then compressing the firing pin spring tightly until it is seared back, fully cocked. The locking sleeve, the rollers of which are guided in the cam slots, is now rotated, unlocking the bolt, at which point the accelerator speeds the bolt rearward. The barrel and locking sleeve are held in a retracted position while the bolt is still recoiling. It carries the empty cartridge case, withdrawn from the chamber by the extractor, until the ejector makes contact with the rim of the case, pivoting it out of the ejection slot in the bottom of the receiver.

Further recoil compresses the driving spring and the bolt hits the buffer, after which an opposite movement begins. As the bolt moves forward, the first round in the magazine is started towards the chamber. At the same time the extractor claw is cammed over the rim of the cartridge. Shortly before the bolt strikes the locking ring, the coupling lever is lifted by an inclined ramp on the bolt body. This releases the retracted barrel and locking ring at the exact instant the bolt lugs are opposite their mating threads in the ring. The rollers in the locking ring follow the camming grooves which rotate the sleeve, quickly locking the entire assembly. If the solenoid is still actuated, the firing lever moves the sear out of engagement with the button on the firing pin. The latter flies forward to fire the propellant charge in the cartridge again.

It was foreseen that the MK 101's tungsten-cored ammunition should be capable of penetrating 70 mm and 100 mm armour at impact angles of 60 degrees and 90 degrees,

This image from an RLM photographic instruction slide shows a 30 mm MK 101 cannon and magazine drum attached to its loading/unloading support frame and fitted into the carrier mount of an Hs 129B-1. The weapon weighed 180 kg. (EN Archive)

respectively, from a range of 300 m. Its muzzle velocity was finally calculated at 960 m per second. Its armour-piercing effectiveness against armour plating of 100 kg/mm² (the equivalent strength of the hull of a Russian T-34 tank) was 70 mm at a 60-degree angle of impact and 100 mm at a 90-degree angle of impact from a distance of 300 m.

By July 1942 supplies of any ammunition for the MK 101 had been curtailed due to manufacturing problems, and once the remaining stock of shells had been used, the cannon were removed from the Henschels and replaced by bomb racks. When a representative of the RLM visited the units at the front to investigate further, he was informed by Hs 129 pilots that when German ground troops made their way through captured areas, no enemy tanks had been found which bore the signs of being hit by aircraft cannon.

Eventually, however, the MK 101 became available in greater numbers and, according to Adolf Galland, by early 1943 all Hs 129s had been fitted with the cannon. When deployed by skilled pilots, it could be a lethal weapon. Hauptmann Franz Oswald was an extremely experienced anti-tank pilot and, successively, *Staffelkapitän* of 5./Schl.G. 1, 8.(Pz.)/Schl.G. 2 and 13.(Pz.)/SG 9. He was awarded the Knight's Cross on 24 October 1944 in recognition of his destruction of more than 50 enemy tanks, plus trucks, anti-aircraft positions, artillery emplacements and horse-drawn transport. He recalled that when attacking enemy tanks with 30 mm cannon:

We would normally open fire at a distance from the target of between 45 and 60 m, and from such a close range, the 30- to 40-degree firing angle brought the aircraft dangerously close to its target. There was always the risk of crashing into the target or striking the ground if the pilot failed to pull out in time, or, if the tank exploded when hit, the attacking aircraft could be badly damaged or even destroyed by blast and flying debris.

There was some degree of misunderstanding among the pilots as a result of a lack of adequate anti-tank training. For example, pilots of II./Schl.G. 1 were of the belief that a tank could only be considered hit effectively if it was set on fire, whereas in reality to critically disable a tank it was necessary for the anti-tank ammunition to penetrate its armour and to kill or disable its crew.

As a preferred tactic, wherever possible, the Hs 129s would avoid approaching enemy tanks from the front, where their armour was the thickest and the line of approach the most dangerous, and attack from the rear where the engine was located. Franz Oswald found the Hs 129 to be a very stable gun platform, allowing him to aim for the cooling grills on the top decking, where the armour was thinnest, and to achieve success in this.

The effect of incendiary shells fired into this area of a tank could produce spectacular results, and if a crew had been unaware that they had been hit, they would continue to drive with flames and smoke streaming from the engine compartment. Eventually, the Russian tank crews became attuned to this method of attack and would rotate their turrets to fire back at the Henschels as they made their approach. Protective grills were also introduced for engines, and these were left open while a tank was moving. Although they could be closed if the tank came under attack, this was not a foolproof form of defence.

Later in the war, Hs 129 pilots would use their cannon in low-level, beam attacks, aiming at the area between a tank's hull and its tracks and wheels. A well-aimed hit with one or two cannon shells would be sufficient to disable a tank. Franz Oswald remembered:

> Sometimes, when a tank exploded, the whole force of the blast went through the bottom of the hull and the earth was scorched in a large circular area all around. As one could never tell in which direction the blast from an exploding tank would go, I would bank away after firing in order to avoid possible blast damage to my aircraft.

Aside from the occasional failures and jamming, the severe conditions of winter faced by Luftwaffe units in Russia inevitably caused serviceability problems. Despite the fairing, the cannon could often become wet, covered in dirt and snow, and frozen.

Hs 129 *Staffeln* known to have been equipped with the MK 101 included 4.(Pz.) and 8.(Pz.)/Schl.G. 1 and 4.(Pz.) and 8.(Pz.)/Schl.G. 2, which were redesignated in October 1943 into 10.(Pz.), 11.(Pz.), 12.(Pz.) and 13.(Pz.)/SG 9, respectively, and 13.(*Panzerjägerstaffel*)/JG 51 which, in October 1943, became 14.(Pz.)/SG 9.

Despite the later introduction of the MK 103, some experienced pilots, such as Hauptmann Rudolf-Heinz Ruffer, *Staffelkapitän* of 10.(Pz.)/SG 9, experienced faults with the newer cannon, and preferred the MK 101 – the weapon with which he achieved most of his kills. At the end of March 1944, Ruffer's score of 63 enemy tanks destroyed made him the most successful Hs 129 pilot, while his *Staffel* had accounted for the destruction of 100 T-34s, six aircraft, 30 assault guns and hundreds of vehicles. Ruffer was awarded a Knight's Cross in July 1944 in recognition of his outstanding record as an anti-tank pilot and marksman. He was known to be able to destroy a tank using a single, short burst of cannon fire.

RHEINMETALL-BORSIG MK 103 30 mm *MASCHINENKANONE*

The development of the 30 mm MK 103 *Maschinenkanone* by Rheinmetall was intended as a faster-firing successor to the MK 101, taking advantage of modern production methods from late 1942 with the least possible need for re-tooling, incorporating stamped sheet metal for the housing, feed and other parts. The RLM armaments specialist Oberst Ingenieur Johannes Mix, the *Abteilungsleiter* of *Abt.* E6 (Aircraft Armament) in the *Technisches Amt*, described its primary role as 'a logical further development in air-to-air defensive action against bombers'. Mix further relates how the development of the MK 103 fell 'chronologically behind' the MK 108 (see page 45), to which 'in deference it had been relegated to the background a number of times'.

Another report prepared by Rheinmetall-Borsig stated that the weapon was 'principally intended for long-range air combat beyond 1,000 m'.

This barrel-recoil-operated, metallic link belt-fed, gas-operated and air-cooled cannon came in two basic forms – an engine-mounted gun assembled from the rear, and a flexible version assembled from the front, yet with the exception of the barrel mounting, the two versions were identical. In the engine-mounted version, a reinforced friction brake and barrel recuperator spring were used, as the weapon was fired without a muzzle brake. With a muzzle brake, the cannon measured 2.3 m and without, 2.03 m. Its barrel was 1.33 m long. It weighed 145 kg, had a muzzle velocity of 900 m per second and was intended to fire high-explosive M-shell ammunition, armour-piercing and incendiary ammunition. Beyond enemy bombers, such ammunition was to be used against armoured ground targets, merchants ships and light armoured vehicles.

A Rheinmetall-Borsig 30 mm belt-fed, gas-operated, air-cooled MK 103 machine-cannon with a 30 mm M-shell alongside. Intended as the replacement for the MK 101, it proved a troublesome weapon. (EN Archive)

Early examples were fitted with a magnetic trigger, while later units had an electro-pneumatic trigger. The claimed rate of fire was 420–450 rounds per minute, but despite attempts to increase the rate of fire of the MK 103 to 600 rounds per minute, the cannon suffered a reputation for generally poor performance. Furthermore, its sheet metal construction made it less strong than the MK 101, and yet it was, initially at least, to use that gun's ammunition.

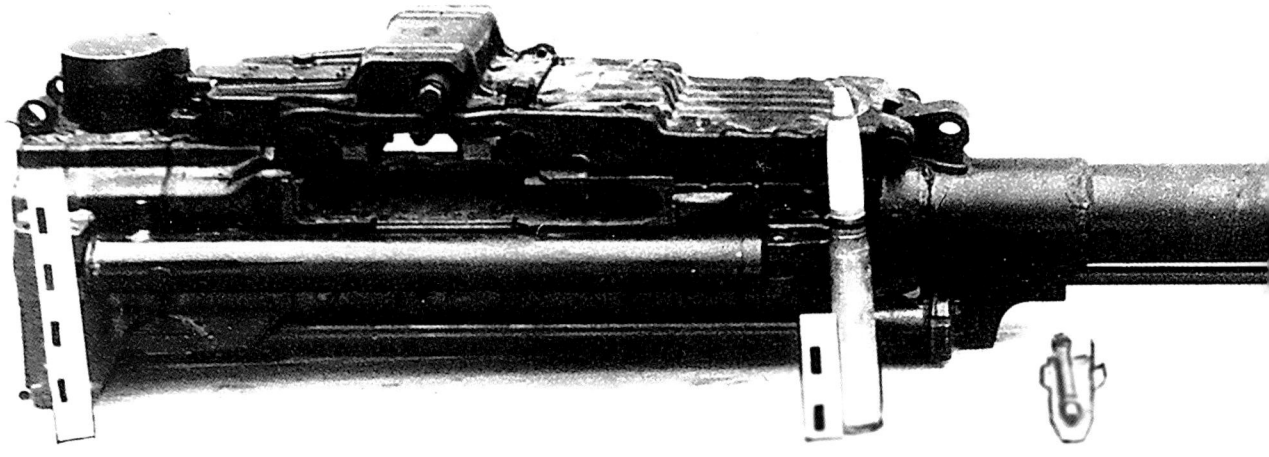

Lt Col George M. Chinn offers a description of the firing process:

The gunner places a loaded belt in the feedway until the first cartridge is behind the belt-holding pawl. Then air is turned into the charging mechanism by actuating the charger valve. The pneumatic effect on the charger's piston-like front and on the driving springs forces the whole assembly rearward, carrying the bolt group with it. The sear engages the bolt and holds it in a cocked position, until another valve releasing air pressure forces the holding device down and allows the bolt to fly forward under tension from the driving spring. On the final movement rearward, the centre pawls in the feed position the first round for stripping.

As the bolt is driven forward, the rammer engages the rim of the indexed round and starts to chamber it. The extractor claw is forced over the lip of the rim at this time. The upper part of the bolt is abruptly stopped as its face strikes the breech. However, a striker on the rear of the piston keeps on to force the two swinging locks out into their locking abutments into the barrel extension, thus locking the bolt securely behind the chambered round.

The cannon now being loaded and ready to fire, pressure on the trigger button closes the circuit and the electric primer in the cartridge is set off. The action remains locked for the first two inches of recoil, but force is transmitted to the feed pawls and they are moved over one space, shoving the cartridge across the spring-loaded rammer, forcing it down. After the projectile has cleared, pressure is brought on a gas piston housed in a cylinder beneath the barrel. Being driven rearward, the piston and slide uncover the two locks, allowing them to move into the bolt body and all recoil together. The extractor pulls the empty cartridge case from the chamber and holds it until the ejector at the rear of the feedway strikes the top of the rim, knocking it down and out of the receiver. Continued recoil fully compresses the driving springs and the final movement ends with the bolt striking the buffer. The first phase of counter recoil places the cartridge-holding pawl in the feeder over the next round. The rest of the forward travel is used to strip and chamber the new round and to lock the action into battery. The cycle is repeated if the electric circuit is still energized.

The MK 103 was too large to fit into the confined spaces of a Bf 109's in-line engine compartment as a *Motorkanone*, but it was earmarked as underwing armament for the radial-engined Fw 190. In the A series of the Focke-Wulf, it was proposed to fit two 30 mm

GENERAL ARRANGEMENT DRAWING OF THE RHEINMETALL-BORSIG 30 mm MK 103

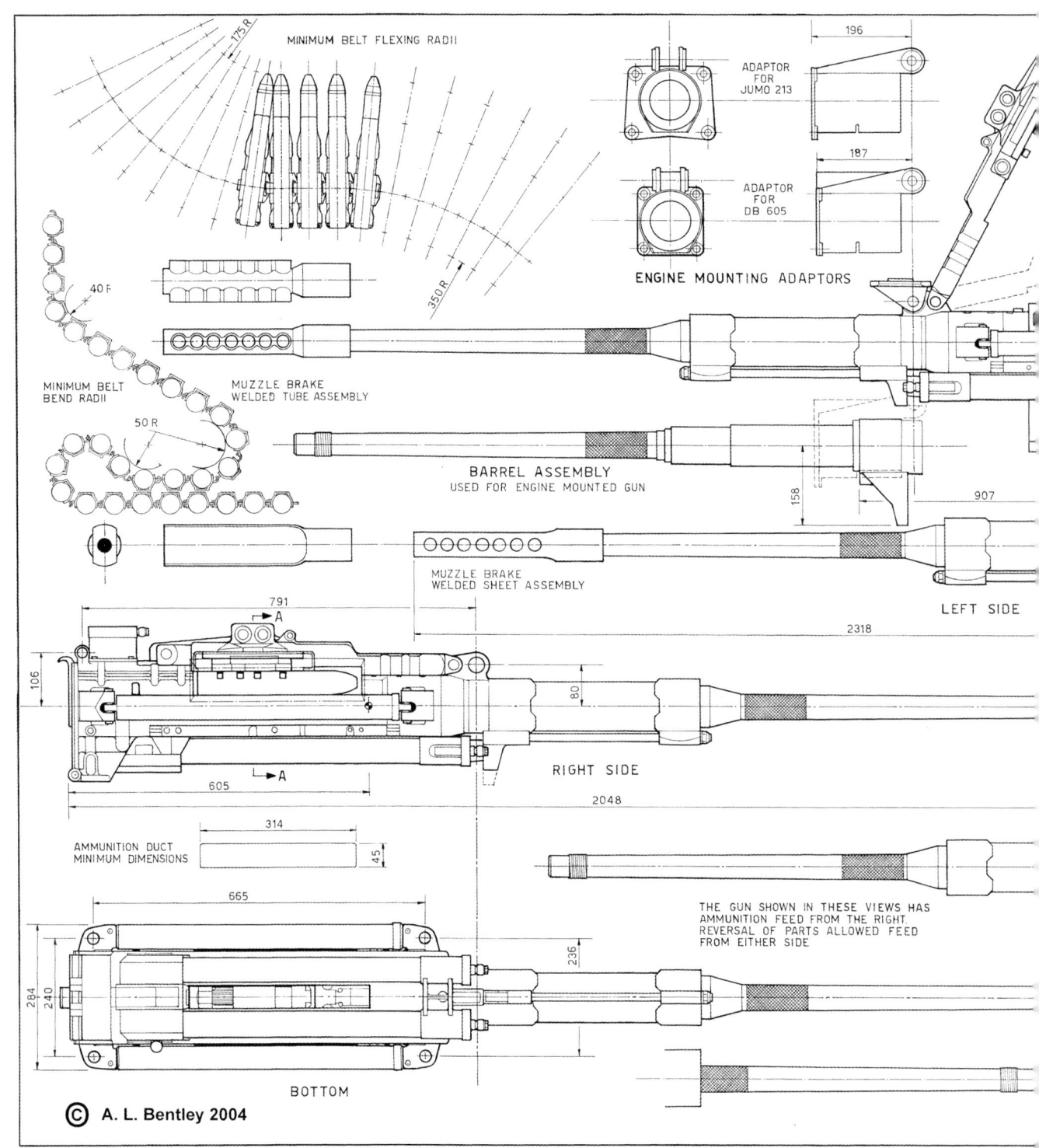

MINIMUM BELT FLEXING RADII

175 R

350 R

40 R

MINIMUM BELT
BEND RADII

50 R

MUZZLE BRAKE
WELDED TUBE ASSEMBLY

196

ADAPTOR
FOR
JUMO 213

187

ADAPTOR
FOR
DB 605

ENGINE MOUNTING ADAPTORS

BARREL ASSEMBLY
USED FOR ENGINE MOUNTED GUN

158

907

LEFT SIDE

MUZZLE BRAKE
WELDED SHEET ASSEMBLY

791

A

106

605

A

2318

80

RIGHT SIDE

2048

AMMUNITION DUCT
MINIMUM DIMENSIONS

314

45

665

284

240

236

THE GUN SHOWN IN THESE VIEWS HAS
AMMUNITION FEED FROM THE RIGHT.
REVERSAL OF PARTS ALLOWED FEED
FROM EITHER SIDE.

BOTTOM

© A. L. Bentley 2004

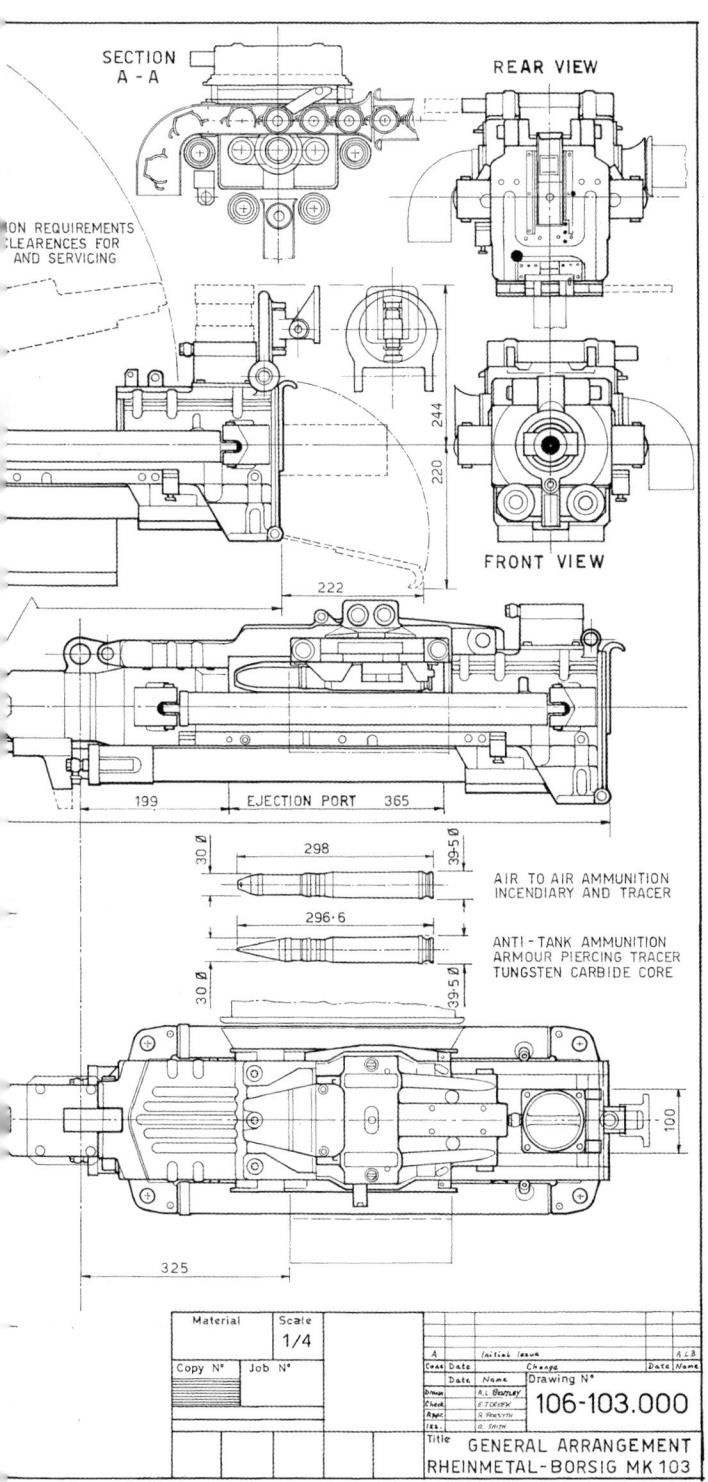

SECTION A-A

REAR VIEW

FRONT VIEW

244

220

222

ON REQUIREMENTS
CLEARANCES FOR
AND SERVICING

199   EJECTION PORT   365

298

296·6

30 Ø

30 Ø

39·5 Ø

39·5 Ø

AIR TO AIR AMMUNITION
INCENDIARY AND TRACER

ANTI-TANK AMMUNITION
ARMOUR PIERCING TRACER
TUNGSTEN CARBIDE CORE

325

100

| Material | Scale 1/4 |
| Copy N° | Job N° |

A	Date	Initial Issue		A.L.B
Cost	Date	Change	Date	Name
	Date	Name	Drawing N°	
Draw		R.L Beamey		
Check		E.Torpix	106-103.000	
Appd		R. Whestyn		
Iss.		R. Smith		
Title		GENERAL ARRANGEMENT		
		RHEINMETAL-BORSIG MK 103		

MK 103 or four MK 108 cannon or two to six 7.92 mm MG 81 machine guns in special containers in the Fw 190A-3/U3 – a *Schlachtflugzeug* (ground-attack aircraft) variant carrying extra armour plate. Similarly, the Fw 190A-5/U11 was to carry a 30 mm MK 103 long-range cannon mounted under each wing outboard of the mainwheel legs. Carrying 32 rounds each, the guns were intended mainly for use against ground targets. The prototype, Wk-Nr 1303 RG+ZA, was tested at Tarnewitz on 19 August 1943 and by E.Kdo 25 on 18 October of that year. This eventually led to the A-6/R3 sub-variant equipped with a MK 103 cannon with 25 rounds per gun beneath each wing.

Two aircraft were completed as Fw 190A-8/R3s featuring two MK 103 cannon with 35 rounds per gun under the outer wing panels, replacing the outboard MG 151s. Two aircraft were tested and a further ten delivered for operational assessment. A specification for an R4 was also considered, likewise intended to carry a pair of MK 103 cannon with 50 rounds to replace both outboard MG 151s, but this variant was abandoned at an early stage. It was proposed that A-9/R3 and A-10/R3 would carry an MK 103 beneath each wing as well, the latter with 40 rounds per gun.

Five prototypes of a dedicated ground-attack variant, the F-8/R3, were built, each fitted with MK 103 cannon with 35 rounds per gun. A production batch of 60 aircraft was ordered.

The MK 103 was also considered for the 'long-nosed' Fw 190D-9, with one such weapon fitted under each wing carrying 40 rounds per gun, but this was never pursued.

In late 1944, the Fw 190 V68 prototype was allocated for purposes connected with the development of the Ta 152 high-altitude interceptor. The aircraft was intended to test the mounting of three long-barrelled MK 103s, one engine-mounted and the others in each wing root, in connection with the proposed Ta 152B-5 *Zerstörer*. Tests carried out at Tarnewitz proved that it was possible to mount an MK 103 between the engine cylinder banks, the barrel of the gun protruding 1,380 mm beyond the front of the spinner. Oberstleutnant Maximillian Bohlan reported that the workshop at the *Erprobungsstelle* was 'heavily overburdened with work in hand, and for this reason the installation of the MK 103 in the wing roots would require two to three months to complete. Focke-Wulf could, however, complete the installation after four weeks.' It is not known whether this was undertaken before the end of the war.

However, although some ground-attack Fw 190s had been fitted with long-barrelled MK 103 cannon for anti-tank missions, the weapon proved incapable of penetrating the armour of most Russian tanks. Furthermore, senior Luftwaffe fighter officers told Allied interrogators that, in their opinion, while the Fw 190 fitted with MK 103s enjoyed the advantage of 'good fire effect', this was outweighed by a speed loss of 60 km/h, a degree of instability while firing, a lowering of flight ceiling and a reduction in manoeuvrability, with the aircraft being noticeably overloaded. Nevertheless, despite this shortcoming, pilots flying the Fw 190 as *Schlachtflieger* generally preferred the weapon for use against rail targets, especially locomotives. Oberst Ingenieur Mix reported:

If the MK 103 was to be used effectively against tanks, there was no alternative but to resort to a projectile with a tungsten core and with a magnesium head to bring about an incendiary effect [at that time tungsten was practically unavailable in Germany]. Fired at a range of 300 m, this special projectile was able to pierce armour plating 70 mm thick and 100 kg/mm² at an impact angle of 60 degrees, and plating 100 mm thick at a vertical impact angle. The armour-piercing incendiary shell designed for use against merchant ships and light armoured vehicles was capable of penetrating marine construction steel (*Schiffsbaustahl*) 20 mm thick at a range of 1,000 m and an impact angle of 60 degrees, and the armour-piercing fragmentation shell, designed primarily for use against ground targets, was able to penetrate steel plating 25 mm thick from a distance of 300 m and at an impact angle of 60 degrees. In their effectiveness and ballistic performance, the cartridges developed for use against air targets corresponded in general to the ammunition types used with the 37 mm anti-aircraft artillery cannon, the *Flak* 18.

After the available supply of MK 101s had been exhausted, a small quantity of this anti-aircraft ammunition had been taken over and, with minor modifications, used for airborne cannon. Only the 37 mm armour-piercing ammunition designed for use against tanks had greater penetration effectiveness (40 mm) than the 30 mm armour-piercing shell. Quite apart from the increase in weight (138 kg per weapon) when the 37 mm anti-aircraft artillery ammunition was used, the 30 mm ammunition was perfectly adequate for use against the Russian T-34, the tank most frequently employed on the Eastern Front. As a result of its threefold greater firing speed [rate of fire], the MK 103, used with the proper ammunition, was also preferable to the 37 mm airborne heavy cannon for use against light armoured vehicles and iron-protected ground targets.

In the harsh operating conditions of an Eastern Front winter, an armourer lends a sense of scale as he checks the breech mechanism of an MK 103 cannon fitted to an Hs 129B-2 of 10.(Pz.)/SG 9 in the spring of 1944. Note the two quick-release bolts hanging from their chain fasteners and the belt of mixed high-explosive and incendiary rounds differentiated by the use of coloured tips. (EN Archive)

In the spring of 1943, after consulting with frontline *Panzerjägerstaffeln*, the Henschel Flugzeug-Werke had commenced work to improve its twin-engined Hs 129B-2 anti-tank aircraft which had been enjoying success, albeit in relatively small numbers, on the Eastern Front. The improvement came in increasing the aircraft's armour protection and the calibre of fixed armament. As the Hs 129C-1, it had been proposed to fit two MK 103 cannon under the fuselage and one 150 kg or four 50 kg bombs under each wing. This concept was later refined to create two separate variants – a bomber and a dedicated anti-tank aircraft with two MK 103s, each with 60 rounds, which were to be placed side-by-side on special mountings with a limited traverse. By May 1944 a model of the C-1 was undergoing wind tunnel tests at the LFA *Hermann Göring* at Völkenrode fitted with two belly cannon turrets.

However, when examples of the Hs 129B-2 arrived at Zaporozhye, in Russia, in the summer of 1943 assigned to 4.(Pz.)/Schl.G. 1 and fitted with a single MK 103, the gun was soon found to be prone to jamming. This was a state of affairs that continued to plague the use of the cannon for the rest of the war. In March 1944, the *Staffelkapitän* of 10.(Pz.)/SG 9, Hauptmann Rudolf-Heinz Ruffer had been credited with 63 enemy tanks destroyed – a total which would have been higher had the MK 103 functioned reliably. Ruffer knew well the problems associated with the cannon: he had once suffered no fewer than eight failures in eight consecutive sorties with an MK 103-equipped Hs 129, hence his preference for the older MK 101. For other pilots, their experience was different.

In February 1945, as Soviet forces attacked the city of Breslau, one Henschel pilot attacked an enemy tank with an MK 103 at low-level above a street. He witnessed the tank blow up and its turret cartwheel into the air and land on the roof of a house some distance away.

Oberst Ingenieur Mix believed the MK 103 to be 'the most modern automatic 30 mm weapon to be introduced in its performance class'. This is given some credence by the fact that the MK 103 was also proposed for one of the Luftwaffe's most modern aircraft, the big, high-speed, all-weather Dornier Do 335 'push/pull' bomber/interceptor/reconnaissance aircraft/nightfighter. In January 1944, Dornier proposed that the Do 335A-0 and A-1 *Schnellbombers* (high-speed bombers) should be armed with two 20 mm MG 151/20 cannon mounted above the engine and synchronised to fire through the propeller arc. A single, long-barrelled MK 103 cannon was mounted between the engine cylinder banks and fired through the spinner. Ammunition for the MK 103 was fed from a 70-round container mounted behind and below the weapon. The first prototype to be fitted with armament was the Do 335 V5 CP+UE, which made its inaugural flight in February 1944, but would not undergo significant testing until August.

In the following Do 335B series, a 310-litre self-sealing fuel tank was to have been mounted in the leading edge of each wing, but this tank was replaced by containers for the 70 rounds of 30 mm MK 103 ammunition for which a single such weapon was installed in each wing. The ammunition containers were protected from the front by a rectangular armour plate.

Meanwhile, at Tarnewitz, the V5's MK 103 cannon proved problematic and was prone to jamming in flight, and although some 1,400 rounds had been test-fired by the end of November, it was still not fully satisfactory. By comparison, the twin MG 151s worked well.

As early as the summer of 1944, while still refining the concept for the Do 335A, Dornier began work on the new B-series aircraft, predominantly as a heavily armed *Zerstörer* that would be able to carry some of the greater calibre weapons being developed for deployment against Allied bombers. It was also hoped that such a fast, yet well-armed, 'destroyer' could avoid or out-perform the increasing numbers of enemy fighter escorts. Eight versions were planned, but it was the B-2 that initiated two prototypes, the first of which was the M(V)13. Powered by DB 603E-1 engines, it flew for the first time on 31 October 1944. The aircraft was pre-armed with two wing-mounted MK 103s, each weapon provided with 70 rounds and resting in extended fairings. To accommodate the cannon, the aircraft's 310-litre wing tanks were removed and replaced by smaller 220-litre tanks in the outer wing sections.

In a further initiative, there were plans to increase the B-series armament even further, to an array of one engine-mounted and two wing-mounted MK 103s as well as two MG 151s in the cowling. If such aircraft had ever taken to the sky in numbers, there is little doubt that their speed and heavy firepower combination would have given the bombers of the USAAF and RAF Bomber Command a battering.

RHEINMETALL-BORSIG MK 108 30 mm *MASCHINENKANONE*

The prime benefit of this weapon, used profusely by the Luftwaffe for close-range anti-bomber work over northwest and southern Europe from early 1942 onwards, lay in its simplicity and economic process of manufacture, the greater part of its components consisting of pressed sheet metal stampings. The gun's operation bore similarities to the Becker-Oerlikon method as used during the closing months of World War I and into the 1930s, and in many ways the MK 108 was considered to have been a masterpiece in weapons engineering, not only saving precious materials but also hundreds of man hours on milling machines and precision grinders.

With the advent of massed American daylight bomber formations bristling with concentrated defensive firepower, the need arose for a long-range, heavy calibre gun with which a German pilot could target specific bombers, expend the least amount of ammunition, score a kill in the shortest possible time and yet stay beyond the range of the defensive guns. It was a virtually impossible requirement, and yet the MK 108 almost achieved this. At first, however, the RLM *Technisches Amt* rejected Rheinmetall's initial proposal on the basis that all German fighter pilots were considered by Udet, the *Generalluftzeugmeister*, to be easily capable of shooting down the heaviest bomber with his preferred 20 mm calibre weapons at extended ranges.

First designed in 1940 by Rheinmetall-Borsig as a private venture, the MK 108 was a blow-back operated, rear-seared, belt-fed cannon, using electric ignition and being charged and triggered by compressed air. Once installed into any aircraft, there was no method of adjustment for harmonisation. One of the most unusual features of the gun was its extremely short barrel, earning it the type-name *Kurzgerät* (lit. 'short apparatus or device') which gave it its low muzzle velocity of between 500–540 m per second, with a maximum rate of fire of 650 rounds per minute. At only half the weight of the MK 103 wing-mounted cannon, two MK 108s represented the same payload, but had a combined rate of fire slightly more than three times that of a single MK 103.

The open breech of the air-cooled, belt-fed, 30 mm MK 108 cannon showing the tray over which an incoming round would pass towards the receiver and the barrel. The cartridge was fired while a heavy bolt housed in the rear breech was moving forward. Clearly seen here is the cannon's pressed metal stamping construction and the wire to which was attached the firing solenoid. (Forsyth)

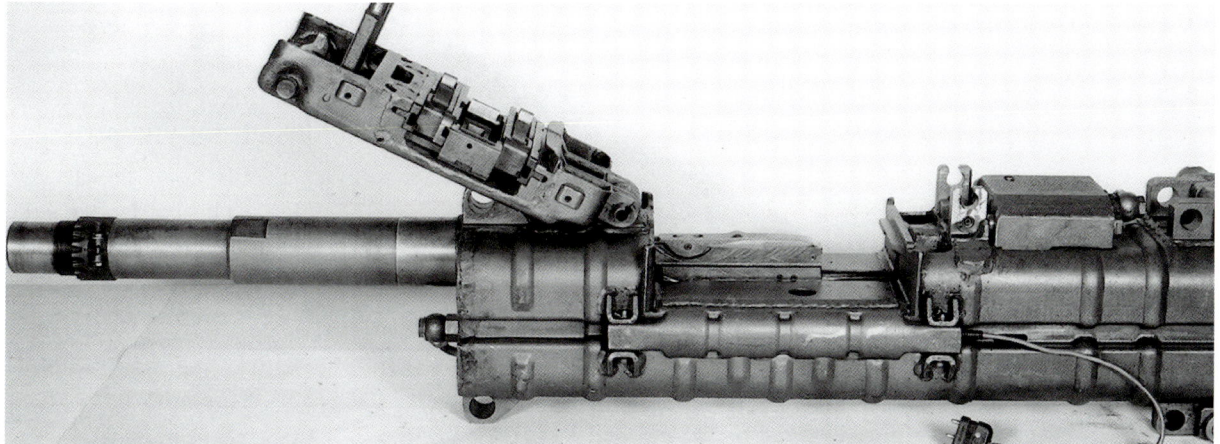

GENERAL ARRANGEMENT DRAWINGS OF THE 30 mm RHEINMETALL-BORSIG MK 108 CANNON

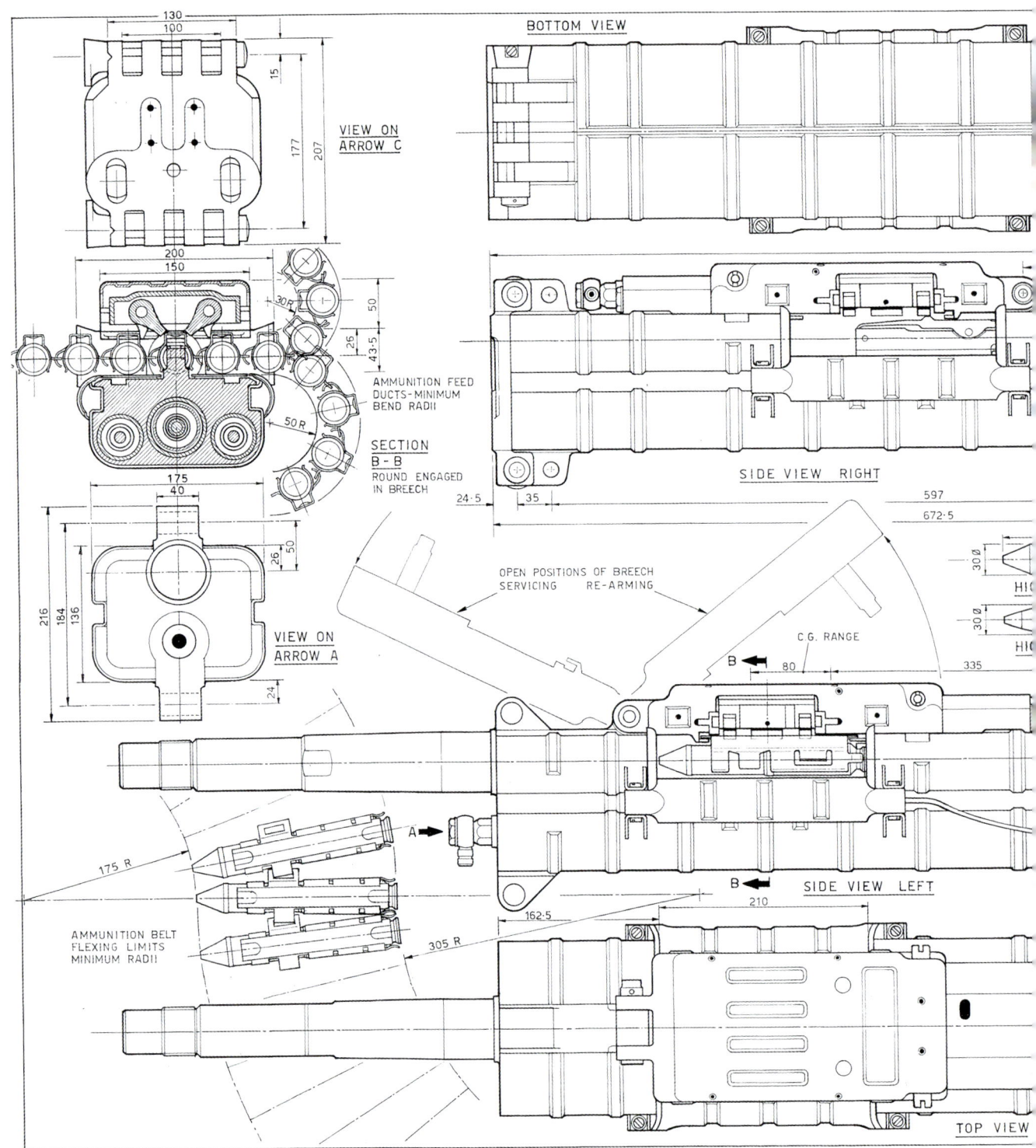

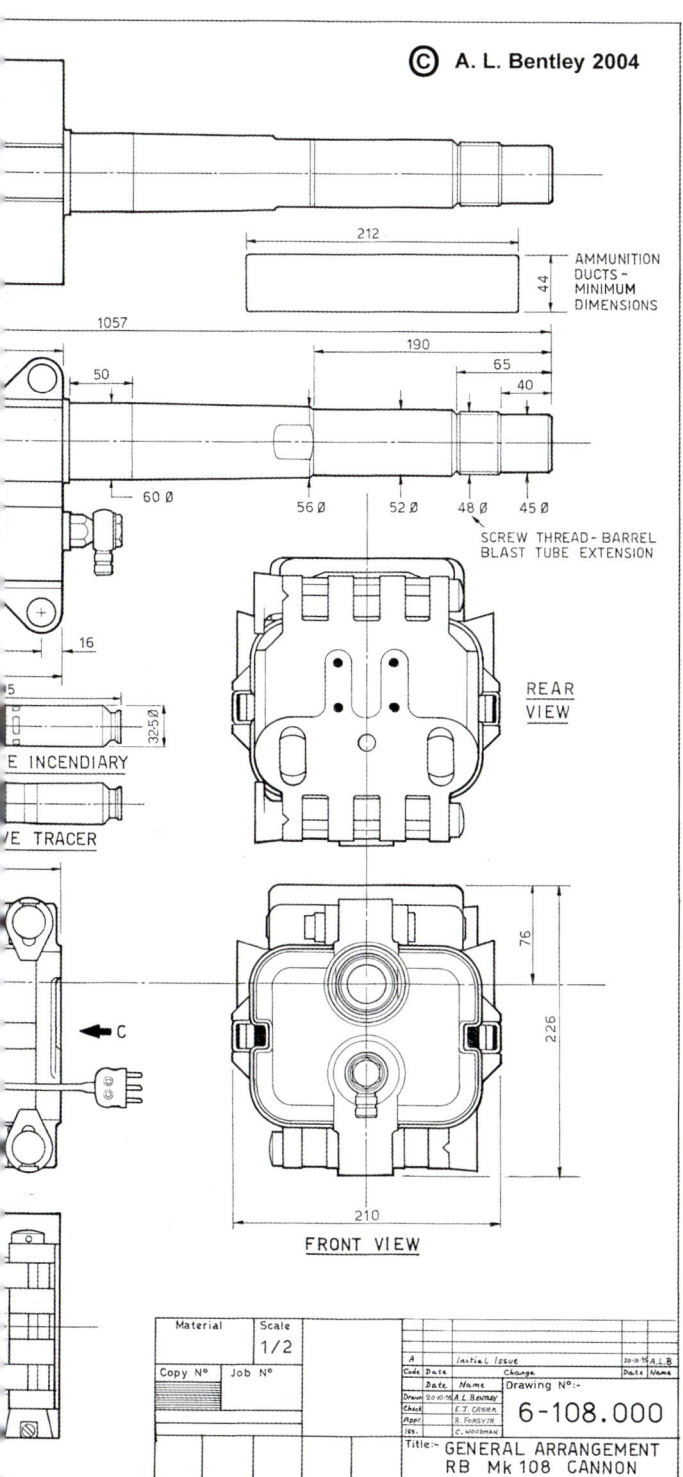

212

AMMUNITION
DUCTS –
MINIMUM
DIMENSIONS

44

1057

190

50

65

40

60 Ø

56 Ø   52 Ø   48 Ø   45 Ø

SCREW THREAD – BARREL
BLAST TUBE EXTENSION

16

32·50

E INCENDIARY

VE TRACER

REAR
VIEW

76

C

226

210

FRONT VIEW

Material	Scale 1/2				
Copy Nº	Job Nº				

A		Initial Issue		20·10·76	A.L.B
Code	Date	Name	Change	Date	Name
Drawn	20·10·76	A.L. Bentley	Drawing Nº:-		
Check		E.J. Croker	**6-108.000**		
Appr		R. Forsyth			
Iss		C. Woodman			

Title:- GENERAL ARRANGEMENT
RB Mk 108 CANNON

A total of 60 rounds were fed by means of a disintegrating belt from an ammunition canister mounted above the gun. On release of the sear, the bolt travelled forward under the action of two driving springs. A projection on top of the bolt then passed through the ring, extracting a round and forcing it into the chamber, before firing the shell while the heavy bolt was still moving forward. After firing, the empty cartridge case re-seated itself in its link. Ejection was performed by means of pawls activated by camming grooves cut into the top of the bolt. The new round then slipped into position using the same process. Neither the barrel nor the receiver moved in recoil, the entire force of the firing process being absorbed by the rearward motion of the bolt against driving springs, which acted as buffers against recoil. Subsequently, no locking mechanism was needed since by the time the fired round had overcome the inertia of the massive bolt, the projectile had left the barrel and the pressure had dropped.

Two basic types of shells could be loaded into the MK 108, namely 30 mm high-explosive M-shells and 30 mm high-explosive incendiary shells.

Oberst Ingenieur Mix described the background to the production process:

Inasmuch as the MK 108 was ordered into production before it was completely finished and before it had been adequately tested, and inasmuch as the ammunition for it was also still in the testing stage, it was extremely difficult to trace certain functional defects apparent during the final testing to their ultimate sources, except in those cases where the reason for failure was clearly evident in the form of damaged parts.

The fact that Rheinmetall-Borsig, whose engineer and workshop personnel were intimately acquainted with the MK 108 and could have carried out whatever modifications might prove necessary and amended their blueprints without delay, was not assigned the task of manufacturing the weapon led to considerable difficulties in the early stages of production in the secondary construction firm, the Deutsche Waffen- und Munitionsfabriken [DWM] in Posen, where plant facilities were designed for mass production. These initial difficulties were overcome to some extent by the assignment of a team from Rheinmetall-Borsig to DWM in Posen. The engineers from Rheinmetall-Borsig were assigned to provide whatever on-the-spot assistance might be needed, and to make certain that the manufacturing specifications in use in the workshops corresponded to the latest stage of development.

The weapon was subsequently integrated into the later variants of the Bf 109. From the summer of 1944, large numbers of Bf 109G-14s built by Wiener Neustädter Flugzeugwerke (WNF), a variant seen as an attempt to standardise following revisions subjected upon the G-6 during production, had their MG 151/20 engine-mounted cannon replaced by an MK 108 with 60 rounds of ammunition to become the Bf 109G-14/U4 sub-variant intended for close-range operations against enemy bombers.

The MK 108 quickly earned a fearsome reputation amongst both German pilots and the Allied bomber crews, the latter dubbing it the 'pneumatic hammer'.

The events of 24 April 1944 bear testimony to the MK 108's destructive power when installed into the Messerschmitt fighter. That day, Hauptmann Hermann Staiger, *Staffelkapitän* of 12./JG 26, claimed his sixth heavy bomber shot down. Staiger had formerly led 7./JG 51 in Russia, where he proved his skill in anti-bomber work when he accounted for the destruction of three SB light bombers in one day on 22 June 1941, with a further four in one day on the 30th of that month, for which he was awarded the Knight's Cross. On 24 April 1944 he led a combined force of aircraft from III./JG 26 and III./JG 3 to attack a formation of 141 B-17s out to bomb Oberpfaffenhofen. Flying a Bf 109 fitted with an MK 108 cannon in the nose, Staiger shot down two B-17s in one minute over Donauwörth. Twenty-five minutes later, he claimed a *Herausschuss* (shooting out a bomber from the protection of its formation) for two more B-17s, before destroying another south of Munich – again all within one minute. Staiger would claim 26 four-engined bombers downed by war's end, making him one of the Luftwaffe's leading specialists in their destruction.

The MK 108 had also been planned as a gondola and nose weapon for the G-6/U4, with the aircraft carrying 60 rounds. However, in spite of its advanced design and incredible destructive power, the cannon began to develop an unfortunate reputation among Bf 109 pilots for being unreliable and prone to jamming.

Nevertheless, from the late autumn of 1944, the same adaptation as the G-14/U4 with the same ammunition capacity took place on WNF-built Bf 109G-10/U4s, as well as with the Bf 109K-4. In the latter, the only 'K' variant to see service before the war's end, an engine-mounted MK 108 was fitted as standard, with 60 rounds of ammunition, along with two MG 131 13 mm machine guns above the engine. It was originally planned for the K-4 to have wooden wings with two integral MK 108 cannon, but after tests in August 1944, this configuration was abandoned.

Not surprisingly, the MK 108 was also fitted into the Fw 190, the fighter favoured by the so-called *Sturm* units that practiced determined close-range tactics deploying massed 'wedges', or *Sturmkeil* formations, of heavily armed and armoured Focke-Wulfs covered by Bf 109s against groups of bombers. The Fw 190A-8/R2 featured two MG 151 cannon installed in the wing roots and two MK 108 cannon in the wings. The cowl-mounted MG 131 machine guns fitted to the standard A-8 were removed to reduce weight, with the empty gun troughs and slots left by their removal covered with armoured plate.

Additionally, a panel of 6 mm armoured glass was mounted on each side of the cockpit canopy, and a sheet of 6 mm armour plate, extending from the lower edge of the cockpit canopy to the wing root, was mounted externally on each side of the fuselage to protect the pilot from lateral fire. Another armoured panel on the underside of the aircraft protected the pilot's seat to a point sufficiently forward to protect the feet and legs.

Within sight of an enemy formation, and some 90–150 m above and 900–1,520 m behind it, the Fw 190s of a *Sturmgruppe* would drop their external tanks, then re-formate from a Vee formation into a line abreast *Angriffsformation* (attack formation) or *Breitkeil*. This was carried out by climbing where necessary and fanning out into a slightly swept-back, line abreast formation of usually more than 20 fighters, either level with, or slightly above, the enemy, with the commander of the *Gruppe* and his deputy flying at its apex. As Major Walther Dahl, the former *Kommodore* of JG 300, recalled for his Allied captors in September 1945:

> Upon sighting the enemy bomber formation, the formation leader gives the signal to attack by rocking his wings, or by radio. The wings of the Vics now pull up until the aircraft are in line abreast, with the formation leader throttling back a bit so the others can catch up. The approach is made from behind and the fighters attack in a line, the formation leader dividing up the target according to the formation of bombers.

Each pilot selected one bomber as his target, closed in to about 360 m and, aiming at the tail gunner, opened fire with his two 20 mm MG 151 wing root cannon. As the *Angriffskeil* closed in to 180 m, each pilot opened fire with his MK 108 cannon, aiming at either of the inboard engines. Should the selected bomber have been damaged or set on fire by a fellow pilot, a *Sturmjäger* would move through the formation and pick another target. Having made their attack, some pilots broke away 45 m from the bomber and side-slipped in the direction of the reassembly area on the basis that this was the quickest and safest way to get clear of defensive fire and debris.

Unteroffizier Oskar Bösch flew MK 108-equipped Fw 190s with *Sturmstaffel* 1 and 11./JG 3. All eight of his confirmed victories in 1944 were *Viermots*, six of them B-17s and two Liberators. He recalled:

> Every day was a struggle to stay alive. We weren't after awards. The best award was to come back at the end of the day. We were outnumbered ten, sometimes twenty-to-one, and we got tired, very tired, but we kept going . . . we had to.

In the beginning, attacking bombers was almost 'easy'. It was exciting. Your adrenalin really pumped.

Everybody had their own tactics, their own tricks, but generally we attacked from behind at about 500 m above the formation, opening fire at 400 m. The air was thin at 7,000 m and often there was turbulence behind the bombers; this sometimes made our approach very difficult. We always went in in line abreast. If you went in singly, all the bombers shot at you with their defensive firepower. You drew fire from the waist gunners. But as an attack *formation*, the psychological effect on the bomber gunners was much greater. First of all you tried to knock out the tail gunner. Then you went for the intersection between wing and fuselage and you just kept at it, watching your hits flare and flare again. It all happened so quick. You gave it all you had. Sometimes, after the first attack, all your energy seemed to go. Your nerves were burnt out.

In the latter half of the war, the MK 108 also became the standard – or intended standard – cannon armament for most of the Luftwaffe's later generation of advanced piston- and jet-engined aircraft such as the Ta 152, Me 163, Me 262, He 162 and Ho 229, as well as the He 219 and Ju 388J nightfighters into which it was planned to incorporate twin MK 108 sets (as the L-108Z) in a *Schräge Musik* (see page 60) installation.

The Me 262A-1a standard jet interceptor was armed with four MK 108s mounted in the nose with a total of 360 rounds. In the relatively brief period that Me 262s were

Standard installation of four 30 mm MK 108 cannon in the nose of an Me 262A-1a jet interceptor. The lower pair of guns was set slightly forward and outboard of the upper pair. (EN Archive)

operational, air combat tactics and formations were fluid, and varied within the two predominant fighter units, JG 7 and *Jagdverband* (JV) 44, as time progressed. It must be stressed that the prime role of the Me 262 was to get to the bombers using their speed (although this in itself meant that getting into an effective attack position at a range of about 1,000 m on a dead-level approach was often challenging) and weight of armament. The latter not only included the MK 108 cannon but also batteries of wing-mounted 55 mm R4M rockets (see page 104) that were to be used specifically against the larger targets of bombers and their formations, although the rockets were not always in plentiful supply.

One pilot to fly the Me 262 in combat was the Knight's Cross-holder Oberleutnant Walter Schuck, *Staffelkapitän* of 3./JG 7, and he experienced the firepower of the MK 108 on several occasions.

On 24 March 1945, just four days after his first flight in an Me 262, Schuck and his wingman engaged an F-5 Lightning reconnaissance aircraft escorted by

Four 30 mm Rheinmetall-Borsig MK 108 cannon were installed into the nose of the Me 262A-1a just above the nosewheel housing. An electric ignition cable can be seen at the rear of the installation, as can the ammunition discharge chutes. While effective when used well, the belt-fed weapons had a slow rate of fire and were prone to jamming. The Me 262 could also be armed with up to 24 55 mm R4M rockets for use against bombers.

two P-51s. It became a draining air battle. Shortly before midday, Schuck had spotted 'three black dots' approximately 120 km southwest of Berlin in the Leipzig-Dresden area. At full speed, the two Me 262s came in behind the three American aircraft and opened fire. Schuck's wingman shot down the F-5, but the Mustang pilots made every attempt to avoid their jet-powered assailants by rolling, swerving and diving. Doggedly, Schuck stayed with them and eventually chose an opportune moment to fire at each. In a formidable example of the power of the Me 262's four MK 108s, Schuck recorded in his memoirs:

> In a fluid, diving curve I positioned myself behind the trailing Mustang of the trio and opened fire. The shells from the four 30 mm cannon smashed with tremendous impact into the enemy machine, which was literally torn apart; the pilot of the Mustang was just able to bail out before it exploded in mid-air.

Schuck then quickly positioned himself onto the tail of the other P-51, whose pilot seemed unaware of the Me 262's presence. Schuck watched as the high-explosive rounds from his MK 108s 'chewed through the right wing of the Mustang'. A moment later the American fighter climbed and then fell back trailing smoke. 'I was hugely impressed by the [Me 262's] devastating firepower,' Schuck recorded.

On 10 April 1945, during a large-scale raid by the USAAF against targets in central Germany, Schuck led seven Me 262s against B-17s over Oranienburg. To avoid the P-51 escort, he brought his formation into attack on a zig-zag course at 10,000 m. With the sight of bombs raining down on the town below, Schuck fired his four MK 108s at a B-17 from 300 m. A wing immediately disintegrated as the German ace flew towards another

Viermot. His second target took hits in its elevator and the crew jumped from the spiralling aircraft. Soon afterwards, two more bombers had exploded under his guns. He had achieved the impressive distinction of shooting down four B-17s within eight minutes. Moments later Schuck was in turn shot down by a P-51, although he was able to bail out.

Adolf Galland had mixed feelings about the MK 108 when installed in the Me 262s of his own JV 44, a small unit comprised mainly of illustrious fighter aces out of favour with the leadership and NCO instructors. JV 44 operated over southern Germany and Austria:

> Firstly, in terms of installation, it was extraordinarily easy to fit four MK 108s into the aircraft. Secondly, it was good to have a gun which solved all our problems; that is to say a gun which had a rapid rate of fire and great destructive effect, although there was the disadvantage of an insufficiently flat trajectory. Then there were other snags; the guns were not much good when you were banking because the centrifugal forces arising from banking ripped the belts. Of course, they didn't rip if you didn't fire! But these teething troubles were easily sorted out by a well-trained groundcrew.

One Me 262, Wk-Nr 112355, was fitted with six MK 108s to become probably the only A-1a/U5 prototype. An extension to the nose of the fighter accommodated the additional cannon. In March 1945 the aircraft was assigned to the Me 262 operational training *Gruppe*, III. *Ergänzungs-Jagdgeschwader* (EJG) 2, at Lechfeld, but when its *Kommandeur*, Oberstleutnant Heinz Bär, was transferred to JV 44 in late April, he took the jet with him to the unit's base at Munich-Riem. It is believed that Bär flew the Me 262A-1a/U5 operationally on 27 April when he engaged USAAF fighters, and shot at least one down.

RHEINMETALL-BORSIG MK 112, MK 114 AND MK 115 55 mm MASCHINENKANONE

Meanwhile, despite the introduction and increasing use of 30 mm weapons, RLM and Luftwaffe ballistics experts recognised that it took a steady stream of *direct* hits by shells of that calibre to bring down a heavy bomber. Despite the efforts of the *Sturmgruppen*, according to Oberst Ingenieur Mix:

> Even the introduction of special offensive tactics – e.g., attacking from the front and aiming at particularly vulnerable points (engines and unprotected fuel tanks, etc.) – had no more than a temporary influence on this unsatisfactory state of affairs. Thus, the development department at the RLM was faced with the task of increasing calibre once more.

Experiments were carried out to determine how much hexogen-aluminium explosive was needed to destroy a large bomber, after which a projectile was to be developed for the required amount of explosive. The blowback effect of the projectile was to be so powerful

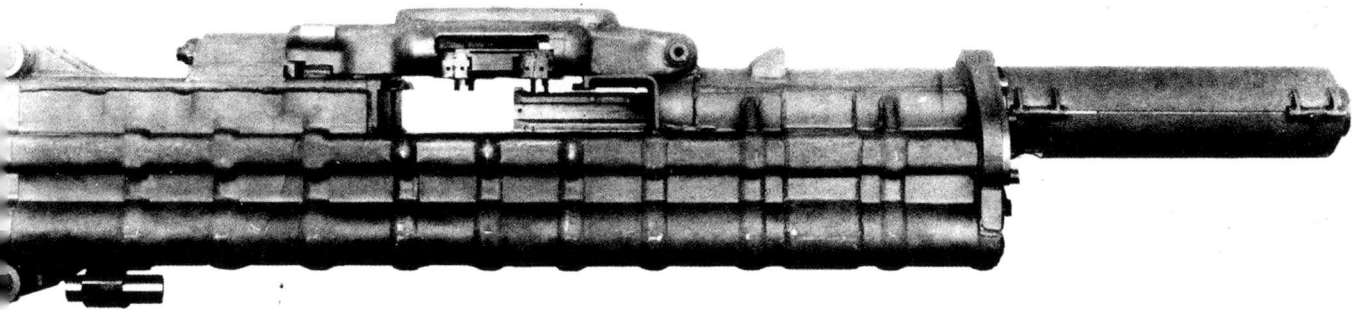

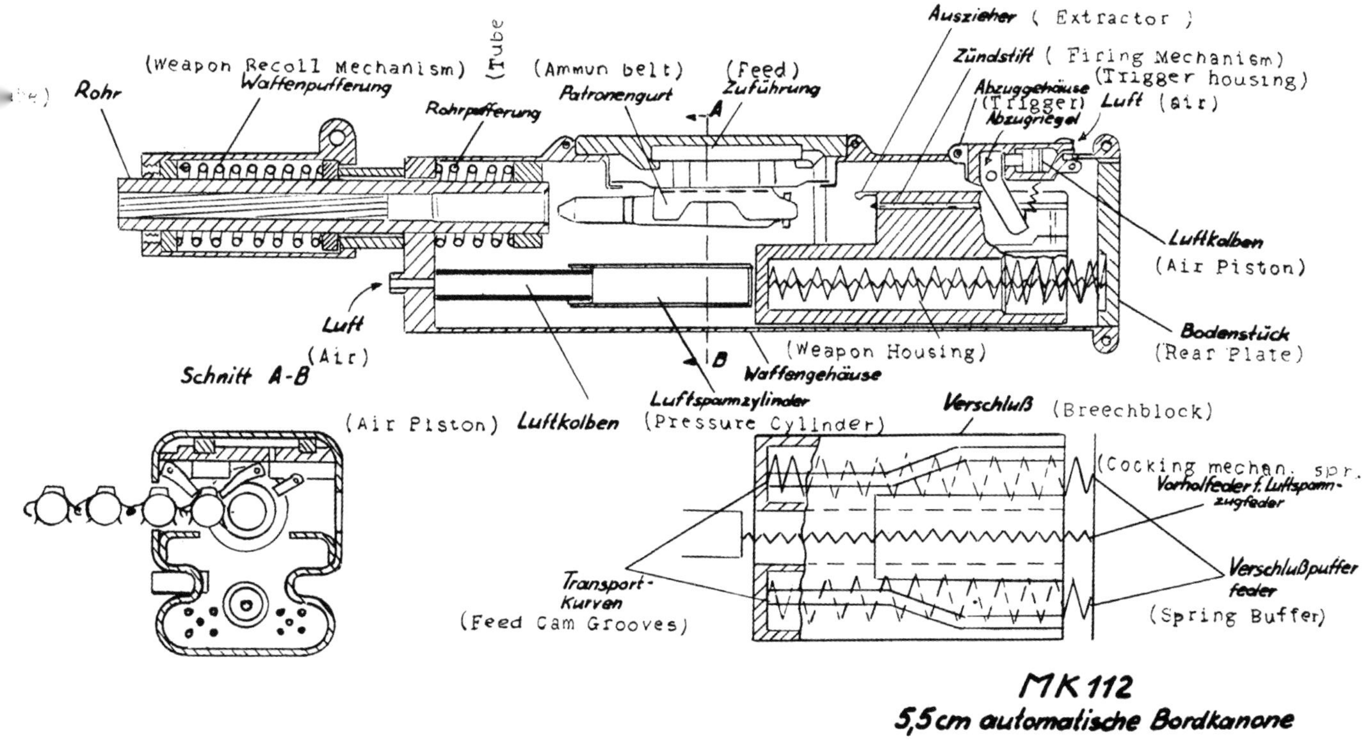

Schnitt A-B

(Air Piston) **Luftkolben**

(Feed Cam Grooves)
Transport-Kurven

Rohr (Tube)   (Weapon Recoll Mechanism) **Waffenpufferung**   **Rohrpufferung**   *(Ammun belt)* **Patronengurt**   *(Feed)* **Zuführung**   *Auszieher* (Extractor)   *Zündstift* (Firing Mechanism)   (Trigger housing) **Abzuggehäuse** (Trigger) **Abzugriegel**   **Luft** (air)   **Luftkolben** (Air Piston)   **Bodenstück** (Rear Plate)

Luft (Air)

(Weapon Housing) **Waffengehäuse**

Luftspannzylinder (Pressure Cylinder)

Verschluß (Breechblock)

(Cocking mechan. spr.) **Vorholfeder f. Luftspann-zugfeder**

Verschlußpuffer feder (Spring Buffer)

MK 112
5,5cm automatische Bordkanone

that only one direct hit would be necessary to guarantee the destruction of a four-engined bomber. The experiments revealed that 420–450 g of explosive were needed to inflict sufficient damage to the fuselage or wings of a large bomber to ensure its destruction, and to produce this, it would be necessary to develop a cartridge of 55 mm.

Subsequently, the RLM issued a requirement for a light-weight, automatic 55 mm aircraft cannon for use up to a range of 1,000 m, with the highest possible rate of fire, simplicity of manufacture, and which could be installed into aircraft with ease. The WKW *Gruppe* at Rheinmetall-Borsig offered the MK 112 and MK 114 machine cannon and the MK 115 recoilless cannon.

Two metres in length, the MK 112 was a belt-fed, recoil-operated weapon with a mass locking breech similar to the MK 108, on which the design was modelled and intended as

Schematic of the belt-fed, recoil-operated Rheinmetall-Borsig 55 mm MK 112 *Maschinenkanone* which bore similarity in construction and design to the MK 108. The weapon proved susceptible to cold weather and was plagued by a problematic electric ignition system. (Forsyth)

a scaled-up version. It had a rate of fire of 300 rounds per minute, it weighed 275 kg, fired a 22.2 cm-long M-shell with a 420 g explosive charge weighing 1.48 kg and had a muzzle velocity of 600 m per second.

An initial batch of ten weapons was manufactured for test purposes, with five more examples held back for further development and improvement based on the test results. These examples had been made with pressed metal housings and heavy, recoil-operated breech-blocks. Extended firing tests were conducted, with bursts of fire up to 20 rounds. However, it is believed that attempts to reduce weight proved challenging, as did cold weather tests conducted at –14.5°C, which revealed weakness in the use of plain carbon steels. It thus became necessary to redesign many parts for use in cold temperatures. Additionally, difficulties were encountered with the electric ignition system and the connections between the firing lever and the ignition bolt in early tests.

It was planned to fit the MK 112 into the Me 262 (including as a combination with the MK 108) and the Do 335.

The heavier, long-range, long-barrelled, gas-operated, electro-pneumatic 55 mm MK 114 came about following a specification issued by the RLM, and it was designed to fire a standard cartridge intended for use by the Flak arm and the Kriegsmarine. The weapon measured 4.2 m in length and was not to weigh more than 1,000 kg, a condition to which it conformed at 700 kg. It had a rate of fire of 150 rounds per minute. The MK 114's 5.3-kg M-shell had an explosive charge of 450 g, and the weapon had a muzzle velocity of 1,050 m per second.

The manufacturing process proved complex, however, with Rheinmetall-Borsig coming up with three different sub-designs, and the use of non-alloyed steel had an adverse effect on design and build quality. Only a small number of guns were completed, with most intended for installation in the Me 262, before the RLM ordered all work to be cancelled in the autumn of 1944.

Also of 55 mm calibre was the MK 115 'jet' cannon, which originated in an order of early 1943 from the *Heereswaffenamt* and the RLM for Oberingenieur Theodor Rakula and his team at Rheinmetall-Borsig to explore the feasibility of a completely recoilless, belt-fed automatic cannon. To achieve no recoil meant that the forces generated by an exploding cartridge had to be equalised. Rheinmetall-Borsig engineers duly devised a method using a short metal base cap to which was secured a heavy, combustible, nitrated paper cartridge case which contained the projectile and the primer – something entirely new in the field of

Rheinmetall-Borsig developed the 3.3 m-long 55 mm MK 115 'jet' cannon with a view to creating a recoilless, belt-fed automatic aircraft weapon. This view shows the feed opening and receiver. Only one example is believed to have been completed. (Forsyth)

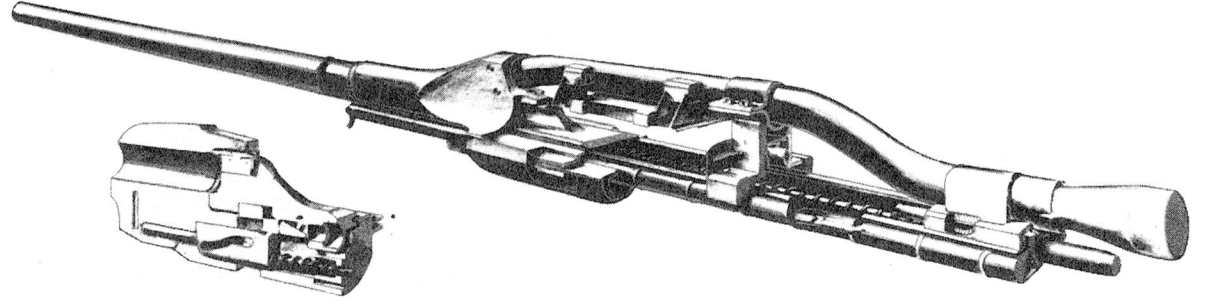

air armament design at the time. When the cartridge was fired, unlike a brass or steel case holding the powder gases, the paper burned, allowing the gases to vent out through a tube behind the weapon and thus equalise the recoil.

The feed opening was located on the top of the receiver on either the left or right-hand side, and the rear portion of the shell case was pulled back into the link and ejected from the side opposite that from which it had been fed.

According to Oberst Ingenieur Mix:

> In my opinion, the basic principle of the jet cannon represents a highly elegant and promising solution to the problem of large-calibre airborne cannon. To be sure, this principle necessitates certain limitations as far as installation is concerned, inasmuch as there has to be a way for flare-back gases and cartridge bits to escape unhindered. In addition, more powder is required, which means that the ammunition weighs approximately twice as much as a normal cartridge of the same capacity. The system of the recoilless barrel, however, has the great advantage that the gun mount can be designed without any regard for recoil reaction. Thus, even a 55 mm weapon of this design would not have been too heavy to preclude installation in the wings of an aircraft.
>
> Provided that a workable jet cannon can be developed for a combat range of between 1,000 and 2,000 m, it seems likely that weapons of this type must be given continued consideration because of their better ballistic characteristics, unless rockets with comparable firing accuracy can be developed.
>
> The MK 115 met the requirements established for the lightweight 55 mm airborne cannon, with one exception – the gas jet had to escape unhindered from the rear.

Just one example of the MK 115 was built for test purposes, which saw the firing of single shots in order to assess the mechanism and a correct nozzle size so that recoil forces could be eliminated. However, the gun's 2.79 kg projectiles with their 0.42 kg explosive charges had a habit of loosening inside their heavy paper cases.

The weapon was 3.3 m in length, weighed 180 kg and had a rate of fire of 300 rounds per minute. It had a muzzle velocity of 600 m per second.

By war's end, development of the MK 115 had reached an advanced stage, but tests had not progressed as far as ascertaining the effects of exhaust blast on an airframe.

MAUSER MK 214 50 mm *MASCHINENKANONE*

Intended for fitment to the Ju 88 and Me 410 as a long-range anti-bomber weapon, the 50 mm MK 214 *Maschinenkanone* originated in a request from the RLM to the Mauser Werke in July 1943. Design and development work went ahead under Ingenieur Nitschke but, ultimately, this long-barrelled cannon was installed for trials in just one Ju 88 and a pair of Me 262A-1s. There were plans for it to be fitted to the Heinkel He 162 *Volksjäger* and the proposed Dornier Do 252/3 advanced nightfighter project, which was to be powered by a pair of Jumo 213J 12-cylinder engines. The Dornier was to have fielded a formidable array of armament comprising two fuselage-mounted 30 mm MK 108 cannon, two further such cannon in the nose and an MK 214 mounted on either side of the lower fuselage.

In a Messerschmitt *Protokoll* of 9 February 1944, an armament option was proposed to equip the Me 262 jet interceptor, then under development, with a single, large-calibre cannon. Two weapons were suggested, the MK 214 and the shorter-range, but faster-firing, MK 112.

Work proceeded slowly, and it was not until New Year's Day 1945 that no less a figure than Adolf Hitler championed the use of a 50 mm gun in the Me 262 when he discussed armament with Hermann Göring. He told the Reichsmarschall:

We must introduce an effective long-range weapon. We must be able to get at the American bombers with a weapon which is so effective at a distance so great that they can't reply. Each hit with a 5 cm projectile will bring down a bomber without question. The bombers are our curse. In my opinion the 5 cm cannon is the least we can use today.

Göring concurred with the *Führer*:

With the cannon, the Me 262 will be able to retain its high speed. I think we shall have success with it.

At an armaments conference a few days later, Hitler ordered the immediate installation of the MK 214 in the Me 262A-1a. The following month discussions were held between representatives from Messerschmitt and the *Erprobungsstelle* at Tarnewitz, as a result of which it was proposed to remove the standard four nose-mounted MK 108s from the jet interceptor and replace them with the breech of the MK 214, with its barrel projecting some two metres forward.

The belt-fed MK 214 was to have a muzzle velocity of 900–1,000 m per second. The repeating mechanism was actuated by a light, automatic part worked by the recoil of the gun. For production reasons the barrel breech and recoil mechanism from the standard 50 mm *Kampfwagenkanone* (KwK) 39 L/60 tank gun were used for trials. The trials were successful, and thus calibre for the planned MK 214 was retained at 50 mm so that existing supplies of KwK 39 components could be used.

Loaded with a 1.8-kg M-shell projectile carrying about 450 g of high-explosive, the cannon would have a muzzle velocity of approximately 900 m per second. The total weight of the shell was 7 kg and recoil was about 4,000 kg. The total built weight, including ammunition, was circa 1,000 kg.

In the Me 262, installation of the MK 214 necessitated a redesign of the nosewheel gear so that it swivelled during retraction to lie flat in the nose. The variant was designated the

Me 262A-1a/U4, and modification of the first sub-variant prototype, Wk-Nr 111899, followed quickly to which the second prototype A-series cannon was fitted.

The aircraft was flown in its U4 configuration for the first time on 19 March 1945, with the Messerschmitt test pilot Karl Baur at the controls. Baur would fly the aircraft on a further 18 occasions, during which he successfully fired 81 rounds while airborne, in addition to the 47 rounds he had fired on the ground. From that point, however, all trials stopped as a result of malfunction and breakage.

In April a second Me 262, Wk-Nr 170083, which had trialled various undercarriage modifications at Lechfeld, was fitted with the third A-series prototype MK 214, but it is not believed to have flown with the cannon mounted before Lechfeld was captured by the Americans.

Meanwhile, following tests at Lechfeld, Wk-Nr 111899 was handed over to the Luftwaffe for service evaluation, this task falling to the diminutive Major Wilhelm Herget. The Luftwaffe's fourth highest-scoring nightfighter ace, he had been appointed *Gruppenkommandeur* of I./NJG 4, equipped with Bf 110s, Do 217s and Ju 88s, in 1942. By June of the following year, when the RAF night offensive against the Ruhr and Hamburg was at its zenith, Herget was awarded the Knight's Cross upon the occasion of his 30th victory. His greatest success came on the night of 20 December 1943, when he downed eight bombers in 50 minutes – one of his victims was claimed using only four rounds of ammunition. The Oak Leaves to the Knight's Cross were awarded by

Me 262A-1a/U4 Wk-Nr 111899 was fitted with a 50 mm MK 214 cannon in its nose. Messerschmitt test pilot Karl Baur is seen in the centre of the three men gathered at the aircraft's side at Lechfeld. Baur flew the aircraft on 19 occasions before Major Willi Herget took over tests for operational assessment. A second Me 262 was also fitted with the weapon for trials. (EN Archive)

THE 50 mm MAUSER MK 214A CANNON

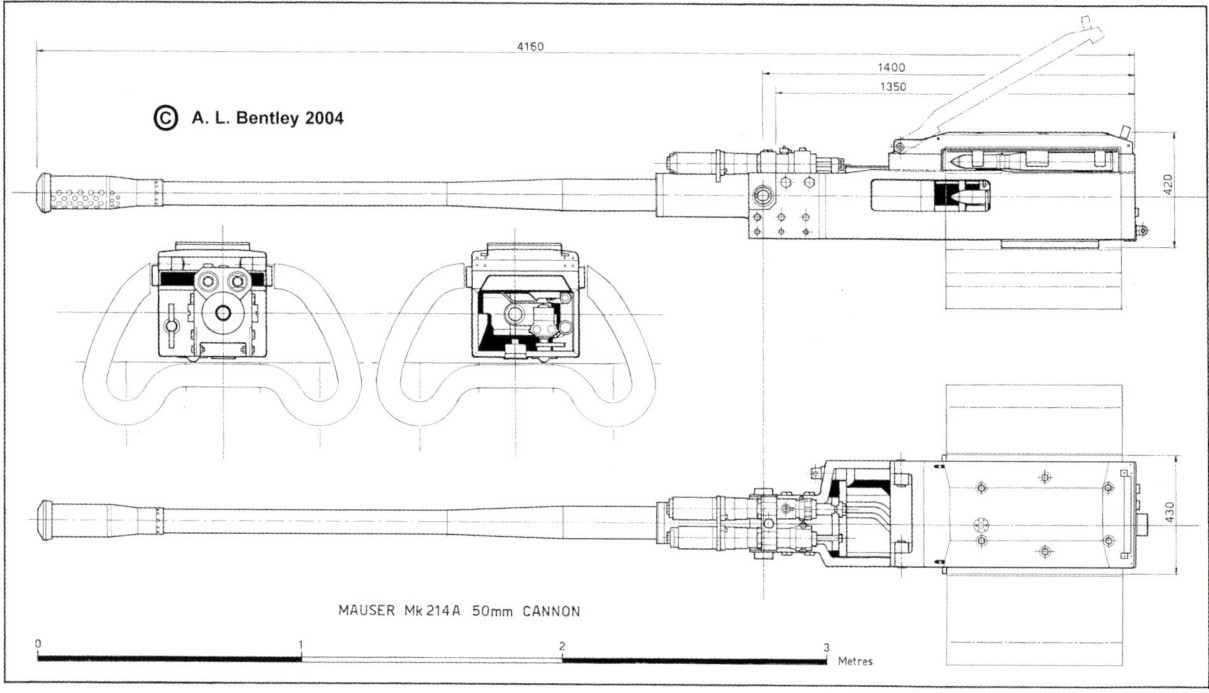

MAUSER Mk 214A 50mm CANNON

© A. L. Bentley 2004

MK 214 Firing Process

The shot is released by initiating the electric ignition current because the weapon is fitted with electric cap ignition. The barrel and the breech recoils and the excessive recoil energy which is not required for actuating the loading mechanism is absorbed by a recoil brake (liquid-filled). During the recoil movement of the barrel, the belt is moved one pitch and the next cartridge is gripped by the claw and moved into the loading tray. Simultaneously, the breech is opened and the empty cartridge ejected. The opened breech is kept in position by claws on the ejector.

The barrel returns to its forward position and during the first part of this movement the loading tray moves behind the barrel and the cartridge is placed in the chamber by a guide combined with the loading tray. During the later part of the barrel movement, the loading tray returns to its original position and after the cartridge has reached the chamber, the breech closes. In the ignition circuit are two contacts, one of which is actuated by the breech wedge

and the other one by the loading tray. The ignition of the cap can therefore only take place when both the breech and the loading tray are in their appropriate position.

In order to load the weapon, it is necessary to introduce the belt by hand and pull on it sufficiently so that the first cartridge detaches itself from the belt and is gripped by the claw. A second cartridge has to be placed by hand into the chamber. If misfire occurs then it is necessary to open the breech by hand and to take the misfiring cartridge from the chamber and replace it with a new one, because no automatic reloading device is provided. The breech is opened by means of a lever which is found on the left-hand side of the weapon.

(Extract from CIOS Target No.2/24: *Report on Visit to Mauser-Werke AG, Oberndorf am Neckar and Mauser personnel at Lager Haiming, Otztal, near Innsbruck*)

Hitler during a much-publicised ceremony on the Obersalzberg in April 1944 to mark his 63rd victory.

On 2 January 1945, Herget had been transferred away from operational duties and attached to a *Sonderkommission* (Special Commission) charged with eliminating bottlenecks in the test-flying and delivery schedules of newly manufactured aircraft. For his part, Herget was appointed head of a mission dealing with the flight-testing of the Me 262, but while touring jet aircraft production and testing facilities in southern Germany he was shocked to discover the extent to which the regional armaments industry was dependent on slave labour. When he complained to local government and SS officials about the conditions in which the workers had to operate, word of Herget's protests got back to Göring and he was severely reprimanded for interfering in matters 'in which he had no concern'. Herget was also forbidden to set foot in any Messerschmitt factory again, with the additional implication that he would be far better off employed on operational duties.

Deeply disillusioned, Herget duly returned to what he was best at, arriving at the Lechfeld test centre on 5 April 1945 to carry out trials with the Me 262A-1a/U4. Taking over testing from Baur once Wk-Nr 111899 had been repaired, he was able to fire another six rounds with the MK 214 against ground targets. Herget made two further flights on 16 April against American bomber formations, but on both occasions the gun jammed yet again. It is likely he then flew the Me 262 to Munich-Riem, since documentary, eye-witness and photographic evidence indicates its presence there later in the month. Riem was also the base of Generalleutnant Galland's JV 44, a unit of Me 262s flown by a mix of 'outcast' fighter aces and instructors which Herget now joined.

During the afternoon of 25 April JV 44 prepared 13 Me 262s in two formations – one to carry out free-ranging patrols, the other to take on B-26 Marauders of the USAAF's Ninth Air Force heading to Erding airfield and a neighbouring ammunition dump. One of these latter Me 262s was the A-1a/U4.

Just before 1750 hrs a single Me 262 was spotted by B-26 crews of the 323rd Bombardment Group (BG) shadowing their formation, way out to the right, just after they had turned away from the target. It is probable that this was the Me 262A-1a/U4. The jet firstly circled around the third box of Marauders to make an attack from behind, closing in to about 500 yards but without opening fire, before diving away to the right. Capt John O. Moench was flying as Group Lead with the 323rd BG, and he remembered:

Suddenly, the intercom came to life with the call out of fighters taking off from the Erding Airdrome below. Almost instantly someone called out a fighter at one o'clock. I looked up, and well out in front of us, swinging around for what looked like a frontal attack, was an Me 262. As the enemy pilot turned, the 50 mm cannon sticking out of the nose of the Me 262 had the appearance of a giant telephone pole. Seconds later, the Me 262 had passed well over the formation without firing a shot, and well

The nightfighter ace Major Wilhelm Herget flew the Me 262A-1a/U4 fitted with a Mauser 50 mm MK 214 cannon, a weapon in which Hitler placed considerable, but misguided faith. It was used in operations by Generalleutnant Adolf Galland's Me 262-equipped JV 44. (EN Archive)

out of range of our 0.50cals. Then we spotted him swinging around in a wide circle, seemingly to get into position to make another frontal pass. Again, he remained out of firing range and disappeared.

Almost instantly, there was a call from the tail gunner that an Me 262 was off the rear. Crossing from the right, the jet swung around and came back from the left. Nineteen of the gunners opened up on the Me 262 at maximum range and, apparently discouraged by the barrage of fire from the Marauders, the pilot broke off.

Once again it seems the 50 mm cannon in the A-1a/U4 had jammed or failed in some way – just as it had done when Herget had attempted to use it earlier in the month. This was the last known occasion that the Me 262 with an MK 214 was used in combat.

With a single MK 214 consisting of around 390 individual parts, in reality the weapon was too complex and costly to be installed in any practical sense in an aircraft. Adolf Hitler's intentions came to naught.

SCHRÄGE MUSIK

In the summer of 1918, Leutnant Gerhard Fieseler, a fighter pilot serving with *Jasta* 38 on the Macedonian front, came up with an inventive idea to supplement the armament on his Fokker D VII fighter. At the time, German machine guns were heavy and unwieldy, and it was a difficult process to remount them, load a new belt of ammunition and make them ready to fire again. After having shot down a Bréguet reconnaissance aircraft, Fieseler, who would later gain fame as an aircraft designer, had one of the enemy's rear-mounted Lewis machine guns removed and mounted in the upper wing indentation of his own aircraft, inclined at an angle of 45 degrees and pointing upwards.

This allowed him to make a firing pass beneath and behind an opponent, and in doing so he stood a good chance of being out of view of his target and able to fire at the enemy machine's most vulnerable area. Fieseler, who would end the war with 19 confirmed victories, enjoyed success with his new weapon, and it was found to rarely miss its target. Indeed, his fellow pilots coined the verb *fieserlieren* ('to fieseler') when referring to his method of shooting at an enemy aircraft from below.

Further development of such a weapon was curtailed by the end of the war, and the concept was largely forgotten until 1938 when Fieseler's former squadron commander in *Jasta* 38, eight-victory ace Fritz Thiede, reminded technical officers in the RLM of the 'below the tail' approach, and how it could be used by a nightfighter when attacking an enemy bomber. Despite the fact that the Japanese had also advised the RLM of their own experiments with such a method, the suggestion generated no interest.

Then, in August 1941, the idea resurfaced when Leutnant Rudolf Schönert, a nightfighter pilot with 4./NJG 1, wrote a report for Generalmajor Josef Kammhuber who, as commander of the 1. *Nachtjagddivision*, bore responsibility for the development, organisation and functioning of the Luftwaffe's still emergent nightfighter arm. Schönert proposed that a single, vertically mounted machine gun could be built into the fuselage of a Do 17 which a pilot could then fly horizontally, below a bomber, with relative ease and out of view of defensive gunners (before the introduction of ventral turrets). But Kammhuber rejected the proposal, having consulted with Oberleutnant Helmut Lent, at the time Schönert's *Staffelkapitän*, and Hauptmann Werner Streib, his *Gruppenkommandeur* in I./NJG 1.

Again the idea lay dormant, but Schönert persisted with his experiments, installing upward-firing 7.9 mm MG 15 machine guns into a Do 17Z-10.

A year later, in the summer of 1942, the *Erprobungsstelle* at Tarnewitz conducted experiments with vertically mounted weapons in a Do 17 and Bf 110, from which they were fired at towed drogues. A negative aspect to these attempts, however, was that a pilot had to crane his head back when aiming a weapon firing at 90 degrees, meaning that his sense of balance and orientation was affected.

Also, around the same time, and following the award of the Knight's Cross in recognition of his 22 victories, Schönert once more approached Kammhuber on the subject of such armament. On this occasion Kammhuber's response was more favourable, and he authorised the installation of vertical weapons in three Do 217Js, at least one of which was so fitted and tested.

From the tests, in the opinion of Oberstleutnant Viktor von Loßberg, a senior staff officer with the *Chef* of the *Technisches Amt* with responsibility for overseeing nightfighter developments, vertically mounted armament could only be effective if a nightfighter and its target were on the same course. Loßberg therefore ordered that instead of being vertical, the guns should be mounted obliquely at various angles between 65 and 70 degrees, so that a targeted aircraft turning at eight degrees per second could be kept in the sights. The results of these tests were favourable enough for three more Do 217s to be fitted with either four or six obliquely mounted 20 mm MG 151 cannon. In early 1943 field tests commenced with 3./NJG 3.

Meanwhile, at the beginning of December 1942, Hauptmann Schönert had been appointed *Kommandeur* of II./NJG 5, equipped with Bf 110s and Do 217s at Parchim, where one of the 5. *Staffel*'s armourers was Oberfeldwebel Paul Mahle. The resourceful Mahle had seen one of the early 'vertical' configurations in a Do 217 at Tarnewitz, and he believed that a similar installation could be fitted into a Bf 110. Utilising parts and equipment that came to hand from stores and workshops, he was able to fit two 20 mm MG FF cannon into the cockpit of a Messerschmitt, with the barrels protruding through the canopy glazing. Schönert then used this aircraft to claim a kill over Berlin in May 1943.

The experimental Do 217s of 3./NJG 3 were also proving successful, and this prompted the development of an oblique armament installation, the R22, intended specifically for the Do 217 and Ju 88C-6 nightfighters.

Rudolf Schönert is seen here as a Major and *Kommodore* of NJG 5 with the Oak Leaves to the Knight's Cross with which he was decorated on 11 April 1944 in recognition of 60 victories claimed at night. Schönert was the driving force behind the Luftwaffe's adoption of obliquely mounted cannon. (EN Archive)

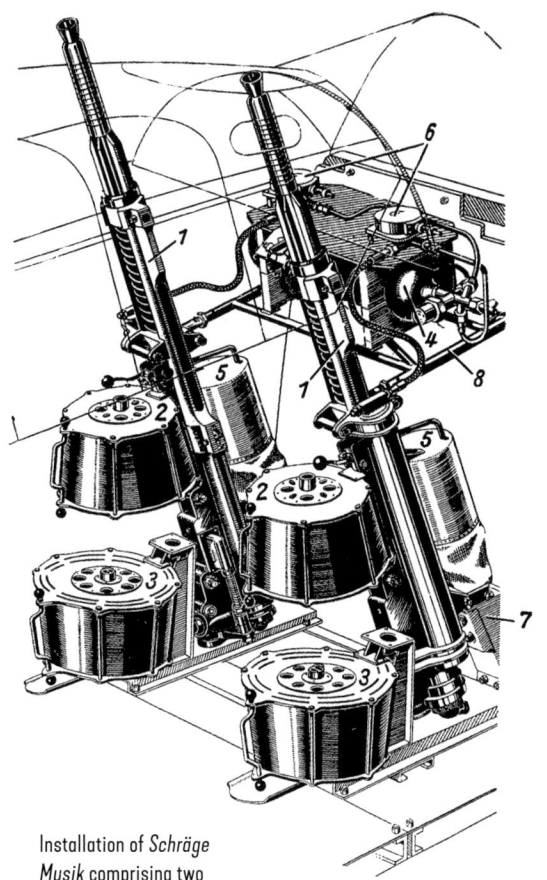

Installation of *Schräge Musik* comprising two 20 mm Oerlikon MG FF cannon in the rear of a Bf 110G-4. Each cannon has a main and reserve spring-driven ammunition drum, while the cylindrical drums numbered 5 in this drawing are the spent case containers. The frame at right beneath the pressure reduction gear acts as the recoil dampener. (EN Archive)

What became known, somewhat oddly, as *Schräge Musik* ('Jazz-' or 'Slanting Music') was not liked universally by the Luftwaffe's nightfighter crews. Some veteran pilots preferred using fixed, forward-firing armament on which they had become proficient. Thus *Schräge Musik* was a weapon which found favour with younger crews who, when using conventional guns, would frequently fire from beyond maximum range, thus failing to strike, betraying their presence and allowing an enemy aircraft to take evasive action. Yet to remain with an enemy aircraft that was diving or weaving evasively was very challenging.

Despite this, from late 1943 the nightfighter schools placed emphasis on instruction using obliquely mounted guns. Trainee pilots were told that when attacking a bomber from below, to aim for the area between the engines and wing roots in order to set fuel tanks on fire. Gun types and calibres varied with aircraft types. Aside from the most common twin 20 mm MG FF installation, as fitted into the Bf 110G-4, the Ju 88G used two 20 mm MG 151/20, while it was proposed to fit four such weapons in the Do 217N. A pair of heavier calibre 30 mm MK 108s were fitted in examples of the He 219A-series nightfighter, and proposed for the Ju 388J, Ta 154 and Me 262B-2.

According to ignition specialist Diplom-Ingenieur Johannes Linke at Rheinmetall-Borsig at Sömmerda:

In an attack from below, the target area of a four-engined bomber is six times greater than the area confronted during an attack from behind. Experiments intended to assess the possibility of attacking against the larger area were carried out using MK 108s. The guns were installed at an elevation angle of 70 degrees in the direction of flight and with a training range of ± 10 degrees in the plane of symmetry of the aircraft and of ± 15 degrees vertically on it.

As a tactic, the fighter attacked by underflying the bomber on a parallel course at a distance of about 800 m and at a speed greater than the bomber. Speed was an important factor in the combat so as to expose the fighter for the shortest possible time to the defensive fire of the bomber. On account of the surplus in speed and the limited training range of 20 degrees, the duration of the combat is very short. With a speed difference of 200 km/h, the attack lasts six seconds. Experiments show, however, that within this six-second period, it is quite possible to obtain the number of hits to ensure the shooting down of the bomber.

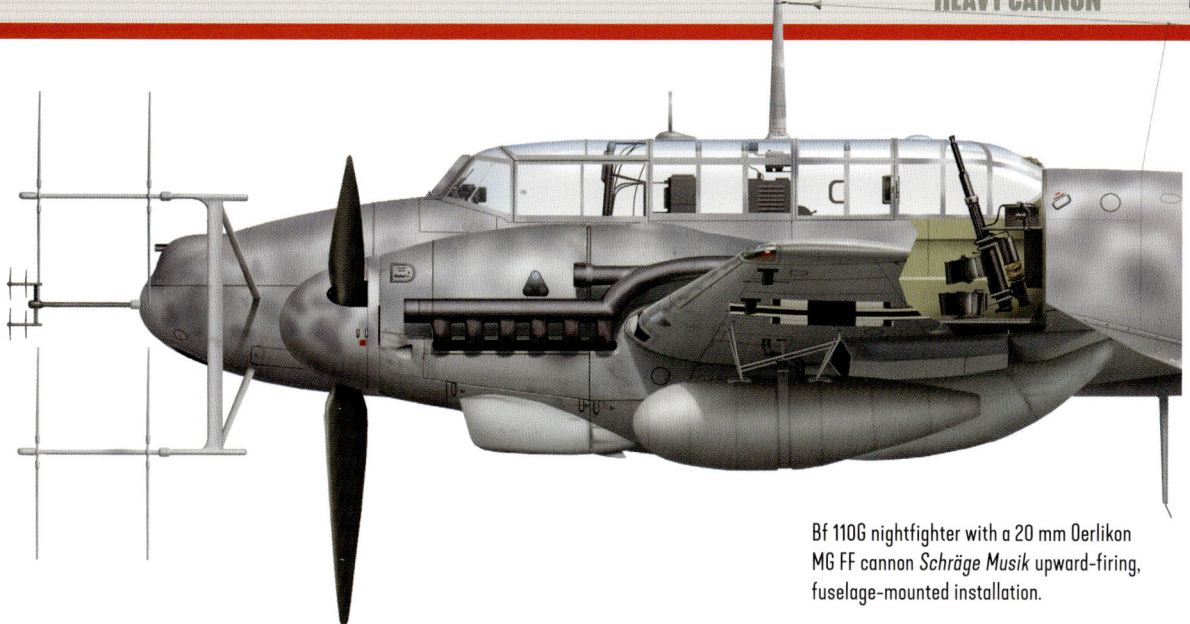

Bf 110G nightfighter with a 20 mm Oerlikon
MG FF cannon *Schräge Musik* upward-firing,
fuselage-mounted installation.

Usual tactics involved a pilot positioning himself directly beneath a bomber at about 50 to 100 m out, before aiming, using an optical sight mounted in the canopy. He would then fire a short burst of non-tracer ammunition in order to conceal his position. Aces such as Heinz Schnaufer of NJGs 1 and 4 (121 night victories), Paul Zorner of NJGs 3 and 5 (59), and Georg Greiner of NJG 1 (47) had notable success using *Schräge Musik* in action against bombers of the RAF.

A British Air Intelligence report of November 1944 based on extensive PoW interrogations noted that:

> The extent to which either type of armament [fixed, forward-firing or upward-firing] is used is generally decided by *Geschwader* or *Gruppe* commanders themselves, and depends largely upon their individual opinions and prejudices.

A Ju 88G of *Stab* or I./NJG 102 fitted with what appears to be a 20 mm MG 151/20 obliquely mounted *Schräge Musik* twin-cannon installation halfway along the fuselage. The aircraft also has a rear-facing 13 mm MG 131 machine gun for defence. On the nose can be seen one dipole antenna for the Telefunken FuG 220 SN-2 *Lichtenstein* AI radar array. (EN Archive)

In June 1943, in another development for obliquely mounted guns, it was proposed to install between six and eight vertical MG 151s into the fuselage of a Do 217 immediately aft of the crew accommodation. The aircraft would also carry *Lichtenstein* air intercept (AI) radar for detecting targets and an upward-facing infrared searchlight for illumination. The Dornier would approach the enemy bomber from 300 m and below, and the searchlight would capture the target to display it on a special sight. This meant an attack could be carried out at a greater range and without the risk of falling debris.

It is believed that in the spring of 1944, trials were carried out at the *E-Stelle* Werneuchen to fit three pairs of upward-firing MG 151s in the Ju 88 angled at about 60 degrees, with proposals for a similar installation in the Ju 188. Each pair of cannon was supplied from specially designed ammunition boxes about a metre in length and 60 cm in height, which were divided at the centre by a partition, with each half holding 60 rounds for each gun.

Some attacks were made using *Schräge Musik* in conjunction with the Telefunken FuG 350 *Naxos Z* detector homing onto H2S navigational and ground-scanning radar emissions from suitably equipped RAF aircraft. In a lecture to crews of I./NJG 4 in November 1944, an officer from the test unit *Nachtjagdgruppe* 10 at Werneuchen said that he had shot down two enemy aircraft with upward-firing cannon using solely *Naxos Z*. Indeed, it was said that a highly skilled operator and a pilot could, with the aid of *Naxos*, close in on an H2S-equipped aircraft sufficiently to obtain a visual for a final attack.

When using FuG 220 *Lichtenstein* SN-2 AI radar in a Ju 88, the pilot and gunner would search visually to port, starboard and upwards, whilst the radio operator watched astern, taking an occasional glance at the SN-2 presentation screen. In a Bf 110, the pilot searched above to port, the radio operator above and to starboard and the gunner astern only.

The point at which the pilot decided to attack with forward-firing armament or with *Schräge Musik* was usually when the radio operator obtained an SN-2 contact. If it was to be with upward-firing cannon, the radio operator attempted to lead the pilot to the target bomber from the fine quarter and to within a distance of not more than 100 m below the contact. An oblique approach was intended to avoid the tail warning device of the bomber, as well as any Window which may have been released, whilst the limit of 100 m difference in height was mainly to avoid entering the H2S scan cone of the bomber until the last possible moment.

In the initial approach, the radio operator concentrated solely upon the SN-2 presentation screen, whilst the pilot gradually adjusted his aircraft's speed to that of the bomber. If the bomber weaved or changed height, the radio operator instructed the pilot to follow its movements based on the reactions on the SN-2 display. The pilot then looked to port, forwards and upwards, while the gunner searched to starboard and upwards for a visual. The radio operator only looked away from the SN-2 when the pilot announced that a visual had been obtained. At that point, the SN-2 was switched off and in the Ju 88, the radio operator moved forward to help with keeping the bomber in view.

The nightfighter closed in slowly at a speed only slightly higher than that of the bomber. The pilot used an additional reflector sight, mounted above and in front of his head, to sight the weapons. If the bomber weaved, the nightfighter would attempt to weave with it, but if the bomber dived to port or starboard, it would often be lost to the nightfighter, and a fresh SN-2 search had to be made. Once the nightfighter was in a position to open fire, speed was reduced to that of the bomber. Careful aim was taken, following the instruction to target the wing tanks. After the attack, during which it was usually found that one burst was sufficient to make a kill, some nightfighter crews would return to base with the trailing aerial of their victim draped around their aircraft.

The effect of a *Schräge Musik* attack was shocking and violent to the crews of RAF Bomber Command. On the night of 24/25 July 1944, for example, Lancasters of No. 300 (Masovian) Sqn were among those targeting Stuttgart. Jim Philpot was a mid-upper gunner in an all-British crew that night. He recalled, 'As we approached the target the welcome was very hostile, to put it mildly.' Shortly after 0210 hrs a twin-engined fighter, probably the Bf 110 piloted by Unteroffizier Heinrich Buhlmann of 11./NJG 5, based at Mainz-Finthen, opened fire from below a Lancaster in a *Schräge Musik* attack.

'It just wasn't our night,' Philpot remembered. 'We were suddenly hit by cannon shells which ripped through the port wing, knocking out both engines on that side. The flames were streaming out alongside my mid-upper turret: the tail gunner and I both saw the fighter make a turn in towards us, so we let fly, only to see our tracers falling short. Another belt of cannon shells came close, but luckily did not score a hit.'

Most likely, as per Luftwaffe tactical directives, the Lancaster had been hit in its tanks. The pilot was unable to maintain control of the aircraft and the order was given to bail out. As he jumped out, Philpot watched the Lancaster 'blazing furiously' in mid-air. Moments later it 'crashed straight into the ground'. Buhlmann claimed a *Viermot* for his first victory at 0225 hrs 80 km northwest of Stuttgart.

On the night of 12/13 September 1944, Lancasters of No. 15 Sqn were amongst a force of 378 to fly a mission to Frankfurt. The crew of wireless operator Bob Kendall were flying their 20th 'op', and as they approached Mannheim they were coned by searchlights and immediately attacked by nightfighters. Kendall's Lancaster took evasive action and went into a violent 'corkscrew' manoeuvre. He recalled:

The Browning guns were chattering away from the rear turret, then suddenly a terrific bang and an explosion occurred under the wireless equipment and the navigator's table. We were consumed by a Dante's Inferno. My brain was numbed by the explosion, cordite fumes seared my throat, and flames were enveloping my body. Flames spurted through the floor and fuse panel as we had been hit smack bang in the bomb-bay, which was loaded with canisters of incendiaries and a 4,000-lb 'cookie'. I snatched up my 'chute from amidst the flames and clipped it onto my harness. I made for the nose. The bomb-aimer and engineer were trying to open the forward escape hatch, and I stood alongside the pilot, who was still skilfully handling the controls.

Kendall's aircraft may have been the second victim of Oberfeldwebel Heinrich Schmidt of Bf 110 and Ju 88-equipped 2./NJG 6, as he claimed a Lancaster east of Mannheim.

Major Heinz-Wolfgang Schnaufer, the highest-scoring German nightfighter ace and an early proponent of *Schräge Musik*, was of the opinion that as much as 50 per cent of nightfighter kills in the latter part of the war were achieved using obliquely mounted cannon. If there was a legacy of *Schräge Musik* beyond its accomplishments in the night war, it was that it prompted German armament experts to explore upward-firing weapons for deployment in the daylight air war.

RHEINMETALL 37 mm *BORDKANONE* 3.7 AUTOMATIC CANNON (*FLAK* 18)

From mid-1942, the increasing quantity and quality of armour deployed by the Russians on the Eastern Front became of growing concern to German commanders, as did the fact that most Soviet tank factories were located well to the east of the front, beyond the range of German bombers. This meant that Russian tanks had to be destroyed on the battlefield, which, in turn, meant that improved anti-tank weapons were needed. That was easier said than done, for Russian factories were turning out ever more tanks with, from 1942, improved versions of the T-34 with its 45-mm-thick sloping armour designed to deflect and minimise enemy fire, and the KV-1 with armour in excess of 75 mm.

As far as the airborne anti-tank response was concerned, larger calibre high-explosive bombs were only found to be effective if dropped within 3.65 m of a target and immediately above ground level, since penetration into the ground channelled blast upwards instead of to the sides. While smaller bombs that could be scattered over enemy tank concentrations were under development, their effectiveness was viewed by many pilots as questionable.

Fortunately, as early as February 1942, *Abt.* E6 at Rechlin had been investigating the practicalities of using 3.7 cm armour-piercing, tungsten carbide-cored, M-shell ammunition. Trials had shown that with a tungsten carbide-cored shell (*Hartemunition*) 85 mm in length and 16 mm in diameter, armoured steel 120 mm thick (equating to a strength of 80 kg/mm^2) could be wholly penetrated from a range of 100 m and at an angle of 60 degrees. This made the prospect of using a 3.7 cm gun as a ground-attack weapon an attractive proposition.

By 1943, the best overall, heavier calibre anti-tank gun deemed for use by Luftwaffe aircraft was the 37 mm *Bordkanone* (BK) 3.7 automatic cannon. This was a development based on the 3.7 cm *Flak* 18 anti-aircraft gun built by Rheinmetall in 1935 which was effective against aircraft flying at altitudes up to 4,200 m. The BK 3.7 measured 3.75 m in length (barrel length 2.10 m) and weighed 275 kg. The cartridge weighed 1.460 kg, to which was added a projectile weight of 0.62 kg, explosive at 0.09 kg and a case weight of 0.61 kg. The rate of fire was 140 rounds per minute at a velocity of 860 m per second. A magazine with six shells weighed 12.5 kg.

Aside from M-shells, the gun could fire ammunition used by Luftwaffe Flak forces such as *Brandsprenggranatpatrone* (incendiary) self-destructing tracer rounds that ignited only when striking a fuel tank, or *Sprenggranatpatrone* (HE) self-destructing tracer. Such ammunition could be alternated with soft-nosed rounds for soft-skinned targets in air-to-ground work.

Among the early considerations for fitting a BK 3.7 in an aircraft was, understandably, the dedicated *Schlachtflugzeug* and *Panzerjäger*, the Hs 129. Studies began in July 1942, with installation commencing in November. In December, Hs 129B-2 Wk-Nr 0280 flew with a BK 3.7 installed, and trials were also carried out in Russia. However, the weight of the cannon when fully installed with mount and fairing was 450 kg, compared to the MK 101 at 231 kg. The weight increased the Hs 129's take-off and landing runs, and also had a significant impact on performance. Further trials were cancelled.

Similar attempts were also made with a Bf 110G (the 3.7 cm-equipped sub-variants would be categorised as Bf 110G-2/R1, R3 and R4) and a Ju 88. In July 1943, one Ju 88, described in a Tarnewitz report as an A-4, had been assigned for trials with two BK 3.7, but the installation required the cockpit and areas of the fuselage to be strengthened by Junkers before such trials could commence. By the middle of the following month, a Ju 88, described as an 'S', again fitted with two BK 3.7, undertook firing trials. The weapons were found to function perfectly, but after 160 rounds had been fired, damage was sustained to the aircraft's engine cowlings. Furthermore, a Tarnewitz weekly report for early August 1943 noted that while the cannon malfunctioned when firing M-shells in low temperatures at high altitude, because of the required 'accelerated transfer' of the aircraft to E.Kdo 25 for operational trials, no further investigation into this could be carried out.

In July 1943, tests were also carried out with the BK 3.7 in the Me 410 V24 as a potential weapon for use against Allied bombers, but the restriction of only 12 rounds made this untenable.

In the early autumn, a Bf 110 fitted with a BK 3.7 carried out tests during which it fired 120 M-shells between temperatures of −35°C to −46°C without any problems. Generally, however, it was felt that when fitted with the cannon, both the Bf 110 and the Ju 88 became unwieldy and vulnerable to ground fire.

Meanwhile, in January 1943, Ju 87D-1 Wk-Nr 2552 had been sent to Rechlin and then to the Eastern Front to trial the fitment of two underwing BK 3.7 weapons, with each gun mounted outboard of the undercarriage leg. Under the supervision of the *Versuchs Kommando für Panzerbekämpfung* (Test Command for Anti-Tank Warfare) led by Oberstleutnant Otto Weiß, tests were carried out in the aircraft at Rechlin and then in Russia by Hauptmann Hans-Karl Stepp and Oberleutnant Hans-Ulrich Rudel (see Chapter One) against captured Russian tanks with encouraging results. But Rudel was circumspect, and wrote later:

> The Ju 87, which is not too fast, now becomes even slower and unfavourably affected by the load of the cannon it carries. Its manoeuvrability is disadvantageously reduced and its landing speed is increased considerably. But now armament potency is a prime consideration over flying performance.

While it was recognised that there was some compromise in handling and performance as a result of the cannon installation, this was not considered adverse enough to prevent further operational development.

Dust swirls in the propeller wash of a Ju 87G Stuka of 10.(Pz.)/SG 2 fitted with 37 mm *Flak* 18 underwing cannon. The white silhouette of a T-34 was applied to a number of such aircraft. Hans-Ulrich Rudel claimed many Russian tanks destroyed with the cannon. (EN Archive)

The Ju 87G-1 emerged initially as a two-seat, specialist tank-destroyer rebuilt from Ju 87D-3 dive-bombers. Most of these short-winged conversions, of which 34 examples were produced, were delivered to the *Vers.Kdo. für Pz.Bekämpfung* in the Bryansk area in February 1943. These were followed by the first series G-1s in April. The Ju 87G was powered by a Junkers Jumo 211J engine that gave the aircraft a maximum speed of 400 km/h and a range of 2,000 km.

In the Ju 87G-1, a BK 3.7 was suspended beneath each wing to fire outside of the propeller arc, thus eliminating the need for synchronisation. The breech mechanism was housed in a streamlined pod fitted to braces just outboard of the wing crank. The pod, which extended approximately three-quarters the width of the wing, aided aerodynamics and gave protection to the breech from dust and combat damage. Above and to the front of the pod, extending slightly over the rear of the barrel, was another, smaller, faired pod that housed the weapon's hydraulic oil heater and air intake.

The cannon were fed by six-round clips, with two clips accommodated in metal hinged trays loaded horizontally into each gun. Each tray extended to either side after fitment, thus giving one aircraft 24 rounds to fire. The empty shell cases were not ejected but fed back into the clip on the opposite side of the gun and then removed after the aircraft had landed.

The low ammunition load meant that missions were generally of short duration and inefficient in terms of fuel consumption. The possibility of welding two ammunition trays together to double firepower was examined but not pursued. When the cannon and magazines were fitted to a Ju 87G-1, all other bomb racks under the fuselage were removed, thus making it a dedicated cannon-armed *Panzerjäger*. Some examples also had the muzzles of their wing-mounted 7.92 mm MG 17 machine guns faired over. The cannon were aimed using a remotely operated pneumatic sighting system.

After introduction in numbers, results with the BK 3.7-equipped Junkers did vary, and some units found the weapon to be unsatisfactory. Thus, they removed the cannon and returned to carrying bombs. But those crews which prevailed began to devise effective short-dive or shallow glide, low-level attack tactics in which an enemy tank was approached in a long, straight run, and fire opened at the closest possible range. Proof of the success of this method came on 5 July 1943 when, despite his earlier reservations, Hauptmann Rudel of St.G. 2 destroyed 12 T-34s, each 'kill' being recorded by a photograph. However, of the introduction of the Ju 87G into service, Rudel remembered:

> The outlook is none too rosy. We are the object of commiseration wherever we appear, and our sympathisers do not predict a long lease of life for any of us. The heavier the Flak, the quicker my tactics develop. It is obvious that we must always carry bombs to deal with the enemy defence. But we cannot carry them on our cannon-carrying aircraft as the bomb load makes them too heavy. Besides, it is no longer possible to go into a dive with a cannon-carrying Ju 87 because the strain on the wings is too great. The practical answer is therefore to have an escort of normal Stukas.

The evolved standard procedures were to fire at a tank's side armour, where the BK 3.7 was effective, or to aim for the thinner armour at the rear of a tank where the engine vents were located. Even if this could not be done, a tank could be immobilised by blowing off a track tread.

Russian tank commanders were known to let off smoke canisters fitted to their tanks in an attempt to simulate destruction, but the more experienced Luftwaffe *Panzerjäger* knew that a genuinely destroyed tank burned with flames. A hit tank would often explode instantaneously if fire broke out close to its store of ammunition, and so for a Ju 87 flying at an altitude of just five to ten metres above it, the situation could be, as Rudel described it, 'uncomfortable'. He experienced personally such a scenario on two occasions in his first few days of flying the Ju 87G operationally.

Under operational conditions it was found that a 3.7 cm cannon round could penetrate 58 mm of armour at a range of 100 m at 60 degrees.

As the Ju 87G-1 went into service, so the Junkers works at Bremen-Lemwerder commenced construction on a run of 208 new G-2 variants based on a conversion from the longer-spanned Ju 87D-5 which gave more strength and stability to the cannon installation. The dive brakes, which were not required for pure ground-attack operations, the wheel fairings and wing-mounted MG 17s and MG 151s were removed to counter drag from the cannon and to enhance aerodynamics. Internally, oxygen equipment

An armourer about to load a clip of six high-explosive 37 mm rounds into the covered outboard metal ammunition tray on the side of the faired breech pod of the BK 3.7 cm gun of a Ju 87G on the Eastern Front. The hinged door to the tray is open and the aircraft's wheel spats have been removed to prevent clogging from dust and clumps of earth. (EN Archive)

Ju 87G-2 Wk-Nr 494320 T6+MU of 10.(Pz)/SG 2 fitted with 37 mm *Flak* BK 18 cannon, Latvia, August 1944.

was excluded to reduce weight, as it was considered unnecessary for the low-level role the aircraft was to perform. But such measures came at a cost in performance, with maximum speed reducing to 270 km/h. Cold weather heaters were later installed in some cannon to enable continued operations in the Russian Winter.

Units known to have operated BK 3.7-equipped aircraft were the *Vers.Kdo. für Pz. Bekämpfung* and the dedicated *Panzer Staffeln* which were assigned as:

- 10.(Pz.) *Staffel* of St.G. 1 (formed from 1./*Vers.Kdo. für Pz.Bekämpfung* – it became 10.(Pz)/SG 77 on 18 October 1943)
- 10.(Pz.) *Staffel* of St.G. 2 (formed from 2./*Vers.Kdo. für Pz.Bekämpfung* – it became 10.(Pz)/SG 2 on 18 October 1943)
- 10.(Pz.) *Staffel* of SG 3 (formed from 4./St.G. 2 in February 1944 – it became 3.(Pz.)/SG 9 on 7 January 1945)
- 10.(Pz.) *Staffel* of SG 77 (formed from 10.(Pz.)/St.G. 1 on 18 October 1943 – it became 10.(Pz.)/SG 1 on 27 January 1944)

'On paper', each *Panzer Staffel* fielded a strength of 12 BK 3.7-armed Ju 87Gs, in addition to four bomb-carrying aircraft intended to suppress defensive fire from local enemy anti-aircraft gun batteries. Ju 87G-1s and G-2s continued to operate in small numbers with the *Panzerstaffeln* until the end of the war, but increasingly at night on harassment missions.

RHEINMETALL BK 5 cm AUTOMATIC AIRCRAFT GUN

In December 1942 the Luftwaffe *Technisches Amt* issued a requirement for a gun with a muzzle velocity of at least 600 m per second and with a rate of fire of 300 rounds per minute. It wanted a weapon capable of causing the almost certain destruction of a bomber with just one shell, but without the attacking aircraft coming within range of the enemy machine's

American armourers prepare a 50 mm BK 5 cannon for movement in a wooden cradle, possibly for shipment to the USA, at the end of the war. Despite its impressive appearance, the BK 5 was a troublesome weapon when mounted in an Me 410. Its weight adversely affected the manoeuvrability of the Messerschmitt *Zerstörer* and Luftwaffe aircrew felt that installation of multiple 30 mm cannon would offer a better prospect for success. (Forsyth)

defensive fire. But the dichotomy was that while there was a requirement for accuracy and extremely low dispersion, due to the heavy-calibre ammunition required, it would be impossible to achieve a high rate of fire. Every shot would have to count. Three aircraft selected to trial such a weapon were the Ju 88, the still embryonic Me 410 *Zerstörer* and, for ground-attack purposes, the Hs 129, while the gun itself was to be built by Rheinmetall-Borsig.

By early 1943, design of the intended weapon had proceeded quickly and to such an extent that construction commenced. The inspiration and basis for the weapon was the 5 cm KwK 39 L/60 tank gun as used in the *Panzerkampfwagen* III medium tank. This gun incorporated the latest armament developments and was available in sufficient quantities. The small number of alterations required to fit the weapon into an aircraft were considered to be worthwhile in view of the fact that it was not necessary to design and develop an entirely new gun. The incorporation of a magazine feed mechanism would make it fully automatic and the gun, its mounting and its feed were to be combined so that the complete weapon was an interchangeable whole.

Among the earliest trials was the installation of KwK 39 L/60 cannon in Hs 129B-2s Wk-Nrs 141291 and 141292 at Henschel Flugzeugwerke's Berlin-Schönefeld plant. It was found that such an installation adversely affected the landing characteristics of the diminutive, twin-engined *Schlachtflugzeug*. This, combined with the fact that the BK 5's projectiles were not sufficient to penetrate the armour of Soviet tanks, brought further work to a stop.

The idea of using a KwK 39 L/60 cannon in the Me 410 was first proposed by officers on the staff of the *General der Jagdflieger* to E.Kdo 25 in July 1943. The BK 5, as it became known, measured 4.348 m, and as a complete weapon weighed 540 kg. It had a rate of fire

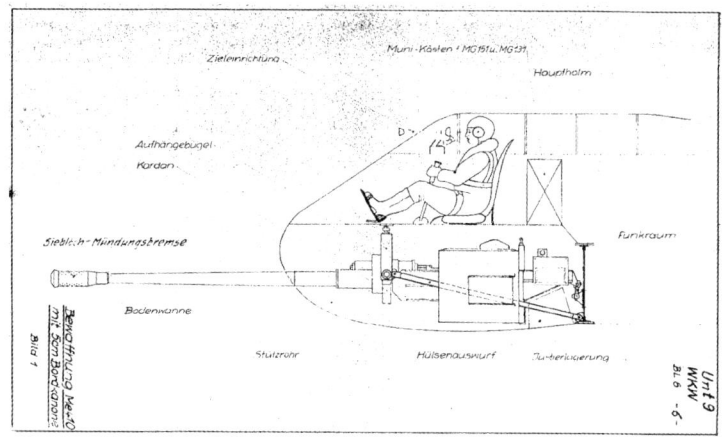

An early, undated sketch produced by Rheinmetall-Borsig at Unterlüss showing planned installation of the BK 5 in an Me 410. The gun is housed in a ventral compartment below the pilot and forward of the radio-operator's compartment. Aiming is performed by a telescopic sight. The gun is supported by a forward bracing tube, vertical bracket with gimbals and a diagonal suspension stirrup on either side. The barrel is fitted with a perforated muzzle-brake. The cartridge ejection chute is at the rear of the breech, behind which is the setting device. (Forsyth)

of 50 rounds per minute and had a muzzle velocity with an M-shell of 965 m per second. The weight of a projectile was 1.52 kg. Recoil with an armour-piercing shell was 2,860 kg and with a high-explosive shell, 2,640 kg. The ring magazine held 21 cartridges, although 22 were loaded, the first round being placed directly into the breech.

The BK 5 was to be used in conjunction with the standard *Revi* 12C/D gun sight or *Zielfernrohr* 4 telescopic sight, although effective use of the latter required considerable gunnery and flying skill, or much practise, as well as the ability to stay in the firing position for a long period of time.

Loading was performed electro-pneumatically, whilst the ammunition was fired by an electric primer. Recoil was taken by two hydro-pneumatic cylinders mounted above the gun cradle.

In the Me 410, the rear part of the gun was fixed to the main spar via an adjusting mechanism on a plate, and the bomb-bay doors were removed and replaced by a ventral fairing. The installation of the gun proved so satisfactory that the outline of the ventral fairing needed to be increased by only some 100 mm at its deepest spot as compared with the original doors. A cartridge ejection opening with an internally mounted chute was cut into the rear of the panel.

By using a perforated muzzle-brake, muzzle blast was so distributed over the airframe that no damage was caused, and therefore no reinforcement necessary. The only problem was the challenge of keeping the gun heated at altitude, and thus heating was provided from the crew compartment.

The gun was hoisted into an Me 410 by means of a large, mobile hydraulic jack known as a *Steinbock*. The muzzle end of the barrel was fastened by means of a rope to the arm of the *Steinbock* which was then pumped up until the rear end of the mounting was inclining upwards. The *Steinbock*, together with the gun, was then run in under the fuselage and the weapon raised until the clamp of the setting mechanism was in line with the plate fixed to the main spar. Two locking bolts were screwed in. A gimbal ring, a pair of lateral bracing tubes and a suspension stirrup were used to secure the mounting of the weapon.

During tests with the gun in an Me 410 at Tarnewitz, at 100 m from the target, the BK 5 gave a pattern of 150 mm x 150 mm for 22 shots, fired either singly or continuously. This was judged to be an extremely good fire pattern, and one which was a necessary condition for the gun's use operationally. Firing the gun during flight had no disturbing influence upon the aircraft or crew. According to one report, the recoil was scarcely noticeable, with no sense of a 'blow', but more of a 'gentle rocking motion'. Nor did the firing have any adverse effect upon the Messerschmitt's controls or its course.

After these first test-firings, the cartridge case ejection opening was enlarged so that a case in its entire length could drop out horizontally.

By late October, it seems Galland's office had revised its opinion of the BK 5 and felt a rapid-fire weapon such as the MK 103 would be better. Nevertheless, when, in November 1943, Hitler watched a demonstration of an Me 410A-1 fitted with a BK 5 cannon, he enthused that it was the 'backbone of the home air defence'. The *Führer* demanded that the Me 410 be committed to Reich defence and that two *Gruppen* or a *Geschwader* be equipped with the aircraft/weapon combination. Göring had to pluck up courage and advise Hitler that only two or three aircraft could be equipped with the cannon since, despite earlier stocks, no further examples were available at that time. His advice falling on stony ground, the Reichsmarschall instructed Generalfeldmarschall Milch on 12 January 1944 to set about equipping two *Gruppen*, each with a strength of 45 cannon-fitted aircraft. Milch wrote to Göring in response two days later:

> The first experimental aircraft have been equipped with a 50 mm gun. Since this gun can no longer be delivered, a conversion for a new 50 mm gun had to be carried out within a very short period of time. This new gun is installed in a different manner. The first delivery of ten guns was to be made in December. This was not possible since the magazine feeding device which we constructed suffered stoppages due to broken belt links whereby the cartridge was hitched on the belt conveyer table. These defects have now been eliminated. There have been difficulties experienced with the high-explosive shell we developed on account of casting defects in the casings and also in respect to the dispersion because of the rotating bands. Alterations are being carried out in the plants. The results have still to be tested.

Once initial investigation on the Me 410 V2 had been carried out at Tarnewitz, adaptation of the gun for aerial use was carried out by Deutsche Lufthansa at Berlin-Staaken. After just over three months of testing, during which various belt-feed and jamming malfunctions had been ironed out, the resulting weapon was fitted to aircraft of II./ZG 26 as the Me 410A-1/U4 from early February 1944, this *Gruppe* moving from Hildesheim to Oberpfaffenhofen to re-equip. By the 8th, 5. *Staffel* had 12 aircraft equipped with the cannon and trial operations commenced over southern Germany and Austria. This, despite the fact that there were prevailing electrics problems, small switches were found to break easily and ammunition belts continued to break. Also, the BK 5's recoil and feed mechanisms were unable to cope with the g-forces of air combat, and it was rare for more than one shell to be fired without the weapon jamming. Cold outside temperatures also caused the recoil brake to operate erratically. These issues were mitigated to some extent by the introduction of an improvised clearing device.

The cartridges cases were manufactured by different companies and came in different grades and, as a result, the cartridge chutes had to be reworked to a certain width. The automatic feeding mechanism was also susceptible to freezing. At first a bag was slung over

An Me 410B-2/U4 fitted with a BK 5 cannon for tests at Tarnewitz. The bulged underside access panel has been removed to show how the gun breech occupies what would have been the bomb-bay. The *Ringmagazin* ('ring magazine' drum) can be seen encircling the rear of the breech. The BK 5 was surprisingly quick to remove or install from the Me 410. (EN Archive)

the rear of the weapon and warm air would be channelled into it from the radiator. This practice was stopped when it was discovered that snow had been thrown into the ejection chutes during take-off, leading to icing. A spring-loaded flap over the chutes solved the problem.

Generally, the employment of such large-calibre guns, forced onto Luftwaffe units by Göring and the high command, proved a fallacy. Many senior officers felt that it would have been better to have employed air-to-air rockets then under development. Nevertheless, installation of the BK 5 in the Me 410 on a broad scale commenced at Auto-Union in Chemnitz and Dornier in Oberpfaffenhofen.

One pilot to fly the Me 410 with the BK 5 was Fritz Buchholz, who joined 6./ZG 26 on 3 February at Hildesheim from 2./*Erg.Zerstörergruppe* at Braunschweig-Waggum, and who recalled:

The Me 410 was a mixed bag. It had good, stable flying characteristics, but it wasn't good in the turn; here the Bf 110 was better. Other than that, in flying the aircraft, generally there were few problems. In air combat, however, it was a different story; for

Me 410A-1/U4 Wk-Nr 420481 3U+LP of 6./ZG 26,
Königsberg-Neumark, Germany, April 1944.

a start, it was easy prey for enemy fighters and in formation it was unwieldy, while it made a nice, big target for the bomber gunners.

Generally, we were ineffective. The *Zerstörer* units were too slow and unmanageable. Our *Gruppe* was frequently the target for enemy fighters, and so we only ever got a short time to deal with the bomber *Pulks*. There was rarely a chance to make a second formation attack because after our first attack, the enemy formation was usually broken up.

Also, the way the Americans staggered their formations was very effective, and it made it very difficult for us to attack. The increasing numbers of escort fighters made a hard job harder, and whenever they appeared we were usually forced to break off because of the risk of being shot down and our lack of adequate defensive armament. Our ground-controllers always tried to get us to the bombers when there were no fighters in the vicinity. Most of our crews just focused on safety and survival; there were few daredevils.

With the BK 5, it was best to attack at an angle of ten degrees to the upper part of the rear of the enemy formation. But because of the rear defensive guns, and our lengthy approach, this was not popular. Because of the trajectory (80 cm to about 1,000 m), we often fired too early, and we were hampered by so many stoppages.

20 February 1944 saw the USAAF launch Operation *Argument*, also known as 'Big Week' – an intense bombing campaign against German aircraft production centres, specifically against the plants responsible for the output of fighter aircraft. On 22 February, during a major raid involving nearly 1,400 bombers from the Eighth and Fifteenth Air Forces hitting targets across Germany, the *Staffelkapitän* of 5./ZG 26, Oberleutnant Fritz Stehle, led ten Me 410A-1/U4s from Oberpfaffenhofen against 183 B-17s and B-24s of the Fifteenth Air Force attacking Regensburg. This would be the first operation on which the BK 5 would be used, but it was still classified as a practice flight.

Shortly before 1300 hrs, Feldwebel Baunicke opened fire with his 5 cm gun and shot down the weapon's first bomber. Around ten minutes later, Stehle claimed another south of Dachau. What this 'practice flight' did tell the crews was that they would need to get within 400 m of the bombers to score a hit, rather than the 800–900 m as claimed by the armament handbook.

This day also brought a serious loss when the *Gruppenkommandeur* of II./ZG 26, Knight's Cross-holder Hauptmann Eduard Tratt, was shot down and killed over Nordhausen. He was credited with 38 aerial victories, making him the highest-scoring *Zerstörer* ace. On 24 February, the crew of Oberfeldwebel Willi Frös and Unteroffizier Gerhard Brandl of 5./ZG 26 shot down a B-17 of the Fifteenth Air Force with the BK 5 at an altitude of 6,000 m over Steyr during an attack on the aircraft plant there.

More missions with the BK 5 would be flown throughout the spring and summer of 1944, but they brought virtually no success. The lack of victories and the operational losses suffered by *Zerstörer* fitted with such armament were disproportionately high in the relatively few missions flown, with their envisaged capability nullified by the aircraft's diminished speed and the defensive fire of enemy bombers. The fact was that most claims made by Me 410 crews were achieved with the Messerschmitt's two MG 151s loaded with 15 mm incendiary ammunition, which accounted for flames observed during combat. This led pilots to question whether the BK 5 was an improvement over what was already available.

A report by Diplom-Ingenieur Kurt Bühler of the *Oberkommando der Luftwaffe* (OKL – High Command of the Air Force) stated:

> The 5 cm cannon is usable in an aircraft but needs careful servicing. Trial missions had shown that 5 cm cannon did not improve the effective firepower of the aircraft's armament since no hits were scored with the cannon. Furthermore, the little tactical advantage the Me 410 possessed was negated by the heavy cannon since the aircraft's manoeuvrability was hampered. The effectiveness of the *Revi* 12C gun sight in combination with the telescopic sight could not be proven, but shows promise for the future.

A necessary tool for long-range shooting was within reach. One of the reasons for the failure of the 5 cm cannon was the 22-shot magazine. This quantity of shells was not enough to bring down a bomber at 1,000 m unless the pilot was an outstanding marksman or very lucky. It has been suggested that the Me 410 armed with four MG 151s or, even better, with four MK 108s, in addition to the normal armament would have considerably more firepower than the cannon-armed aircraft.

Consider this comparison chart:

1 x 5 cm BK	45 rounds/min	250 g explosive charge
4 x 20 mm MG 151/20	700 rounds/min	840 g explosive charge
4 x 30 mm MK 108	600 rounds/min	3,200 g explosive charge

From March 1944, P-51 Mustangs of the Eighth Air Force were able to accompany the bombers on deep penetration raids, ranging from the west as far as Stettin, Berlin and Munich, while Vienna came within reach of the Fifteenth Air Force. As a consequence, losses among the Me 410 units began to escalate, often because the single-engined fighters assigned as their escorts were drawn into self-defence.

The BK 5 was a small attempt by the Luftwaffe to solve a massive strategic crisis – as was so often the case.

RHEINMETALL-BORSIG BK 75 75 mm GUN

From 1942, beyond the development and limited deployment of large calibre guns such as the BK 3.7 and BK 5, the Luftwaffe continued to search for an even more powerful weapon with which to impact Soviet armour. This was particularly the case with regard to the Hs 129, where it was realised that the BK 5 would not be adequate. Thus, eyes turned towards the Pak 40.

The 75 mm Pak (*Panzerabwehrkanone*) 40 was the superlative, standard anti-tank gun in the Wehrmacht's arsenal. More than 23,000 would be produced by war's end, and it would prove itself capable of being effective against just about every type of Allied tank. The gun was light and very powerful, and could penetrate, head-on, 133 mm of armour at a distance of 915 m with its 3 kg tungsten carbide core projectiles which were fired at a velocity of 930 m per second. At 2,286 m, the velocity remained powerful enough to penetrate armour of 83 mm thickness head-on. However, when hit at an angle of 30 degrees, penetration decreased to 96 mm at 915 m and 53 mm at 2,286 m.

The issue with adapting the Pak 40 for an aircraft was, not surprisingly, its weight at 1,500 kg. To overcome this, Rheinmetall engineers removed the ground carriage and introduced a new muzzle brake, while the breech was altered for semi-automatic loading. In such a configuration, the gun was renamed the BK 7.5. Production was restricted to a run of just 20 weapons on orders of the *Führer*, who wanted to see how it performed operationally before allowing any further production. It was viewed as a weapon 'for combatting tanks that had broken through'.

As the Luftwaffe's principle, dedicated *Schlacht* aircraft, the Hs 129B-2 was considered initially as a carrier for the BK 7.5, but even though Rheinmetall had incorporated a semi-automatic breechblock, this still meant that the 7.5 cm projectiles would have to be fed into the breech manually, which necessitated either a second crew-member or the development of a fully automatic mechanism. However, provision of the latter was expected to take too long, and the Henschel could not provide accommodation for a second crewman.

A plausible alternative to the Hs 129 lay in the Ju 88A-4 with its space for a crew of four. This meant that one crew-member could handle the loading of the semi-automatic mechanism. The Junkers also had a strong airframe able to absorb the recoil generated from the BK 7.5 and the additional engineering required for installation would be minimal. In due course, a BK 7.5 was fitted into a Ju 88A-4, adapted with a solid nose, which subsequently became the P-1 variant, although it is believed some later Ju 88 prototypes were used for tests.

One of the few Ju 88P-1s of the *Versuchs Kommando für Panzerbekämpfung* fitted with a 75 mm BK 7.5 cannon, possibly photograped at Rechlin. Note the large faired housing under the fuselage and the long, perforated muzzle brake. The aircraft carries an emblem used by the unit believed to depict a dagger piercing the shell of a tortoise. (EN Archive)

The gun was laid in a large, purpose-built pod which replaced the aircraft's standard ventral gondola (*Bodenwanne*). The gun and its mounting weighed 1,439 kg, and each of its 12 rounds weighed 11.8 kg. The shells were housed in a drum magazine. The weapons were installed at a downward angle of four degrees from the longitudinal axis and offset 16 cm to the left of the fuselage centreline. The barrel projected 1.83 m from the aircraft's nose. There was accommodation for a crew of three. At the rear of the new faired pod for aft defence was a twin 7.9 mm MG 81Z (Z for *Zwilling* – 'twin') gun set which augmented the MG 81Z mounted in the rear of the cockpit and the single, forward-firing MG 81 used by the pilot, mainly for aiming.

Initial ground and air-firing trials were carried out at the Junkers plant at Bernburg and Roggentin airfield, a part of the *E-Stelle* Rechlin. At the latter site, the gun was fired at captured Russian tanks, two rounds being customary during one firing pass, with the first being fired from a height of 300 m and the second from 75 m. These tests proved encouraging, despite the considerable weight of the weapon. Crews also quickly became accustomed to the gun's unique characteristics, specifically the recoil, noise and vibration that accompanied each shot. Indeed, very satisfactory results were achieved in firing against practice targets and vehicles.

One negative was that when the BK 7.5 was fired, the jets of gas ejected from the muzzle brake and, on some examples, from vents in the side of the gun pod caused damage to the Ju 88's cockpit area, engine cowlings and propellers. The solution was to strengthen the cockpit and cowlings with armour. The shell ejection process was also reworked, but the risk of damage to propellers remained, compelling replacement after the firing of every 100–150 rounds. Following these test-firings, a small number of P-1 conversions was ordered, each example having a loaded weight of 11,068 kg.

In March 1943, some examples of the Ju 88P-1 were assigned to the 3. *Staffel* of Oberstleutnant Otto Weiß's *Versuchs Kommando für Panzerbekämpfung* (see Chapter One) at Poltava in southern Russia. Their deployment was not auspicious: the crews comprised bomber men who were learning to undertake anti-tank missions with a ferocious weapon, something for which they were ill-prepared and about which they were decidedly uncomfortable. The situation was not helped by a raft of technical problems and breakdowns associated with both armament and engines. The Ju 88s also suffered from a lack of manoeuvrability and a low maximum speed of only 393 km/h.

On one mission it took 75 minutes for four Ju 88P-1s to take-off in response to a call to attack tanks, and even after that, only one or two aircraft got as far as the battle area, some 60 km away. Over the front, the BK 7.5-equipped Ju 88 presented too large a target and its liquid-cooled engines were more vulnerable to ground fire than air-cooled ones. If one engine was damaged, the aircraft did not have sufficient power to fly unless the gun was jettisoned and therefore lost. Furthermore, only a small number of rounds could be fired during an attack approach. The Ju 88P-1s carried out just four sorties before they returned to Germany, where they were transferred to III./KG 1, joining that *Gruppe*'s Ju 88s and He 177s under Hauptmann Werner Kanther.

The ceasing of BK 7.5 activity with the Ju 88 initiated a return to investigating the gun in the Hs 129, and despite the development undertaken with the Ju 88, Henschel had not terminated its own development work. In August 1943, the *E-Stelle* Tarnewitz reported that both static and air trials had been carried out under the supervision of Rheinmetall-Borsig with an Hs 129B-2 fitted with a BK 7.5, albeit without an automatic loading mechanism. Eight rounds of 2.5 kg projectiles had been fired satisfactorily. Rheinmetall-Borsig then proposed fitting the gun with a simple ten- to twelve-round magazine along with an automatic loader that had been designed by Henschel. These took the form of a rotating cylinder and an electro-pneumatic feed.

On 10 December 1943, in something of a marked turnaround in opinion, Oberstleutnant Viktor von Loßberg of *Abt.* E2 in the RLM's *Technisches Amt* wrote to Oberst Hubertus Hitschhold, the *General der Schlachtflieger*, regarding the proposed installation of the BK 7.5:

Hs 129B-2/B-3 Wk-Nr 140494 DO+XG fitted with BK 7.5 cm cannon, *Erprobungsstelle* Tarnewitz, Germany, July 1944.

Now the Ju 88P has proved to be insufficient for the use of this heavy weapon, and consequently not suitable for use at the front, GL C [*Generalluftzeugmeister/Abteilung* C] has returned to the earlier use of the Hs 129 with the 7.5 cm BK. The firing characteristics of the Hs 129 with cannon are particularly praised by the *Erprobungsstellen*. No deterioration in flight characteristics or performance has occurred, according to the unanimous judgment of the *Erprobungsstellen* and frontline officers who have flown with such installations. As a result of the fuselage construction as an armoured aircraft, the damage to the cockpit that occurred in the Ju 88 was not shown in the Hs 129. The use of the 7.5 cm BK for anti-tank warfare from the air appears vital in order to achieve penetration of armour with the use of tungsten ammunition. The capability of an aircraft type suitable for this purpose, as experience has shown, currently exists only in the Hs 129, and it is therefore recommended to equip it with this weapon.

The multi-loading device for ten to twelve rounds, which has been developed in the meantime, has been simplified in such a way that functional difficulties in this regard are not to be expected beyond what would be considered normal. The manufacture of this loading device can be carried out mainly by the Henschel company.

In order to advance the presently stopped development work in such a way that the Hs 129 with 7.5 cm BK can be delivered in January/February 1944, an experience report is attached, which shows the high stability of the Hs 129 after firing with Flak ammunition.

An order for six Hs 129s fitted with the BK 7.5, to be known as the Hs 129B-3, then followed from the RLM in January 1944. The gun was to include an electrically firing breechblock, a heater to warm the recoil mechanism fluid in cold weather and a redesigned muzzle brake.

The gun was suspended in a frame below the Hs 129B-3's fuselage, but with the magazine and feed mechanism inside the fuselage. The entire mounting was housed in a large, faired pod, with the gun's barrel protruding about one metre ahead of the aircraft's nose. In an improvement on the Ju 88P-1, gas jets were channelled in a way that only the armoured nose would be affected.

To give the Hs 129B-3 sufficient power to operate with the 1.6-ton weapon and 12 shells, its 700 hp Gnome-Rhône 14M 14-cylinder air-cooled radial engines were replaced by 850 hp 14M 38s. However, like the Ju 88, in the event that the Henschel lost one of its engines, the whole gun, breechblock, magazine and pod would have to be jettisoned.

While awaiting delivery of the automatic mechanism, single-round firing trials were carried out which caused some damage to the left engine cowling and wing leading edge of one test aircraft, as well as jamming to a sliding canopy. When the mechanism arrived, air trials proceeded which revealed that the Hs 129 had a rate of fire of 40 rounds per minute, comparing very favourably with the Ju 88P-1's 24. The Henschel could also fire a round every 1.5 seconds, meaning that four shots could be fired at a target in an approach from 1,000 to 200 m.

On 25 August 1944, Hs 129B-3s Wk-Nrs 162033, 162034 and 162035 were delivered to E.Kdo 26 (see Chapter One) for operational assessment. Because of a lingering problem associated with shell case ejection, these aircraft were recommended for use on a limited scale only. Nevertheless, at the end of October, they were transferred to 13.(Pz.)/SG 9 at Schippenbeil for testing under fully operational conditions.

Despite all the testing, however, ultimately, the few Hs 129B-3s which reached the frontline *Staffeln* proved unreliable. When fired, muzzle blast and recoil caused a brief

Armament handbook drawing showing the installation of a BK 7.5 cannon in an Hs 129B-3, which is also fitted with a ZFR 3B gunsight. Seen here are the height adjustment screws just behind the framework over the gun's breech, and at rear the large shell drum, beneath which are the return spring and locking lever. The pipe running from the two cylinders on top of the drum lead to the external compressed air point. (EN Archive)

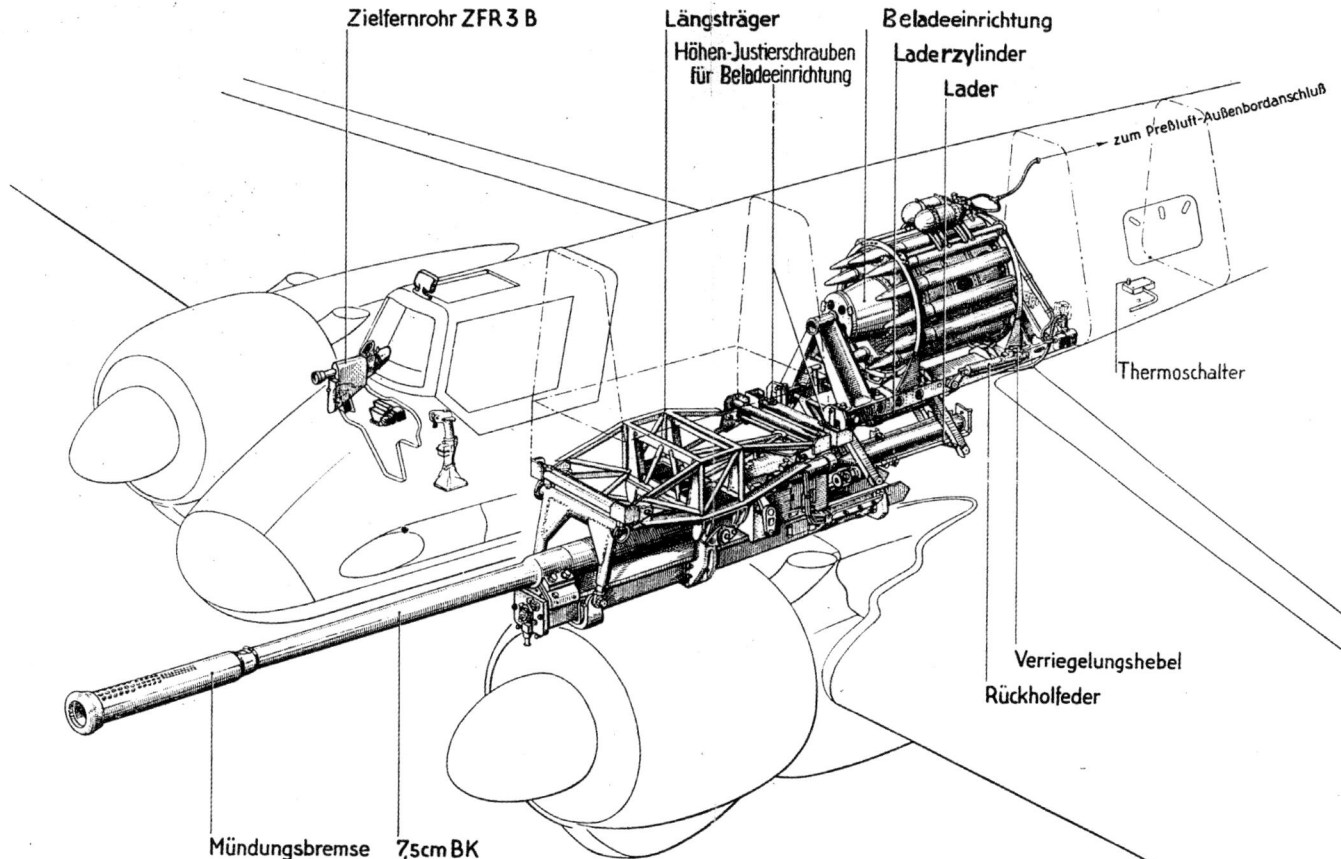

Zielfernrohr ZFR 3 B

Längsträger
Höhen-Justierschrauben
für Beladeeinrichtung

Beladeeinrichtung
Laderzylinder
Lader

zum Preßluft-Außenbordanschluß

Thermoschalter

Verriegelungshebel
Rückholfeder

Mündungsbremse   7,5 cm BK

10 km/h loss of speed. The electrical system on the magazine was also prone to malfunction. This meant that for the greater part of November 1944, during which time seven BK 7.5-equipped Henschels arrived with 13.(Pz.)/SG 9, technicians from Rheinmetall-Borsig and Tarnewitz grappled with the problems. The situation was compounded when aircraft that had fired the weapon operationally returned with missing nose panels or damaged areas around the cockpit.

In January 1945, both 13.(Pz.) and 12.(Pz.)/SG 9 (which had also taken delivery of some B-3s for a brief time) switched to Fw 190s intended for equipping with *Panzerblitz* anti-tank rockets (see Chapter Four). The available, operationally active B-3s were then passed to 10.(Pz.) and 14.(Pz.)/SG 9. Generally, the pilots of these units viewed the aircraft somewhat more favourably, and were impressed by the increase in firepower the BK 7.5

Hs 129B-2 Wk-Nr 140494 was modified to B-3 specification for testing by the RLM at the *Erprobungsstellen* at Rechlin and Tarnewitz in 1944. It was one of the few aircraft fitted with a 1.6-ton BK 7.5 cannon. (EN Archive)

presented. The commander of 10.(Pz.)/SG 9, Oberleutnant Gebhard Weber, recounted how one round from a BK 7.5 was sufficient to blow a one-metre-diameter hole in the turret of a T-34 or, in another case, cause a tank to explode. 'The effect of the 7.5 cm cannon was extraordinary,' he recalled.

The small numbers of Hs 129B-3s with SG 9 fought on as long as their cannon remained serviceable and there was ammunition to fire, but ultimately the *Schlachtflieger* became overwhelmed along with all German forces as the Eastern Front collapsed in the spring of 1945.

CHAPTER THREE

AIR-TO-AIR WEAPONS

RHEINMETALL-BORSIG *RAUCHZYLINDER* 65 AIR-TO-AIR ROCKET

Soon after taking command of E.Kdo 25, Hauptmann Horst Geyer (see Chapter One) investigated the use of wing-mounted rockets and mortars by single-engined fighters against four-engined bombers. As a first trial, two Fw 190s were fitted with a pair of external, wing-mounted 'firing frames' each built to carry four 65 mm spin-stabilised RZ (*Rauchzylinder* – 'smoke cylinder') 65 rockets.

First designed by the *E-Stelle* Tarnewitz in January 1936 and then produced by Rheinmetall a year later under Luftwaffe direction, the RZ 65 was intended to be launched from tubes on

This Fw 190A-5, believed to be from E.Kdo 25, is fitted with RZ 65 rocket launchers mounted internally in the wing, with the firing ports visible along the leading edges. The RZ 65 proved so disappointing in the limited operational trials conducted during the summer of 1943 that the weapon was dropped from the *Jagdwaffe*'s inventory later that year. (EN Archive)

an externally mounted rack or from a *Föhn* 'honeycomb' barrel of tubes. The rockets were charged at the front of the tube and a terminal contact block was fixed at the back of the tube for the connecting of ignition wires. Originally loaded with diethylene glycol dinitrate as a propellant charge, described in reports as a 'compressed black powder', this was later replaced by a more efficient, smokeless powder and activated by an electric threaded primer. Weighing 3.15 kg, in its initial configuration, including a warhead of 840 g, the rocket had a nominal velocity of 300–380 m per second and was fitted with an electric and mechanical percussion fuse. It was to be fired at a maximum range of 300 m from a target, and once launched, the rocket would spin at 19,700 rpm with a maximum thrust of 200–220 kg.

The flight path of the RZ 65 was erratic and followed a corkscrew trajectory, easily visible with the naked eye after launch, which was due to unequal pressure distribution from the combustion chamber to the nozzles. This problem resulted in a sighting error of up to 7 per cent (70 mil), where 1 mil is 1/1,000th of a radian.[1] Since a sighting error of 2.5 per cent (25 mil) had been stipulated by the Luftwaffe, the project was dropped in 1941. Nevertheless, for the next two years Dr. Klein at Rheinmetall continued to work on it, attempting to improve the rocket's ballistic performance.

The stabilising spin of the rocket shell was induced by an *Antriebsturbine* (turbine motor), which consisted of a series of tangentially angled nozzles in the baseplate. In an attempt to improve the accuracy of the shell, a wide range of nozzle angles was tried from eight to 37 degrees from the longitudinal axis of the shell. With a tangential nozzle angle of eight degrees, it had a rotational acceleration of 15,700 to 19,700 revolutions per minute. The rocket shell could hit a target within a rectangle of 1.65 x 2.0 m at a range of 105 m (approximately 15–20 mil), which was well within the Luftwaffe's requirement. The refined warhead weighed 0.238 kg, with an explosive charge of 0.14 kg. The final design had an electrical firing device and an impact detonator.

The RZ 65 is believed to have first seen operational deployment with E.Kdo 25 when two Fw 190s took off as part of a four-aircraft *Schwarm* to intercept raids by the Eighth Air Force on Bremen and Kiel – targets close to the *Kommando*'s base at Wittmundhafen – on 13 June 1943. Horst Geyer led one of the two-aircraft *Rotte*, each machine carrying RZ 65s, while the other *Rotte* was led by Oberleutnant Erwin Hardtke, who had joined E.Kdo 25 from Schl.G. 1. Geyer recalled:

This mission was to see my first *Abschuss* [victory] as *Kommandoführer* of E.Kdo 25. Scattered bomber units were making their way home after their raid on the ports.

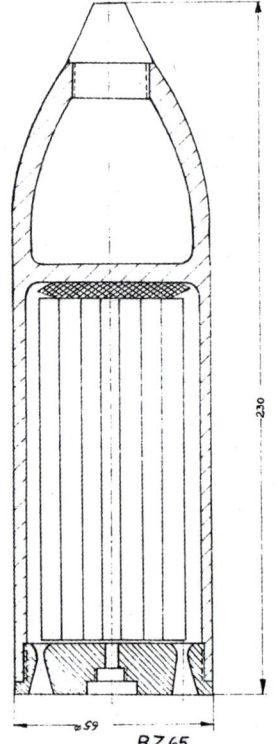

Rheinmetall-Borsig sketch of a 65 mm RZ 65 rocket showing its internal structure. (Forsyth)

---

1   The gunnery unit of dispersion of fire is termed the 'mil' (or 'mille') and refers to the angle subtended by one metre at a range of 1,000 m. One mil is almost exactly equivalent to an angle of one milliradian (mr). One mr equals 0.057°.

Close-up of the RZ 65 installation in the starboard wing of an Fw 190, with recoil vents towards the trailing edge. (EN Archive)

An armourer loads and charges a 65 mm RZ 65 rocket into a fairing on the port wing of a Bf 109. (EN Archive)

There were no escort fighters in sight, so I attacked two B-17s that were flying close together. I fired all eight RZ 65s and after the two bombers were forced to separate, I was able to wreak havoc on the machine flying lowest and to the right with several bursts from my MK 108 cannon. From about 2,000 m, I observed two parachutes fall out while the B-17 was evidently trying to go for an emergency landing somewhere. Meanwhile I had lost contact with my three comrades, but they all landed back at Wittmundhafen without damage. What was key here was that the rockets had weakened the bombers' defensive fire, shocked the crews and enabled me to get in close to make my shoot-down.

RIGHT Experimental installation of four RZ 65 rocket tubes fixed to the underside of a Bf 109F-2's port wing. Note the proximity of the wheel well to the nearest tube. (EN Archive)

BELOW Rheinmetall-Borsig sketch for a planned revolving 65 mm RZ 65 *Rohrblock*, with the option for volley-shot or single-round firing. (Forsyth)

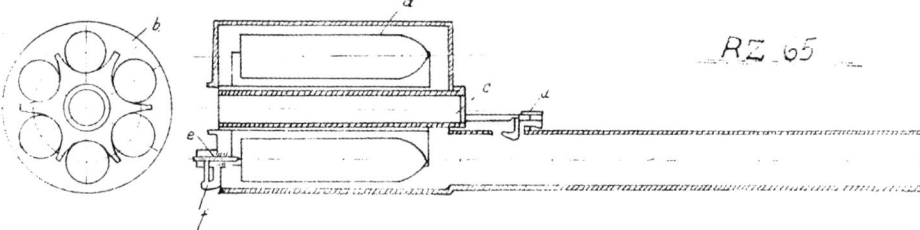

a.) Projectile
b.) Drum
c.) Catch
d.) Catch bolt
e.) firing mechanism
f.) bolt for single-round firing

In a later adaptation, the wing leading edges of at least one of the *Kommando*'s Fw 190s had launch tubes for six RZ 65s installed internally, with three tubes built into each wing, but this brought little result.

Despite isolated successes, deployment of the RZ 65 by E.Kdo 25 proved largely unsatisfactory due to technical problems. A week after Geyer's attack on the bombers he reported that:

> The use of the RZ 65 has shown that impact on a target at a key point cannot be seen. The calibre [of the rocket] is too small when compared to the size of a four-engined bomber. In addition to the extraordinary stability of these aircraft, it has recently been discovered that crews (or at least the pilots) are provided with chain mail armour in addition to the normal armour plating. This provides excellent protection against small shell splinters. Even when fired in mass, there is little chance of success due to the poor ballistics.

Indirectly, Geyer's success on 13 June was principally due to his flying skills, and further trials with the rocket on day fighters were dropped. According to a post-war report prepared for the Allies by two former Rheinmetall-Borsig ballistics engineers:

The use of the RZ 65 rocket was doubtful because of the poor dispersion. In spite of development over several years, it was not possible to achieve satisfactory accuracy with the rocket, which was stabilized by rifling. Apart from this, the installation as a *Zusatzwaffe* [an external or additional weapon to standard armament] was not very favourable because of the large diameter and the small charge. The rocket had to be shot from barrels. On one hand they were very bad and in insufficient number to install under the wings because of the large number of rockets required and their size. On the other, they had a strong drag, which meant a high loss of efficiency.

A Bf 110 nightfighter was also fitted with two RZ 65-based eight-round 'drum devices' in 1943, but nothing is known of any results achieved with the installation.

RZ 65 (latest known specification)	
Calibre	65 mm
Total Weight	2.5 kg
Weight of explosive charge	130 g
Weight of propellant charge	400–700 g
Weight of warhead	840 g
Discharge	Electric
Fuse	Percussion fuse
Maximum thrust	200–220 kg
Duration of thrust	1/3 sec
vmax	273 m per second after flying distance of 150 m

WERFER-GRANATE 21 AIR-TO-AIR MORTAR

What had been learned by Hauptmann Geyer and the pilots of E.Kdo 25 from the tests with the RZ 65 was the tactical value in *dispersing* the bomber *Pulks* so that their defensive fire could be weakened and confusion caused within a formation to enable single, isolated bombers to be targeted more easily. At that point, the fighters could engage more closely and use their cannon and machine guns to bring bombers down. What was needed to do this, however, was a more powerful weapon that could cause ever greater breakdown of an enemy formation.

One option lay in the form of an army infantry weapon designed for use in ground warfare – the 21 cm *Nebelwerfer* 42 mortar. In a lecture given to a committee of air technicians on 27 October 1944, Flieger Oberst Stabsingenieur Günther Voss, an aircraft armament theoretician, explained the concept behind the use of an adapted long-range mortar:

Any armament which would make it possible for a fighter to carry out an attack against a large formation of bombers from the rear would have to be able to open fire

Lille-Vendeville, May 1944 – a classic view of armourers from *Stab/JG 26* loading a 21 cm *Werfer-Granate* into a launch tube suspended from the port wing of an Fw 190A-8/R6 that is also fitted with a centreline 300-litre drop tank for longer-range missions. The man kneeling at the rear of the tube is ready to connect the ignition wires and check the screw bolt that holds the shell secure. (EN Archive)

from outside the effective range of the enemy's defensive armament. In spite of this, it should not impair the flying characteristics of an aircraft through the excessive weight of the armament. Such heavy armament intended for fighters operating against bombers should be able to be jettisoned so that combat with escorting fighters can be taken up at any time using the standard weapons of the fighter. This long-range firing would also be of advantage in attacks against large formations of [enemy] fighters because in such an instance, it would be unnecessary to pass through the enemy's formation. In consequence, the danger of splitting up our formation is removed. This would, obviously, have exposed our own formations to attacks from the enemy's escort fighters. A first experiment for putting into practice such armament was the use of the Army 21 cm mortar shell.

As hopes fell for the effectiveness of the RZ 65 in June 1943, so a consignment of 30 21 cm mortar tubes, together with 200 shells from the Wehrmacht munitions storage facility at Lübeck-Gestringen, was delivered to the Fw 190-equipped I./JG 1 at Schiphol, with a further 34 tubes and 200 shells going to II./JG 26 in France, also flying Focke-Wulfs, where trials were placed under the supervision of Leutnant Otto Hummel of 5. *Staffel*.

It was at this point that Hauptmann Tratt, who in the meantime had been appointed to command the *Zerstörerstaffel* of E.Kdo 25, was assigned temporarily to I./JG 1, where he formed the *Erprobungskommando/*JG 1, equipped with four Fw 190A-4s, specifically to carry out tests with the mortar. Firing took place over the North Sea, and as early as 13 June, three B-17s were claimed by mortars over the German Bight, while on the 22nd, Oberfeldwebel Hans Laun and Oberfeldwebel Günter Fick of I./JG 1 claimed a further two *Viermots* shot down and two damaged. These initial results proved sufficiently satisfactory for trials to continue using aircraft of both JG 1 and JG 26, as well as E.Kdo 25 and the weapons testing centre at Tarnewitz.

Underside of Fw 190A-8/R2 fitted with 21 cm W.Gr. 21 air-to-air mortar tubes.

Geyer remembered:

Unlike other missiles, the 21 cm *Werfer*, which came to us from the Army, was not equipped with fins or stabilisers. Rather, this weapon was stabilised by its own spin which, in turn, was created by the blast from initial ignition and the subsequent velocity. The 21 cm shell turned two or three times per second after leaving its launch tube, but speed increased rapidly thereafter.

We observed that the shell did not run straight to its intended target, but rather it spiralled and therefore often missed the target. To overcome this, the manufacturer built in a time fuse intended to detonate the shell at a pre-set time. We usually fired the weapon from a range of 400 m, and from our experience with it, we were able to set the fuse correctly, compensating, of course, for the approach speed of the target. However, the closer to the target you were, so the greater the blast and the success of the weapon.

The W.Gr.21 rifled mortar launching tube took the form of an open cylinder constructed of rolled steel sheet 85 mm thick, butt-welded down its length. The external diameter was 250 mm

The view from the rear of a rolled, steel sheet W.Gr.21 mortar launching tube fitted to the wing underside of an Fw 190. Visible is the screw bolt to hold in the shell when loaded, ignition wires, four bracing struts and the central suspension hook and lug. On the wing surface can be seen the connection plate. In an emergency, the launch tube could be jettisoned by activating an electrically primed explosive charge that severed the central hook. (EN Archive)

Fw 190A-5 Wk-Nr 1372 'White 4' of E.Kdo 25 was jacked up and rested on wheel chocks to ground-test arrays of single, double and triple sets of 21 cm mortar launchers at Barth on the Baltic coast. The aircraft was pointed towards the waters of the Barther Bodden, and in this photograph the mortars appear to have been loaded for a rearward-firing trial. (EN Archive)

and the length 1.3 m. Inside the tube were three symmetrically placed guides riveted to the tube, of extruded L-section, which projected 338 mm into the tube. The guides acted as runners for the projectile and, simultaneously, served as stiffeners.

One tube was suspended from beneath each underside wing surface of an Fw 190A-4/R6 by means of four bracing lugs and a central hook with a suspension bracket. Three retaining springs, located near the rear end of the tube, held the 112 kg shell with its 40 kg warhead in place and a screw bolt, also at the rear end of the tube, prevented the shell from sliding out. In an emergency, the launching tube could be jettisoned by activating an electrically primed explosive charge which severed the central hook.

The mortars were controlled from a cockpit armament panel containing two armament switches and a *Revi* 16B reflector sight. Two spin-stabilised shells were fired simultaneously when the pilot depressed a button on his control column. As Geyer states, the mortar shells were fitted with a time fuse which was of the S 30 standard clockwork type, capable of giving a delay of up to 30 seconds. It was pre-set prior to delivery to an operational unit and not subsequently adjusted. In theory, the firing range was therefore invariable, and the weapon's low velocity meant that to be effective it had to be aimed 60 m above its target and a shell had to detonate within 28 m of a bomber.

Fw 190A-5 Wk-Nr 1372 'White 4' fitted for trials with three 21 cm W.Gr.21 mortars, *Erprobungskommando* 25, Barth, early 1944.

In this dramatic image, the tail of Fw 190 'White 4' has been lowered to the ground as a mortar is remotely launched forward. Note the camera at left that is filming the process and the spare launch tube on the trolley in the foreground. (EN Archive)

The weapon, or 'stovepipe' as the Germans came to call them, was used in numbers for the first time operationally on 28 July 1943 during USAAF raids to Kassel and Oschersleben, and results were acceptable in as much as fragmentation from blast did break up the bombers and a number were claimed destroyed as an indirect result. In a report prepared in late August 1943, the headquarters of the Eighth Air Force warned that the mortar appeared 'to be the most dangerous single obstacle in the path of our bomber offensive.'

TOP RIGHT Accidents will happen: the damaged trailing edge and control surfaces of the starboard wing of 'White 4' following the launch of a 21 cm mortar shell. The lack of the securing bolt at the end of the tube, the angle of the tube and the area of damage suggest that this was as a result of a rearward-firing test. (EN Archive)

BELOW This B-17F was lucky to survive and return to base after having been hit by a 21 cm mortar shell during a mission in July 1943. The damage to the bomber's radio room and navigation station was probably caused by close blast effect from an exploding shell. (Forsyth)

Due to the limited ground organisation at Wittmundhafen, E.Kdo 25 relocated 150 km south to Achmer in early October 1943, and from there, Geyer claimed one more victory with the *Kommando* when he shot down a B-24 from a group belonging to the 2nd Bomb Division (BD) between Münster and Osnabrück on the 8th. As the enemy formation made its return from a raid on Vegesack, Geyer took off from Achmer leading a flight of three Fw 190s each fitted with two W.Gr.21 mortars. Attacking the Liberators from the rear, Geyer fired his mortars at one bomber, but realised only one of his launch tubes was functioning. As he jettisoned both tubes, he noticed

Armourers heave a 21 cm mortar shell into the outer launch tube of a pair fitted to the underside port wing of a Bf 110, their efforts being closely watched by Hauptmann Johannes Kiel, *Gruppenkommandeur* of II./ZG 76. The shell weighed 112 kg, so a 'battery' of four such mortars added considerable weight to a Bf 110 or Me 410, greatly affecting speed and manouevrability in types that were already outclassed in the air by Allied escort fighters. Johannes Kiel was awarded the Knight's Cross on 18 March 1942, and he is believed to have scored around 20 victories. Kiel was killed in action on 29 January 1944 in air combat with aircraft of the USAAF. (EN Archive)

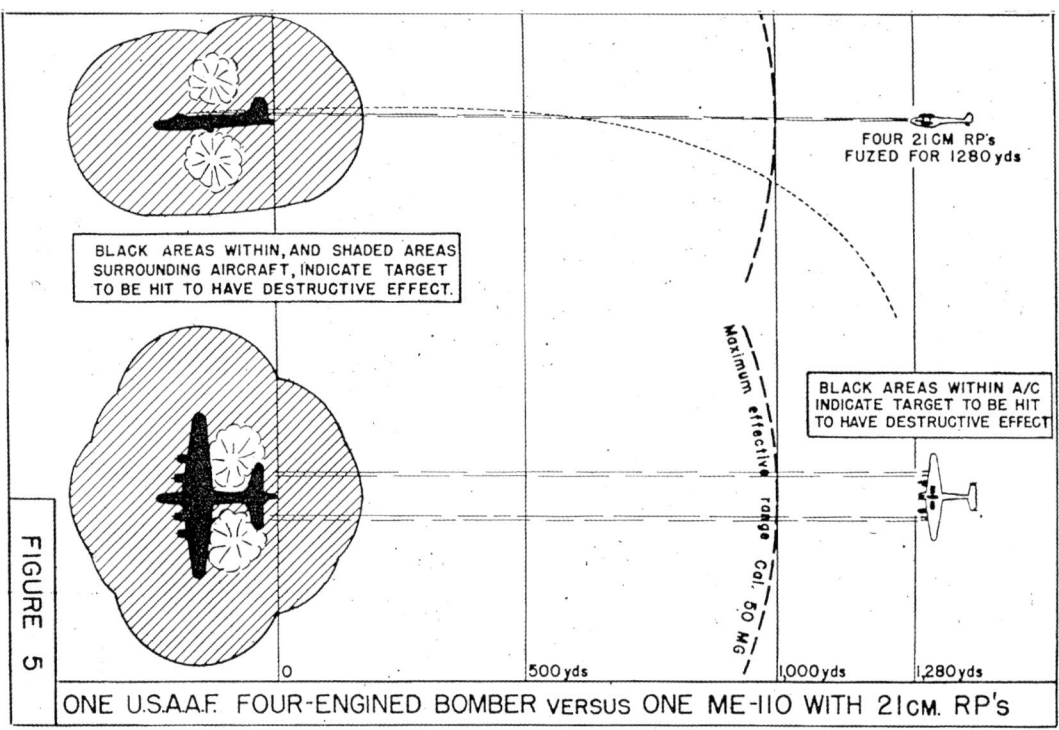

FOUR 21CM RP's
FUZED FOR 1280 yds

BLACK AREAS WITHIN, AND SHADED AREAS
SURROUNDING AIRCRAFT, INDICATE TARGET
TO BE HIT TO HAVE DESTRUCTIVE EFFECT.

BLACK AREAS WITHIN A/C
INDICATE TARGET TO BE HIT
TO HAVE DESTRUCTIVE EFFECT

Maximum effective range   Cal. 50 MG

FIGURE 5

0   500 yds   1,000 yds   1,280 yds

ONE U.S.A.F. FOUR-ENGINED BOMBER versus ONE ME-110 WITH 21CM. RP's

Instruction diagram for USAAF bomber crews illustrating the potential destructive area surrounding a B-17 by four 21 cm projectiles when fired by a Bf 110 at a range of 1,280 yards. The diagram also illustrates how the Bf 110 would have been beyond the range of the bomber's 0.50cal machine guns. (Forsyth)

that the B-24 had tipped away from its formation and was falling through the sky. Geyer pursued it and opened fire with several long bursts from his 20 mm MG 151/20 cannon, following which he observed:

> . . . considerable damage to the fin assembly and heavy smoke coming from the inner starboard engine. But right then several Mustangs suddenly rushed down on us and I gave the order to evade. One of my wingmen had also succeeded in shooting down a B-24 using his mortars.

However, the mortar was perhaps used to its greatest effect against the infamous American mission to Schweinfurt on 14 October 1943, when 62 *Viermots* were shot down, many as a result of being dispersed from their formations by the use of the mortar.

'Stovepipes' were also fitted to Bf 109G-6s of IV./JG 3, I., II. and III./JG 53, I. and III./JG 77 and I./JG 5, and were used to varying effect in the Mediterranean and in Rumania from August 1943 until early 1944. Other Bf 109s of 7./JG 3, 5./JG 11, 2./JG 27 and 6./JG 51, so-equipped, operated in the *Reichsluftverteidigung*, and a number of Bf 110G-2/R-3 *Zerstörer* of ZG 76 and Me 410As of ZG 26 carried pairs of twin mortar sets specially assembled by the Maschinenfabrik in Donauwörth, in addition to an array of cannon and machine guns, to operate as heavily armed bomber-destroyers.

On 10 October Major Karl Boehm-Tettelbach led the Bf 110s of ZG 76, together with Me 410s of III./ZG 1, against B-17s of the 3rd BD during an attack on the marshalling

Me 210A-1 Wk-Nr 2102251 2N+DD of ZG 26 has been fitted with four 21 cm air-to-air mortar tubes intended for breaking up and dispersing USAAF bomber *Pulks*. The tail fin and rudder have been painted white, probably for tactical and recognition purposes during operations in the air. (EN Archive)

yards at Münster. The Division left England without escort due to bad weather and had already been mauled by single-engined fighters, but the 14th Bomb Wing was particularly badly hit near Münster when the mortar-armed *Zerstörer* made a mass attack from the rear, inflicting considerable damage. As the American post-mission synopsis recorded:

> The fighters appeared to stay out of range, Me 110s firing at formation with long-range weapons slung under each wing and lobbing explosive . . . attacked from 800–1,000 yards firing rockets from under each wing (two distinct puffs were seen from each ship). Their formation resembled our defensive formation.

Furthermore, Bf 110s were seen to:

> . . . hit a B-17 by rocket, tail came off, plane broke in two. It then collided with another B-17 near Saerbeck. Both went down. No chutes.

No fewer than nine *Zerstörer* were also lost during the Münster raid, but this was exceptional. Losses usually ran at 5 to 10 per cent per mission and success levels were considered good, not just if bombers were shot down, but also where formations were scattered and disorganised, leaving them prey to the single-engined fighter units.

Aircraft from E.Kdo 25 were in the air to respond to the USAAF raid on Bremen on 29 November 1943 when 154 B-17s dropped 410 tons of bombs on the city. Two Bf 110s of the *Zerstörerstaffel* were involved, each fitted with four 21 cm mortars. Success was elusive and the *Kommando* advised Milch and Galland on 1 December:

> In operations to date, it has been shown that operations using a special weapon are purposeless when under *Schwarm* strength as smaller units are rendered unfit to engage by the strong defence even before making their own attack.

However, although further intensive trials continued under E.Kdo 25 until mid-1944 with the aim of improving the W.Gr.21 in terms of strength, weight, functioning and operational longevity, ultimately it was found that the launch tubes robbed German fighters – particularly the heavier *Zerstörer* – of their performance and made them vulnerable to Allied fighters. Senior Luftwaffe fighter commanders recognised the psychological effect of the mortars on bomber crews, but equally that when fitted to the Fw 190, a loss in speed of 40–50 km/h was incurred, as well as a loss of ceiling and manoeuvrability.

There was also a lack of a range-measuring device and therefore an inability to control the point of detonation. Over the Italian front on 30 January 1944, the *Staffelkapitän* of 2./JG 77, Hauptmann Armin Köhler, flying a Bf 109, recorded how on one mission against US bombers over Udine, 'I take hits in the starboard wing and the [W.Gr.21] tube is shot away.' Next day, when the Allied bombers returned, Köhler complained that, 'The mortars overshoot.'

Irrespective of what was happening in the Mediterranean, back in the Reich, Horst Geyer was overseeing another development for the W.Gr.21 – a rearward-firing version. In February 1944, following a suggestion made by Stabsingenieur Reyle of the RLM's Technical Office, Geyer noted that initial ground tests with a rearward-firing mortar had brought 'positive results', although further, presumably airborne, tests still had to be carried out. 'This installation is ready to go,' he recorded, 'and it is expected that with little effort, the W.Gr.21 used this way will be much more advantageous compared to the previous attack methods used.'

The intention was that a pilot would fire the mortar, known as the *Krebsgerät* (Crab Device), after he had made a firing pass using forward armament against a bomber formation and was in the process of passing through the enemy *Pulk*. The fuse would be set to detonate at 1.5–2 seconds after the weapon was fired, giving sufficient time for the carrying fighter to fly ahead and clear. There was a plan to make the tube jettisonable after firing but it is not thought this was ever followed through. It was hoped that a rearward-firing mortar would achieve surprise in the manner of a 'Parthian shot'.

In May 1944, Galland ordered that 20 Fw 190A-8s were to be fitted with the *Krebsgerät*, while Oberst Hannes Trautloft, the Inspector of Day Fighters, required one Me 410 to be installed with the rearward-firing mortar for trials with the *Zerstörerstaffel* of E.Kdo 25. On the one occasion the weapon was fired, the Me 410 suffered from a strong blowback, thick smoke filled the cockpit and its hydraulic system was very badly damaged. Despite this, by 15 July, it was planned to have 60 *Krebsgeräte* ready for installation into Fw 190s, with 16 fighters of E.Kdo 25 fully fitted out by 15 August. A new automatic, optically controlled firing mechanism, known as the *Wurzen*, was also being worked on by the Hugo Schneider Aktiengesellschaft Metallwarenfabrik (HASAG) company in Leipzig. By the end of August, however, Geyer recorded that only one Fw 190 had been fitted with the automatic device.

In early February 1944, at least one Me 410, believed to have been B-1 variant Wk-Nr 420416, was trialled at Tarnewitz with a rotating set of six 21 cm launch tubes. The tube 'battery' was installed in the Messerschmitt's underside nose compartment usually reserved

Fw 190A-8/R2 'Yellow 17' fitted with rear-firing 21 cm mortar and flown by Unteroffizier Willi Unger of 12./JG 3, Barth, Germany, May 1944.

Unteroffizier Willi Unger of 12./JG 3 photographed with his Fw 190A-8/R2 'Yellow 17' at Barth during trials with the *Krebsgerät*, a fuselage-mounted, rearward-firing 21 cm mortar. The intention was that a pilot would make a frontal attack against a bomber *Pulk*, and as he passed overhead, would fire the mortar. The weight of the device adversely affected speed and manoeuvrability, and posed a risk to other Luftwaffe fighters following behind. The weapon was abandoned. (Unger/Forsyth)

for bombs or the breeches of heavy cannon. However, the test-firing proved a failure when the airframe suffered significant damage and the idea was abandoned.

In the meantime, in May 1944, pilots of the Fw 190-equipped 12./JG 3, while based briefly at Barth, had attempted trials with a single rearward-firing 21 cm mortar tube. Just four of the *Staffel's* aircraft were installed with a form of the *Krebsgeräte* fitted beneath their centre sections, but they proved unreliable mechanically. The additional armour already added to the unit's Fw 190A-8 *Sturmjäger* affected performance, and at least one pilot who tested the weapon in combat, Unteroffizier Willi Unger, reported that the *Krebsgerät* simply caused a further deterioration in the fighter's speed and manoeuvrability. By late September, trials with both E.Kdo 25 and other operational units seem to have petered out.

An experimental installation carried by Me 410B-1 Wk-Nr 425416, comprising a revolving *Werferrohrbatterie* of six 21 cm mortar tubes. It was tested in February 1944 but the aircraft sustained severe damage in the process and there were no further firing attempts. (EN Archive)

In a report prepared for the Allies, one former German ballistics engineer who worked with E.Kdo 25 on the development of the W.Gr.21 gave his somewhat jaundiced opinion:

> The first experimental attack with the mortars was a success. But this was a one-off, lucky, direct hit on a bomber that probably caused its bombload to explode, which, in turn, caused the downing of two further bombers flying next to it. Encouraged by this chance success, the *Erprobungskommando* recommended the installation of the mortars. In spite of repeated warnings from the *Technisches Amt* over premature optimism with regard to the strike and success possibilities, the OKL ordered the installation. Many fighter units installed them by their own means. But no further success could be reported next time they were used by the squadrons. The experiences of the pilots and the personal experiences of the author of this report with the fighter units can be summarised as follows with regard to the installation of the 21 cm mortar:

Successful shoot-downs were possible only through the occasional direct hit. A shell which detonated at a distance of ten metres from a bomber was without effect.

The installation was generally rejected by the fighter pilots because its use was without success. Apart from this, our fighters were much more exposed to the enemy escort fighters as a result of a loss in performance. Only the development of the special R4M rocket removed the misgivings of the *Technisches Amt* and the squadrons.

While dispersion (see RZ 65) had caused difficulties with the W.Gr.21, the dominant error lay in the fact that firing was usually carried out at a much greater range than that for which the time fuse was set – i.e., ranges of 2,000 to 3,000 m instead of the nominal range of 1,200 m.

In his lecture of October 1944, Flieger Oberst Stabsingenieur Voss damned the use of the 21 cm mortar, while also admitting its value in inspiring future weapon development:

> This experiment was bound to fail because it was carried through with entirely inadequate resources, quite apart from the fact that high-explosive shells of heavy calibre with time fuses are, on principle, to be considered as unsuitable for combat against air targets and would be of little success even when effectively delivered. Nevertheless, these experiments proved that combat against air targets with heavy calibre *rockets* [author's emphasis] launched from aircraft is, in principle, possible, and that success can be expected from this kind of armament if such rockets are suitably constructed.

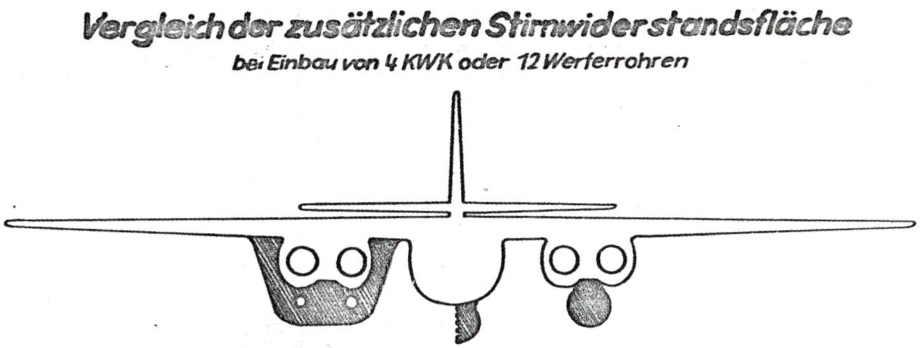

Vergleich der zusätzlichen Stirnwiderstandsfläche
bei Einbau von 4 KWK oder 12 Werferrohren

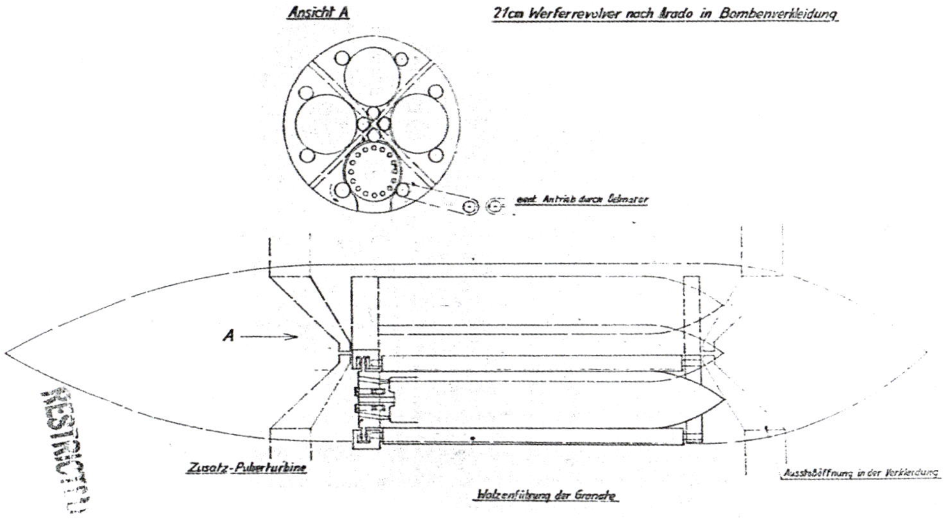

Ansicht A

21cm Werferrevolver nach Arado in Bombenverkleidung

Zusatz-Pulverturbine

Walzenführung der Granate

Ausstoßöffnung in der Verkleidung

In the summer of 1944, the Arado firm explored the possibility of mounting revolving batteries of four 21 cm *Werfer-Granate*, in what were termed *Werferrohren*, on to its new Ar 234 jets. The batteries were to be housed in faired casings which resembled bombs or drop tanks. It was proposed that the shells could be released singly or in salvos, but no development work is known to have been undertaken beyond the drawing board.

By late 1944 the W.Gr.21 had all but disappeared from use, although in March 1945 a small number of Me 262A-1a jet fighters of the *Stabsstaffel* and III./JG 7 were fitted with mortars in a brief and ultimately fruitless experiment.

HENSCHEL Hs 298 AIR-TO-AIR ROCKET

Designed by Professor Herbert Wagner (see Chapter Five, Hs 293), the Hs 298 was intended as a rail-launched, air-to-air rocket for deployment against enemy bombers with a 50 kg high-explosive warhead that was to be guided visually to a target and radio-controlled by two operators in the carrier aircraft.

Built of aluminium and magnesium alloy, the mid-wing Hs 298 had tapered, swept-back wings and a single horizontal stabiliser mounted high on the tail and twin vertical fins mounted outboard of the stabiliser. It was powered by a two-stage Schmidding 109-543 rocket motor, with solid fuel propellant manufactured by the Westfälisch-Anhaltische Sprengstoff-Aktiengesellschaft (WASAG) giving a burn time of 25–30 seconds. The first stage gave acceleration to push it ahead of the launch aircraft, while the second stage gave the rocket sufficient thrust to maintain flight and course at a constant speed until it reached its target. The warhead was to be fitted with a *Fox* radio *Zielabstandszünder* (proximity fuse) made by Allgemeine Electricitäts Gesellschaft (AEG) of Berlin or a Ruhrstahl *Kranich* acoustic proximity fuse that was effective within ten metres, and there was an impact fuse in case of direct hits.

A panel in the nose section carried the radio receiver, filter network, gyroscope, proximity fuse, generator and the propeller used for driving the generator. The main body of the rocket held the explosive charge and propulsion unit, which was supported by two bulkheads.

It was planned to launch the rocket from Do 217 or Fw 190 carrier aircraft. The Do 217 was to have been loaded with two rockets under each wing and a fifth on the fuselage centreline, while the Fw 190 would carry one under each wing. The rocket was held to its launch rail by an electrically operated bolt, and a rail provided a run at launch of around 1.8 m. Launches could be carried out in good visibility or at night in periods of moonlight.

The Hs 298 carried two spools of wire on its wingtips. One end of each wire was attached to the guidance controls on the carrier aircraft. As the rocket flew towards its target, the wires unwound from the spool, with electrical impulses for control sent along the wire. It was recognised that this system was susceptible to enemy countermeasures.

Control of the rocket was maintained by the Telefunken FuG 203/230 *Kehl-Strassburg* radio-controlled guidance system for guided/glide bombs. One operator on board the carrier aircraft aimed at the target using a reflector sight while the other used a telescope locked to the reflector sight by a servo system and guided the rocket by a joystick. The pilot had to manoeuvre the aircraft so as to keep the target ahead and to starboard as the aiming device was mounted on the starboard side.

The Hs 298 air-to-air missile suspended for testing beneath a Do 217M bomber fitted with flame dampers. (EN Archive)

During the rocket's final stages of development, it was decided to fit the Hs 298 with the advanced high-frequency, electromagnetically operated *Kakadu Zielabstandszünder* manufactured by the Donaulandische Apparatus Gesellschaft in Vienna. The *Kakadu* used the Doppler principle, with separate receiving and transmitting circuits for the purpose of increasing anti-jamming characteristics. The range of action was adjustable and had a maximum value of 25 m. A folded dipole arrangement feeding into a tuned capacity chamber was used. The antenna radiation had side lobes and was considered suitable for use against aircraft.

The average speed of an Hs 298 for a horizontal attack and an attack from above at 1,820 m was estimated at 865 km/h, while an attack mounted from below from 915 m was 725 km/h.

Test launches were carried out on 22 December 1944 using a Ju 88G as the carrier, with three Hs 298s being fired. One detonated during flight, one crashed and one remained fixed to the launch rail. Details of any further tests are not known. On 7 January 1945, a Ju 88G piloted by Flugbaumeister Diplom-Ingenieur Max Mayer, test pilot for unmanned, automatic and remotely controlled missiles at the Luftwaffe's *Versuchsstelle* at Peenemünde-West, carried out a test-firing of the Hs 298 V2, which had a higher-thrust rocket motor. On launch, the rocket cleared its rail but exploded just a short distance ahead of the Ju 88. 'A massive piece of missile debris became snagged in the Ju 88's port engine propeller,' recalled Mayer, 'and was slammed into the cockpit, causing me severe head injuries.' Mayer was knocked out for a few seconds but regained consciousness. He persuaded his crew to remain with the aircraft and managed a successful landing on one engine with blood pouring down his face.

In reality, it is doubtful whether the Hs 298 could have been effectively aimed at single bombers, but it may have had some use in dispersing *Pulks*. Production of the Hs 298 was to have commenced on 1 January 1945, but all work had ceased by 6 February. However, Henschel continued to work on 135 examples of the improved, increased range Hs 298, based on the V2, of which more than 100 in an almost complete condition were destroyed during a Soviet attack on the production plant at Wansdorf to the northwest of Berlin.

Henschel Hs 298	
Length	2.003 m
Span	1.290 m
Wing area	0.42 m²
Tail unit span	0.570 m
Fuselage Width	0.205 m
Weight	95 kg
Explosive weight	25 kg
Rocket Engine Weight	29 kg
Fuel	25 kg
Maximum speed	234 m per second
Range	1,600 m optimum (but between 500 –2,500 m)

The DWM 55 mm R4M *Orkan* air-to-air rocket with locked stabilising fins. Visible on the main body is the suspension lug. (Forsyth)

DEUTSCHES WAFFEN UND MUNITIONS FABRIK R4M *ORKAN* 55 mm AIR-TO-AIR ROCKET

Since the summer of 1943 German ballistics engineers had recognised that the installation of rockets would become 'indispensable' as the possibility of introducing greater ranges of fixed armament into a single fighter aircraft grew more difficult, combined with the corresponding increase in the defensive firepower of Allied bomber formations.

Throughout the latter half of 1943 and into 1944, following the mixed success of the W.Gr. 21 cm air mortar, German armaments experts came to the conclusion that the only plausible alternative was for a fighter formation to attack a bomber *Pulk* simultaneously, firing batteries of rockets carried either in underwing racks or in nose-mounted 'honeycombs'. These weapons, as it was described, could create 'a dense fire-chain' that would be impossible for the bombers to avoid. In June 1944, a requirement was put forward by the TLR for an electrically fired, fin-stabilised weapon, the warhead of which would contain sufficient explosive to destroy a four-engined bomber in one hit. Four weeks later, a consortium of companies comprising the DWM Research Institute of Lübeck, WASAG at Reinsdorf, Rheinmetall-Borsig and the Luftfahrtgerätewerk at Hakenfelde, each with individual responsibility for different components, was formed and led by DWM.

The resulting solution was an elongated rocket projectile guided by eight fins that extended automatically once the weapon had left its launching device. Extensive studies and experimental work indicated that if 400 g of explosive from such a missile penetrated any part of a bomber it would result in its destruction. The consortium actually settled on 500 g in case larger bombers with improved construction were manufactured. Like the MK 108 M-shell, the rocket depended primarily on blast for its effectiveness.

This consortium ultimately presented a proposal for an 814 mm-long 55 mm calibre rocket with a warhead containing 520 g of HTA explosive, ignited by an AZR 2 detonator, all bearing a weight of 3.5 kg. The rocket was intended to be launched against aerial targets from a range of 800 m and be stabilised by eight fins that would open automatically by aerodynamic drag immediately after launching.

The proposal was received favourably and the designation 'R4M' (*Rakete 4 kg Minenkopf*) applied to the project. Firing trials took place at the end of October 1944 on the Strehla firing range at the Westin works of Brünn AG and at Curt Heber at Osterode. However, the *Erprobungsstellen* at Rechlin (which had conducted the first air launches in December 1944) and Tarnewitz both judged that the missile was still unsatisfactory as a result of the poor standard of manufacture of some individual parts. By the end of January 1945, once some initial burn-out problems had been solved, a general re-working of the rocket, incorporating various aerodynamic and warhead refinements, was conducted.

In its final form, the R4M appeared as an unrotated, rail or tube-launched, single venturi, solid fuel-propelled, multi-fin stabilised missile, with the warhead contained in an exceptionally thin 1 mm sheet steel case enclosed in two pressed steel sections welded together and holding the Hexogen high-explosive charge. The missile bore a high charge

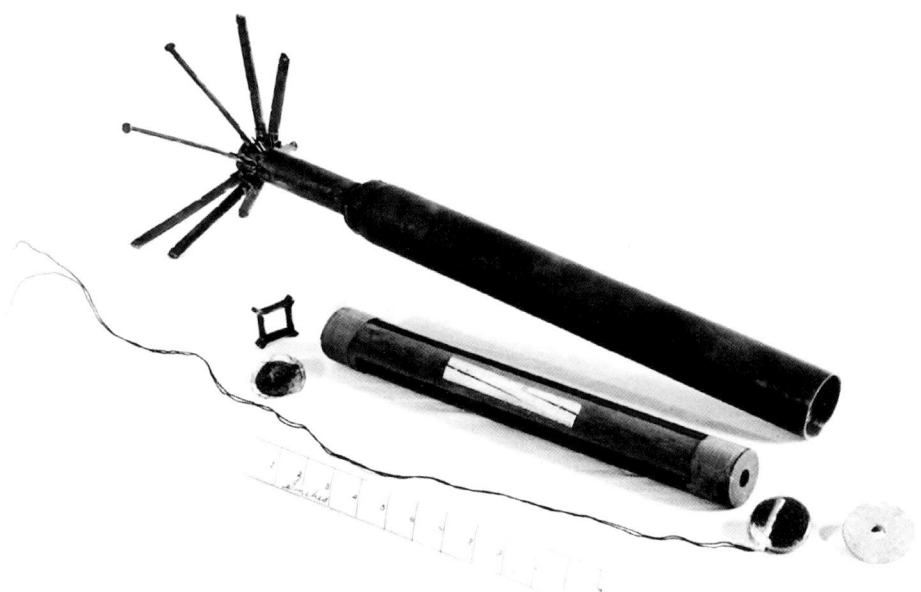

The component parts of the R4M rocket less the fuse and warhead, but showing the extended stabilising fins, ignition wire and the removed inner propellant chamber. (Forsyth)

weight to case weight ratio. The fuse was designed to discriminate between thin skin and main aircraft structure and to penetrate 60 to 100 cm into a target aircraft before detonation to give maximum blast effect.

Curt Heber was also developing another weapon in which R4Ms would be arranged in hexagonal tubes in a 'honeycomb' arrangement. A cluster of six tubes were to have a bore width of 56 mm and a sheet thickness of 0.75 mm. The tube length was equal to the length of the R4M. The carrier could be adapted in accordance with the number of projectiles required and the type of aircraft to which it was loaded. Due to the fact that the tail fins of the R4M were spring-loaded, the sequence of firing the projectiles was staggered using a special switching installation so that the rockets did not interfere with each other as they exited the canister.

For the Luftwaffe, the hope was that the rocket could be used by the new, high-speed Me 262 jet interceptor, which had entered service in the summer of 1944.

From the Me 262, it was intended to launch the R4M from wooden underwing racks, mounted by four screws and positioned outside of the engines, with the connections between the launch rack and the wing surface faired in to counteract the possibility of air eddies as much as possible. The standard launch rack – known as the EG.-R4M – measured approximately 700 mm in length, with each rocket being fitted with sliding lugs so that it could hang freely from the guide rails.

Prior to loading into the rack, seven of the R4M's eight fins were held in a folded-down position by binding them with spring-steel wire made with spherical or similarly thickened ends. The wire ends were then crossed and the eighth (free) fin pressed down to hold the other seven in place. Each rocket was then loaded from the rear of the rack, with the eighth fin held in place by the rail securing the wire binding. The rocket was pushed along the guide rail until the rear sliding lug was arrested by a notch in the rail. At the back of each rail was a terminal contact block connecting the ignition wires that

ASSEMBLY OF THE DWM 55 mm R4M *ORKAN* AIR-TO-AIR ROCKET

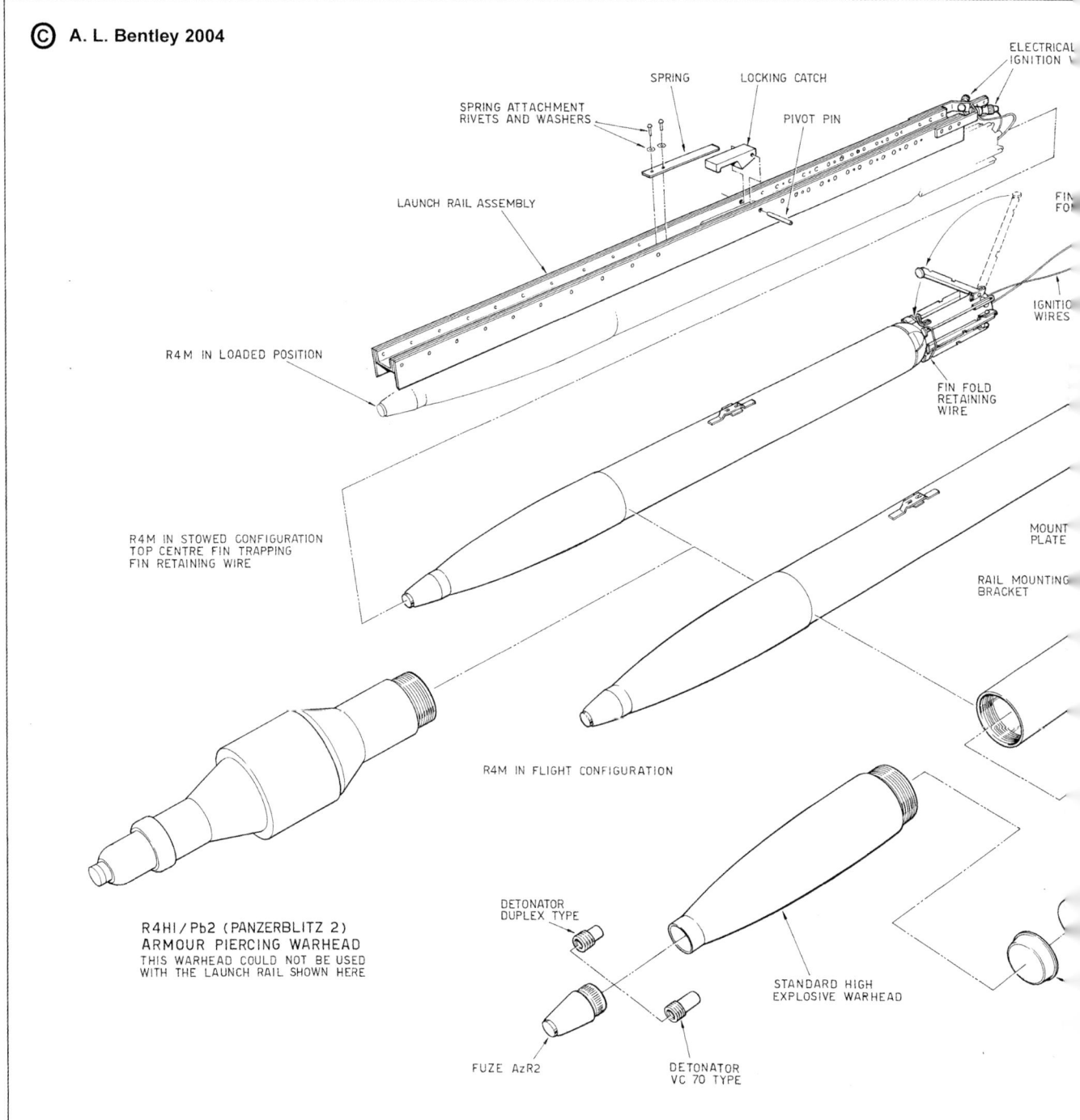

© A. L. Bentley 2004

SPRING

LOCKING CATCH

ELECTRICAL IGNITION

SPRING ATTACHMENT RIVETS AND WASHERS

PIVOT PIN

LAUNCH RAIL ASSEMBLY

FIN FO

IGNITIO WIRES

R4M IN LOADED POSITION

FIN FOLD RETAINING WIRE

R4M IN STOWED CONFIGURATION TOP CENTRE FIN TRAPPING FIN RETAINING WIRE

MOUNT PLATE

RAIL MOUNTING BRACKET

R4M IN FLIGHT CONFIGURATION

R4HI/Pb2 (PANZERBLITZ 2) ARMOUR PIERCING WARHEAD
THIS WARHEAD COULD NOT BE USED WITH THE LAUNCH RAIL SHOWN HERE

DETONATOR DUPLEX TYPE

STANDARD HIGH EXPLOSIVE WARHEAD

FUZE AzR2

DETONATOR VC 70 TYPE

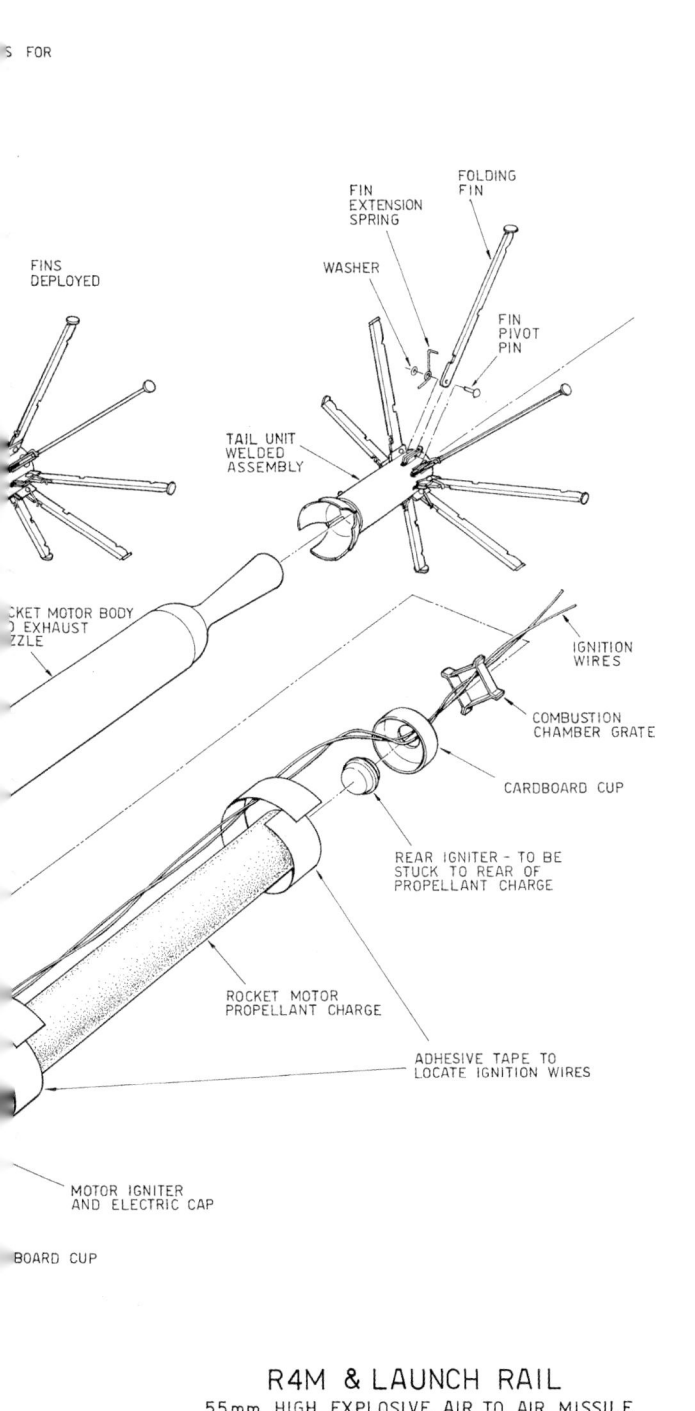

R4M & LAUNCH RAIL
55mm HIGH EXPLOSIVE AIR TO AIR MISSILE

FOLDING
FIN

FIN
EXTENSION
SPRING

WASHER

FINS
DEPLOYED

FIN
PIVOT
PIN

TAIL UNIT
WELDED
ASSEMBLY

S FOR

CKET MOTOR BODY
EXHAUST
ZZLE

IGNITION
WIRES

COMBUSTION
CHAMBER GRATE

CARDBOARD CUP

REAR IGNITER - TO BE
STUCK TO REAR OF
PROPELLANT CHARGE

ROCKET MOTOR
PROPELLANT CHARGE

ADHESIVE TAPE TO
LOCATE IGNITION WIRES

MOTOR IGNITER
AND ELECTRIC CAP

BOARD CUP

hung down close to the socket. Once fired, the eighth fin was designed to spring free, which, in turn, released the binding wire, thus allowing the remaining seven fins to open – a process which commenced at about 400 mm from the rail and finished once the rocket had flown approximately 2.5 m.

As many rails as desired could be fitted together to make one launch rack by means of transverse connection, with a gap of 65 mm between each rail. It was usual to carry a maximum load of 12 R4Ms under each wing of the Me 262, using a 21 kg rack. It was calculated that the loss of speed incurred to an Me 262 as a result of a Heber launch rack being fitted was approximately 16 km/h. It had been expected to obtain an 80 per cent kill score at 500 to 600 m.

Operational evaluation of the R4M was conducted by Major Georg Christl's *Jagdgruppe* 10 based at Redlin. His unit had been assigned the task of testing various types of experimental weapons systems produced by manufacturers and which were intended specifically as advanced fighter armament in the war against the bombers.

On 21 February 1945, a total of 200 practice missiles (the R4M-Gb) was delivered to the unit, but still defects were observed including corrosion in the combustion chambers and the warheads, which DWM diagnosed as a 'non-homogeneous mixing of powder'. These problems were eventually rectified and a pyrotechnic trials session followed in which the rockets 'smoked around in wild curves', having been fired from a static Fw 190. Christl and his technical officer, Hauptmann Karl Kiefer, then developed their own wooden underwing launching rack to carry 12 rockets. Still, the pattern of fire dispersal was far too wide, and so Christl and Kiefer attempted to fire the missiles at intervals using a bomb release switch taken from an He 177. Eventually it was found that control could be gained if six rockets were launched in two salvos.

Christl recommended that the first consignment of R4Ms be dispatched to 11./JG 7 at Parchim following eventual trouble-free launching on 15 March 1945. JG 7 was the first Luftwaffe fighter wing to be slated for full Me 262 deployment. Under the supervision of Willi Langhammer of Messerschmitt Augsburg, twin launch racks, each capable of carrying 12 R4Ms, were fitted to the Me 262 of Leutnant Karl Schnörrer of 11./JG 7. The jet was then immediately flown by Messerschmitt test pilot Fritz Wendel on the request of III./JG 7's *Kommandeur*, Major Rudolf Sinner. Schnörrer also made two subsequent flights in the aircraft before the weapon mountings were pronounced acceptable. Wendel reported:

> Without further ado, Wendel tested the converted machine. No adverse changes in flying qualities were discovered, except for perhaps a minor loss of airspeed in the climb. On 8 March, Leutnant Schnörrer carried out the first firing test. This did not go quite as planned as several rockets failed to leave their rails or burned out on the rack, fortunately without exploding. A second attempt by Schnörrer went so smoothly that his 9. *Staffel* immediately began to convert the other aircraft, and steps were taken to convert the other *Staffeln* plus the I. *Gruppe*.

Two days later, the Inspector of Day Fighters, Oberstleutnant Walter Dahl, visited Parchim and watched Oberleutnant Günther Wegmann fly a demonstration in one of the newly armed Me 262s. Using an old Savoia transport aircraft parked on the edge of the airfield as a target, Wegmann made a gentle descent and fired all his rockets. The Savoia was completely destroyed.

On 18 March nearly 1,200 American heavy bombers attacked railway and armaments factories in the Berlin area. They were escorted by 426 fighters. 9./JG 7 put up six aircraft, each fitted with two underwing batteries of 12 of the new R4M rockets. The jets intercepted the *Viermots* over Rathenow and a total of 144 rockets was fired into the American formation from distances of between 400 and 600 m. Pilots reported astonishing amounts

Twelve R4M rockets have been loaded onto the launch rack carried by this Me 262A-1a of 9./JG 7. Visible is the running fox emblem of the *Geschwader*. (Forsyth)

of debris and aluminium fragments – pieces of wing, engines and cockpits flew through the air from aircraft hit by the missiles.

Oberfähnrich Walter Windisch, who had two victories to his credit by the time he joined JG 7 from JG 52, was one of the first pilots of the *Geschwader* to experience the effect of the R4M in operational conditions:

Flying the Me 262 was like a kind of 'life insurance', but I was on that first sortie on 18 March during which R4M rockets were used and I experienced something beyond my conception. The destructive effect against the targets was immense. It almost gave me a feeling of being invincible. However, the launching grids for the rockets were not of optimum design – they were still too rough and ready, and compared with conventionally powered aircraft, when you went into a turn with the Me 262, flying became a lot more difficult because the trimming was not too good.

Based at Munich-Riem in southern Germany since early April 1945, JV 44, under the command of Generalleutnant Adolf Galland, was also equipped with Me 262s flown by a number of fighter commanders who had fallen foul of Göring, as well as several highly qualified and experienced Luftwaffe flying instructors. Adolf Galland recounted:

On the Me 262 we could mount the R4M outside of the turbines under the wings, 12 on each side, with little aerodynamic disturbance. They were fired over a switch relay in 0.03 seconds of one another and aimed in exactly the same way as the MK 108, with a natural dispersion of about 35 square metres. But on account of the arrangement of the rockets, a shotgun-like pattern was made creating a rectangle around the bomber. One hit – any hit – no matter where scored, sufficed to destroy a four-engined bomber. The loss of speed from the Me 262 as a result of mounting the R4M was insignificant. The projectiles were mounted with an upward inclination of eight degrees and fired at 600 m, at which range they had the same ballistics as the MK 108. When you fired them, you just heard 'ssssshhhh'. Just a whisper.

Johannes Steinhoff, the 176-victory ace who flew with Galland in JV 44, recalled even greater range capability:

The great advantage of the rockets was that though their speed only slightly exceeded that of sound, they could be let off 1,100 m away from the target – and from this range continued until they represented a field of fire of over 30 m x 14 m. This meant that by releasing all his rockets at once against a close formation of bombers, a pilot couldn't miss. We had, at last, the means not only of combating these hitherto almost unassailable formations, but of destroying them. But – and it is a big 'but' – it was 'five minutes to twelve'. In other words, early April, before we got the rocket armament, and then only enough to equip a few aircraft.

TOP RIGHT The wooden EG.-R4M launch rack, with channels for 12 R4Ms, fitted to the underside starboard wing of an Me 262A-1a of JG 7. Weighing 20 kg, more than one rack could be carried but as far as is known, this never happened. (EN Archive)

BELOW View of a loaded EG.-R4M rack carrying 12 R4Ms as fitted to an Me 262A-1a of JG 7. (Forsyth)

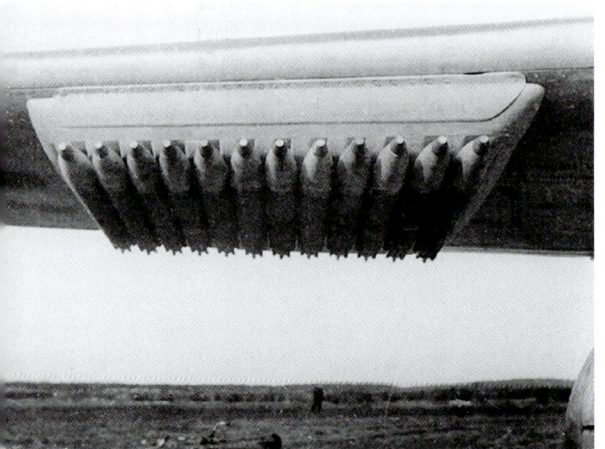

An example of the devastating effect of the R4M can be illustrated by JV 44's attack on a formation of Marauders from the Ninth Air Force's 323rd BG, which was one of three B-26 groups sent to bomb marshalling yards at Memmingen on 20 April 1945. Flying with the 455th Bombardment Squadron that day was B-26F-1 *Ugly Duckling*, flown by 1Lt James L. Vining. The Marauders of the 323rd BG had just reached the target area in clear skies when they were intercepted by around 15 Me 262s of JV 44, several armed with R4Ms, flying west in line astern in loose *Ketten* at between 3,500 and 4,000 m. Closing in, the jet pilots fired their salvos of rockets into the USAAF bombers. Vining recalled:

I tucked my wing closer to No 4, and at that instant a terrific blast went off below my knees and the plane rolled to the right. Sensing that my left leg was gone, I looked toward my co-pilot, and while ordering him to take the controls, I noted that the right engine was at idle speed. So in one swift arcing motion with my right hand I hit the feathering button, moved to the overhead rudder trim crank and trimmed the plane for single-engine operation, and, just as rapidly, pressed the intercom button to order the bombardier to jettison the two tons of bombs. We were losing altitude at 2,000 ft per minute.

Despite the loss of Vining's leg from the R4M attack and the severe damage sustained to the B-26, *Ugly Duckling* managed to crash-land in friendly territory, where the armourer/ gunner was killed and another gunner sustained broken legs and back injuries when the bomb-bay collapsed upon landing and the aircraft split into four pieces.

Also in the autumn of 1944, ballistics specialists at Rheinmetall-Borsig at Unterlüss conducted feasibility studies into developing 'honeycomb' launch pods, loaded with R4Ms, under the wings of the He 162 *Volksjäger*. A post-war report prepared by Rheinmetall-Borsig on the feasibility of mounting such a system onto the He 162 stated:

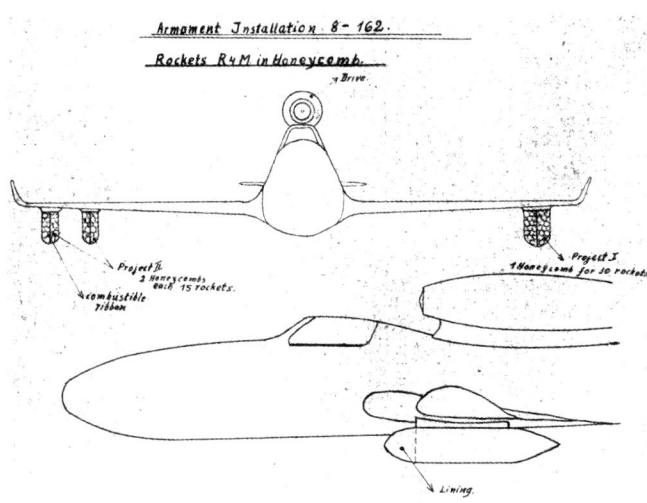

A drawing copied by technicians at the Halstead Exploiting Centre in southern England from a German original, showing proposed installations of 'honeycomb' batteries of R4M rockets in the He 162 *Volksjäger*. (Forsyth)

The R4M leaves behind a considerable stream of gas flows which means it cannot be installed in front of other fuselage components, and consequently the only arrangement which can be considered is under the wings.

It was planned to accommodate 30 rockets under each wing, packed tightly together and loaded into a hexagonal honeycomb container. With such an arrangement, only two rounds can be fired at a time, one round from each honeycomb. There must be an interval of 70 ms [milliseconds] between rounds in order to prevent succeeding projectiles being obstructed by the preceding ones. A rate of fire of 1,700 rounds per minute is therefore achievable.

By sub-dividing the 30-cell honeycomb into two combs with 15 rounds each and suspending them under the wing at a distance of approximately 500 mm apart, it is possible at any time to discharge four rounds simultaneously, thereby raising the cyclic rate to 3,400 rounds per minute. Since the R4M has a calibre of 5 cm, this is a considerable improvement in performance compared with other armament installations.

The firing of a burst of four rounds is effected by means of an electrical cylinder switch, such as has been designed for the SG 119 [see elsewhere in this chapter]. It is possible in this way to fire continuous bursts, as well as release a pre-determined number of bursts.

The honeycomb magazines are enclosed in an aerodynamic cover, which is discarded by the pilot before the attack by igniting a combustible tape.

The weight of the complete weapon installation including the ammunition is approximately 250 kg. An advantage here is that after firing the rockets, the aircraft has only 40 kg to carry, and this can also be jettisoned if desired. In other weapon installations high dead weights or the armament remains with the aircraft.

A similar installation was planned for the Bachem Ba 349 *Natter* vertically launched rocket interceptor in which grids of 24, 32 or 48 R4Ms would be accommodated in the nose behind a jettisonable cover. The fundamental purpose of the *Natter* was to destroy at least one bomber during its mission, and in this regard, a 'honeycomb' launcher would 'fit' perfectly within its nose. However, there were difficulties in accommodating such a weapons system in the *Natter* – the exhaust gases could not be discharged to the rear, but

had to be carried laterally past the cockpit. This caused comparatively sharp curves in the outlet channels so that the gas pressures reacted on the entire structure at almost full strength. The first experimental model of the *Natter* nose with this armament was completely destroyed on the first firing.

The 48-rocket honeycomb was planned as the solution to the *Natter*'s armament problems by circumventing the problem of the unacceptably long firing time of the single-discharge 32-round battery. The R4Ms would still be housed in a honeycomb installation, but the rounds would be fired in three salvos of 16. The passage of the first salvo down the hexagonal tubes would cause the next salvo to be fired, and so on. All the rounds could be fired in 0.3 seconds. The order of firing would allow adequate lateral separation of the rounds to avoid mutual interference.

According to a report written by R. Riecker and Dr. Kokott of Rheinmetall-Borsig:

> To use the superior speed of jet fighters to best advantage, to break the enemy fighter screen and to launch surprise attacks on bombers from any desired position, development changed to the arrangement of a *Schrottschussbewaffnung*, or multi-barrel armament. This arrangement consisted of a concentration of 48 R4M rockets in a honeycomb guide in the nose.

LEFT It was proposed to fit a battery of R4Ms in the nose of the Bachem Ba 349 *Natter* VTO rocket-powered interceptor. Here, the M22, fitted with a Walter rocket motor and boosters, and crewed by a dummy pilot, is made ready for a test-VTO at Heuberg, near Stetten am kalten Markt, on 22 February 1945. The take-off was successful, with the dummy pilot falling clear after separation of the craft's nose. However, residual fuel caused the landed fuselage to catch fire on the ground. Trials continued, even with one, unsuccessful, brief manned flight, but ultimately the *Natter* never entered service. (EN Archive)

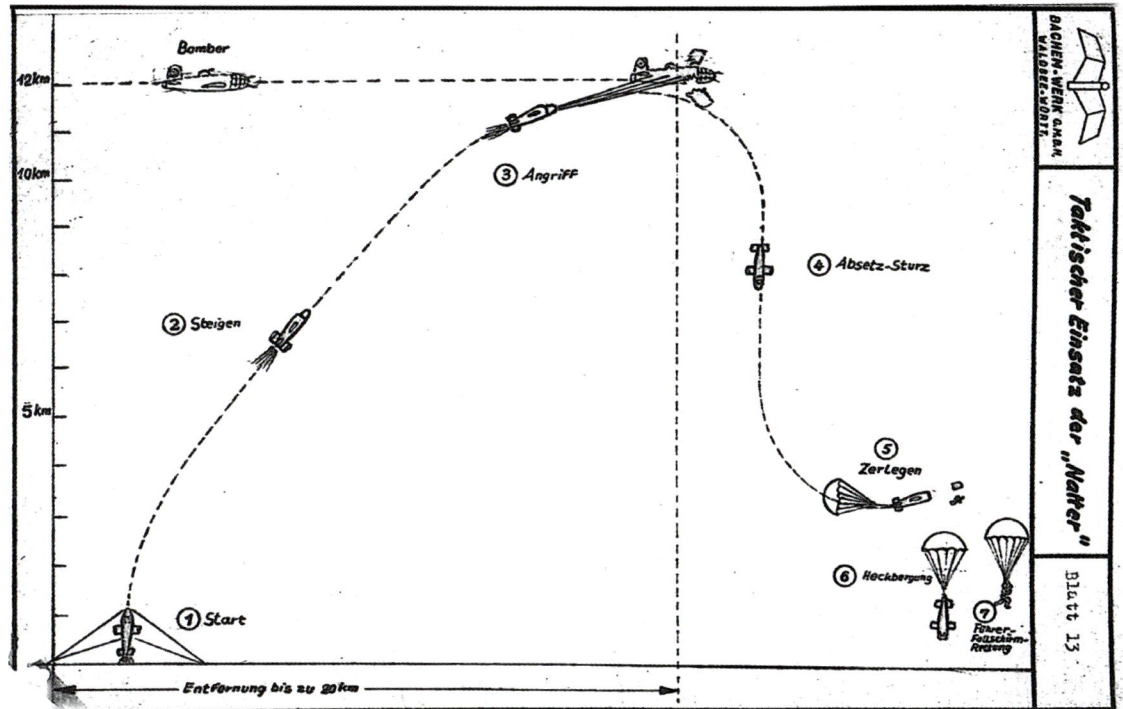

ABOVE A diagram prepared by the Bachem-Werk issued on 27 November 1944 showing the proposed 'Tactical Operation of the "*Natter*"' against enemy bombers. The Ba 349 *Natter* was to climb rapidly from its launch pole to an altitude of 11,000 m, where it would attack a bomber with a salvo of rockets. The *Natter* would then glide back to earth, and in the process its pilot would bail out as the craft's nose and main body separated. 1 – Take-Off; 2 – Climb; 3 – Attack; 4 – Return descent; 5 – Separation of components; 6 – Descent and recovery of rear section; 7 – Pilot parachutes and is recovered. (Gooden)

The concentrated number of projectiles is fired by the pilot and its dispersal is arranged to cover aiming and sighting errors and to secure a hit. The main idea of this armament was to give the fighter pilot, whose superior speed was his greatest combat asset, the possibility of using his firepower like a blow at the shortest possible distance which he could reach tactically and which in any case, for an attack, he would have to reach.

Concentrated discharge, fired from a number of barrels, has the advantage that any required suitable dispersion diagram can be controlled and registered. Its disadvantage, however, is that its action is limited to short range only, as at greater distances, owing to the limited quantity of ammunition, neither the required area of the dispersion diagram, nor sufficient concentration of shots can be obtained. Another disadvantage, which must not be underrated, is the effect on the morale of the fighter pilot, in so far as he is forced, for a certain time, to stay in the defensive zone of the bomber without being able to defend himself by firing bursts from his own weapons.

Despite the initial operational success with rack-mounted R4Ms installed on the Me 262s of JG 7 and JV 44, it was not until April 1945, two-and-a-half weeks before the end of the war, that DWM received an order to produce, immediately, 25,000 R4Ms. But it was too late. At the cessation of hostilities, the DWM plant was found cluttered with rockets in the process of production, but practically none had been completed.

R4M Air-to-Air Rocket	
Length	810 mm
Diameter	55 mm
Span of fins (extended)	240 mm
Launch weight	3.5 kg
Propellant weight	0.9 kg
Weight of explosive	500 g
Time of burn	0.8 sec
Velocity at end of burn	550 m/sec
Burn distance	180 m
Propellant grain	Single tubular
Fuse type	Pre-set impact
Manufacturer	DWM, Lübeck

RHEINMETALL-BORSIG R 100/BS 21 cm AIR-TO-AIR ROCKET

In 1943, Rheinmetall-Borsig at Berlin-Marienfelde submitted a proposal to the TLR for a *Bordtorpedo* ('aerial torpedo') for combatting enemy bomber formations. It was a proposal which, after some consideration, found favour, and TLR/*Flak* E6 issued a development order for an aerial torpedo 'similar to the Ruhrstahl X4 and the Hs 298' that was to be used at ranges of around 1,000–1,200 m from a targeted bomber *Pulk*, and which was to have, as it was described, the maximum 'shrapnel effect'.

Rheinmetall proceeded to work on the development of a 21 cm rocket with a warhead packed with 400 small, thermite *Brandsplitter* (incendiaries) each of 56 g and each fitted with a cutting ring. The rocket was 1.84 m in length and weighed 110 kg. It was adapted from an earlier rocket, the R 100/M, which had a *Minen* (high-explosive) warhead, but following tests it was decided that the blast produced from such a warhead was not as effective for attacking bomber formations as the composite incendiary warhead of the R 100/BS.

Although the late 1944 test-firings of the BS rocket suffered from interruptions caused by the worsening war situation, the first order for 500 rockets was eventually made in January 1945. By that time, the R 100/BS (the 'BS' suffix denoted *Brandsplitter*) was seen as a quick armament provision for the Me 410, which was to have carried six rockets, the Fw 190, which would have carried two, the Me 262, which would have carried five, and the Ar 234, which would have carried up to nine.

The R 100/BS was to have been fired from a Rheinmetall-Borsig-manufactured *Abschussschiene* (AG) 140 discharge rail which, in turn, was suspended from a standard ETC 50 bomb rack. All wires from the AG 140 were connected through one plug to the ETC 50. The AG 140 was 1,800 mm in length and had three sliding lugs, one in front and two at the rear. The front sliding lug had a hole through which projected an insulated pin. After ignition of the propelling charge, once the rocket had moved forward, the pin tipped over a bend in the AG 140 and pressed gently into the R 100/BS, cutting through the short circuit wire of the primer.

After another short distance the pin tripped over another bend (a contact curve), charging the primer in the warhead. This was done since if the primer was made live simultaneously with the rocket, both aircraft and crew would have been in danger in the event of a failure in the rocket's propelling charge. In such a case, the rocket remained on the discharge rail with a live primer, which detonated shortly afterwards according to the setting of the time fuse. This danger was avoided by use of the pin. Once the rocket had moved far enough for the pin to be completely pushed in the propelling charge, the rocket was discharged. The time fuse in the rocket detonated the warhead after an accurately pre-determined time.

As mentioned, the R 100/BS was intended for firing at approximately 1,000 m while the carrier aircraft made a frontal attack. The carrier aircraft's crew were able to compute the range and firing point by aid of an 'Oberon' clock, which took into account the speeds of approach; that is to say the speed of the crew's own aircraft and that of the target. Manufactured by Goertz of Vienna, the clock formed one element of the *Oberon* system, which was made up of a FuG 217 or 218 radar for range determination, the *Elfe* computer manufactured by the Siemens *Wernerwerk*, the Askania EZ 42 gyroscopic gunsight and an *Anstellwinkelgeber* device made by Zeiss Ikon in Dresden for measuring the attitude angle of the attacking aircraft.

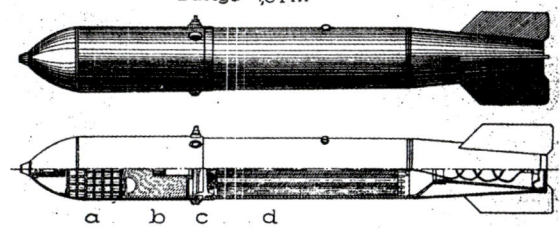

Die Bordrakete R 100 BS

Gesamtgewicht 110 Kg
Durchmesser 21 cm
Länge 1,84 m

a   b   c   d

a   400 Brandsplitter zu 56g Gewicht.
b   Ausstossladung
c   Einstellbarer Zeitzünder
d   Treibladung

ABOVE A Rheinmetall-Borsig drawing of the R 100/BS 21 cm air-to-air rocket. a – 400 Incendiaries each weighing 56 g; b – Primer; c – Adjustable time fuse; d – Propellant (Forsyth)

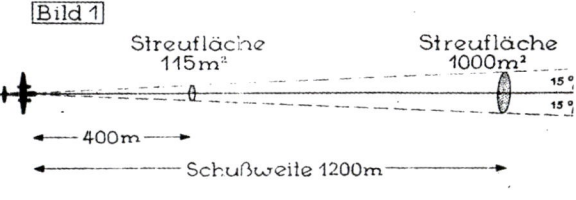

Bild 1

Streufläche 115 m²   Streufläche 1000 m²

15°
15°

400 m

Schußweite 1200 m

Vergrößerte Streufläche und erhöhter Munitionsaufwand bei großen Schußweiten

Bild 2

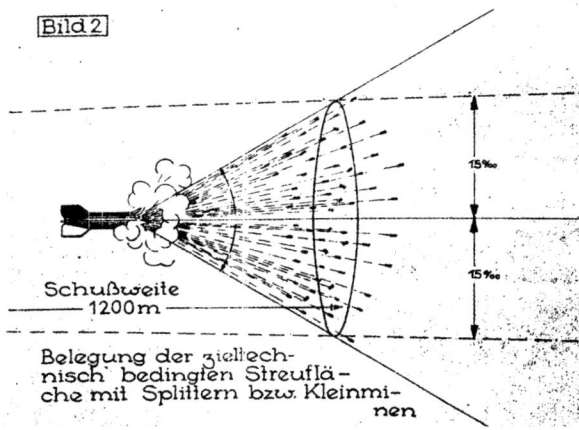

Schußweite 1200 m

15‰

15‰

Belegung der zieltechnisch bedingten Streufläche mit Splittern bzw. Kleinminen

ABOVE The first diagram shows how the greater the distance at which the R 100/BS was fired, the greater the area of dispersal covered by the incendiaries. The second diagram illustrates how one 56 g incendiary, once blown out from the R 100/BS, would penetrate the outer metal skin of an aircraft, trailing flaming splinters, before breaking into a fuel tank and igniting both the spilling fuel and the fuel still in the tank. (Forsyth)

The pilot firstly selected the fusing time delay from tables correlating closing speed, range and fusing delay. In pre-setting the time fuse, the speed of approach and the range were taken as a basis for calculation and the detonation was timed so that it occurred at a short distance before the target. The firing range was set in the *Oberon* clock and the rockets discharged at a distance of 1,000 m from the target.

Knowing his intended method of attack, and choosing a firing range of approximately 1,000–1,200 m, the pilot set an appropriate fusing time delay on a control knob either before take-off or during flight. The fighter then approached the bomber formation, the pilot followed the target using the EZ 42 and at a suitable point the rocket was fired, fused to explode at the target. Apart from setting the fusing time delay and tracking with the EZ 42, the pilot played no further part in the attack since the automatic devices based on the chosen fusing delay time made the necessary range determination, allowed for the aiming deflection angle, aircraft attitude angle and closing speed, and fired the rocket when the correct point was reached.

Upon detonation, the R 100/BS blew apart to release 400 incendiary splinters in a cone-shaped blast by means of a propelling charge. The incendiary splinters had such force that they were able to penetrate any part of a bomber's fuselage and/or fuel tanks, setting the fuselage or any type of fuel in the tanks on fire with their incendiary properties.

In a report dated 6 December 1944, a Rheinmetall engineer stated:

It can be assumed that when the R 100/BS is used for the first time, about five to six rockets will be required to shoot down an enemy aircraft. This not very favourable result is likely to last only the first few weeks of use, after which the completion of a useable warhead, which is now necessary, will probably enable a more balanced device [warhead, attack method] which can be used later. The effort required per shoot-down of an enemy should then drop to about three rockets.

   In other words, in the first few weeks there will probably be a need to deploy three Fw 190s (with two rockets each) to achieve one shoot-down. However, once the development has been perfected, two enemy aircraft can be expected to be shot down for each Ar 234 (equipped with six to nine missiles each) used.

Despite the advanced technical development work, however, it is believed that no more than 25 rockets were completed by the end of the war, and even those remained unused.

RUHRSTAHL X4 AIR-LAUNCHED ROCKET

In December 1944, a committee headed by Generalmajor Walter Dornberger, the commander of the *Heer*'s A4 (V2) rocket organisation, was set up to coordinate new anti-aircraft missile development for the defence of the Reich. The committee's main task was to cancel projects which were believed to duplicate others or which were too slow or complex, and to award contracts to those projects deemed to be workable and realistic.

TOP LEFT An Fw 190F-8 at the *E-Stelle* Karlshagen in the autumn of 1944 fitted with a Ruhrstahl X4 rocket under each wing suspended from ETC 71 racks. (Forsyth)

BELOW LEFT Close-up of a Ruhrstahl X4 rocket suspended from its ETC 71 carrying rack. Visible are the missile's insulated guide wires extending from the outrigger at the tip of the swept-back stabilising fin. Note the 'V71' applied to the lower fin, suggesting that this was just the 71st example to be built and still regarded as a 'prototype'. (Forsyth)

One of the few projects falling into the latter category was the Ruhrstahl X4, a remotely controlled, wire-guided, rocket-propelled, fin-stabilised, air-to-air missile fitted with a combination proximity/impact/self-destroying fused warhead intended for launching by fast fighters against formations of heavy bombers. Originally conceived in June 1943 by Dr. Max Kramer, a scientist and aerodynamicist who was employed by Ruhrstahl, and inspired by his SD 1400 *Fritz X* guided anti-ship glide bomb, the missile was manufactured by the Ruhrstahl AG Presswerke at Brackwede and Brinkmann & Co. in nearby Hövelhof.

The Dornberger Committee felt that the weapon offered sufficient promise to merit further development, and it was planned, initially, for carriage by the Fw 190 and Me 262, usually in batteries of four missiles per aircraft.

The X4 had a 20 kg warhead carrying a dinitroglycol-based explosive. It was cigar-shaped, with the warhead section made of steel. This was attached to the central section of the missile by wooden screws, together with an adaptor for the fuse attached to the nose in the same way.

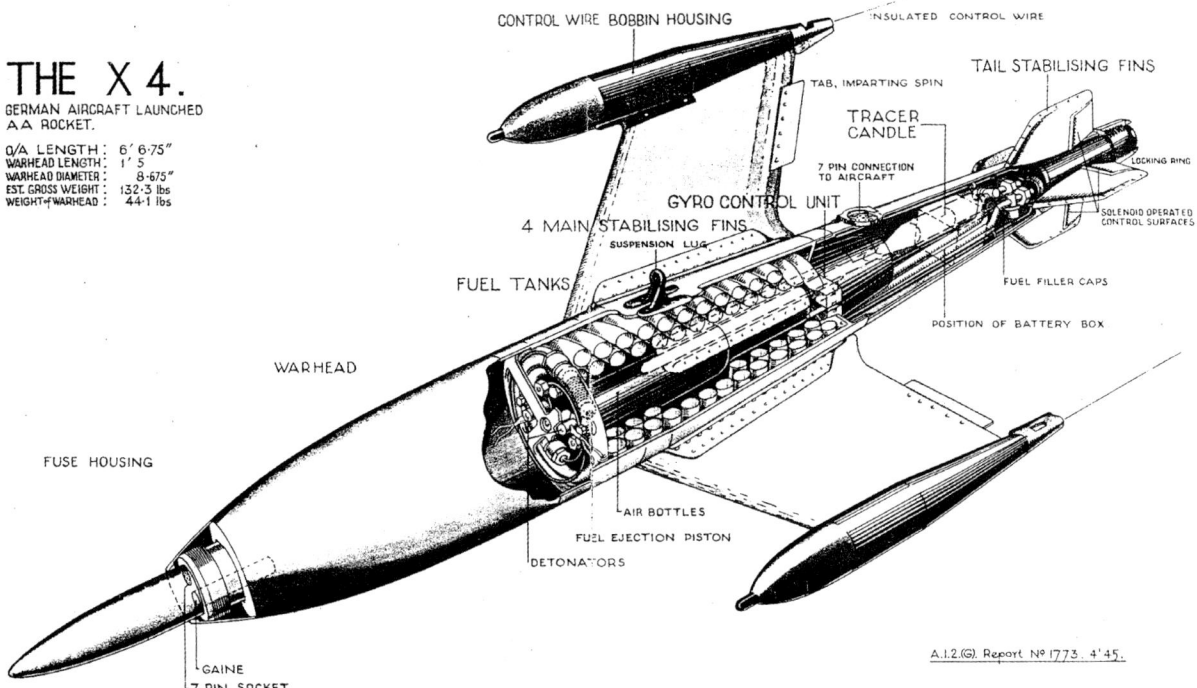

THE X 4.

GERMAN AIRCRAFT LAUNCHED
A.A ROCKET.

O/A LENGTH : 6' 6.75"
WARHEAD LENGTH : 1' 5
WARHEAD DIAMETER : 8.675"
EST. GROSS WEIGHT : 132.3 lbs
WEIGHT of WARHEAD : 44.1 lbs

CONTROL WIRE BOBBIN HOUSING
INSULATED CONTROL WIRE
TAB, IMPARTING SPIN
TAIL STABILISING FINS
TRACER CANDLE
7 PIN CONNECTION TO AIRCRAFT
LOCKING RING
SOLENOID OPERATED CONTROL SURFACES
GYRO CONTROL UNIT
4 MAIN STABILISING FINS
SUSPENSION LUG
FUEL FILLER CAPS
FUEL TANKS
POSITION OF BATTERY BOX
WARHEAD
AIR BOTTLES
FUEL EJECTION PISTON
FUSE HOUSING
DETONATORS
GAINE
7 PIN SOCKET

A.I.2.(G) Report No 1773. 4'45.

Detailed cutaway of the Ruhrstahl X4 produced by British Air Technical Intelligence. (Forsyth)

The central body of the missile, made of cast, machined aluminium, and the tail section of four pieces of thin aluminium sheets were welded together. The central section housed the helical aluminium tube fuel tanks and a combined, two-compartment steel air bottle. Through this section passed the electrical services to the fuse and air bottle opening valve.

The 'wing' fins and tail fins were cruciform in appearance. Four sharply swept-back plywood main stabilising fins were fitted symmetrically to the body of the missile by angled aluminium brackets. A steel suspension lug for hook-type suspension was riveted to the body mid-way between two of the fins. Two diametrically opposite fins carried torpedo-shaped containers, each housing a spool wound with approximately six kilometres of 0.2 mm-diameter insulated steel wire, to the end of which was fitted a plug to enable the wire to be connected to the carrying aircraft's control equipment. The other two fins were fitted with a tracer candle at their tips and with smaller imparting tabs than those fitted to their companion fins.

Four symmetrically placed fins were fitted to the aluminium tail unit, of aerofoil section set at 45 degrees in relation to the main stabilising fins. Each tail fin was fitted with a double-acting solenoid-operated control spoiler, the spoiler surfaces being in the form of small rakes. These vibrated five times per second, moving the full amount of travel to either side with each vibration. The tail section housed the gyro control unit, nine-volt battery and box, and fuel leads from the tanks to the fuel combustion chamber that was located at the extreme rear of the tail section.

The X4 was to be fitted to an underwing or under-fuselage 70 kg ETC 70 A1 or ETC 71 C1 launch rack. The latter had two 'outriggers' to which the spools of guide wire

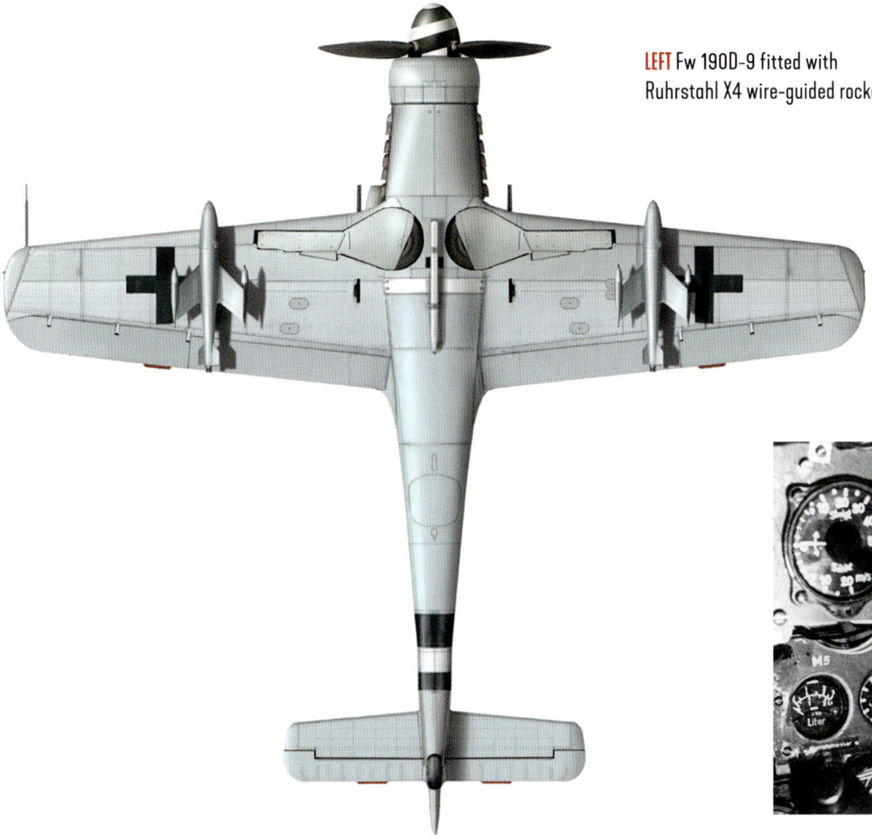

LEFT Fw 190D-9 fitted with Ruhrstahl X4 wire-guided rockets.

BELOW The X4 control toggle installed to the right of the instrument panel of an Fw 190. Its fitment forced the removal and resetting of some instruments, but immediately to its left is the aircraft's rate of climb/descent indicator. Visible at bottom left is the gun button on top of the pilot's control column. (Forsyth)

leads were fitted to the starboard and port sides. These outriggers were fitted with cap devices to jettison the ends of the wire which were still attached to them after the missiles were fired. The wires were fitted, during guiding, to the X4 through 'banana' plugs. The cap operation could be carried out separately for each ETC rack by means of a jettison lever. In an Fw 190 fighter, the starboard ETC jettison lever was labelled *Flächenlast* ('wing load'), while the port rack had a modified lever that was attached to the port inner wall of the cockpit.

Once fired, ideally from a range of 1.5–2.5 km, the X4 was controlled by the pilot using visual observation in conjunction with signalled corrections applied to the Telefunken FuG 510 *Düsseldorf* transmitter and FuG 238 *Detmold* receiver from a small control unit in the aircraft. This measured 10 cm x 10 cm and was connected to the missile's controls by two insulated wires. The pilot would move the control unit's *Knirps* joystick fore and aft for elevation correction and side-to-side for azimuth correction. Separate switches were used for pre-selection of any one of the four missiles carried and for starting the stabilising gyro prior to launching.

Testing the firing release mechanism on the ETC 70 rack beneath the starboard wing of an Fw 190. (Forsyth)

GENERAL ARRANGEMENT DRAWINGS OF RUHRSTAHL X4

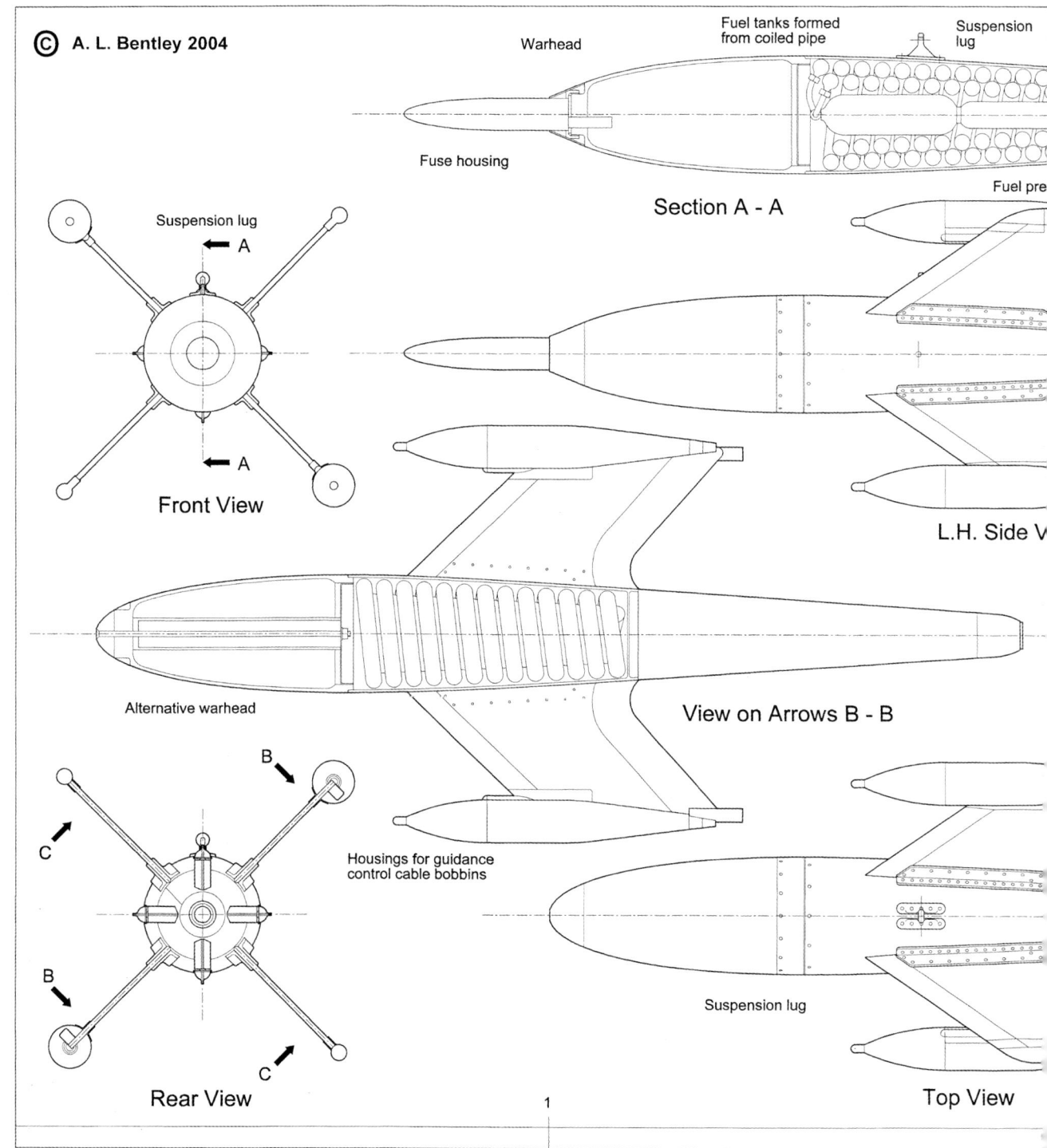

© A. L. Bentley 2004

Warhead

Fuel tanks formed from coiled pipe

Suspension lug

Fuse housing

Section A - A

Fuel pre

Suspension lug

← A

← A

Front View

L.H. Side V

Alternative warhead

View on Arrows B - B

B

C

Housings for guidance control cable bobbins

B

C

Suspension lug

Rear View

1

Top View

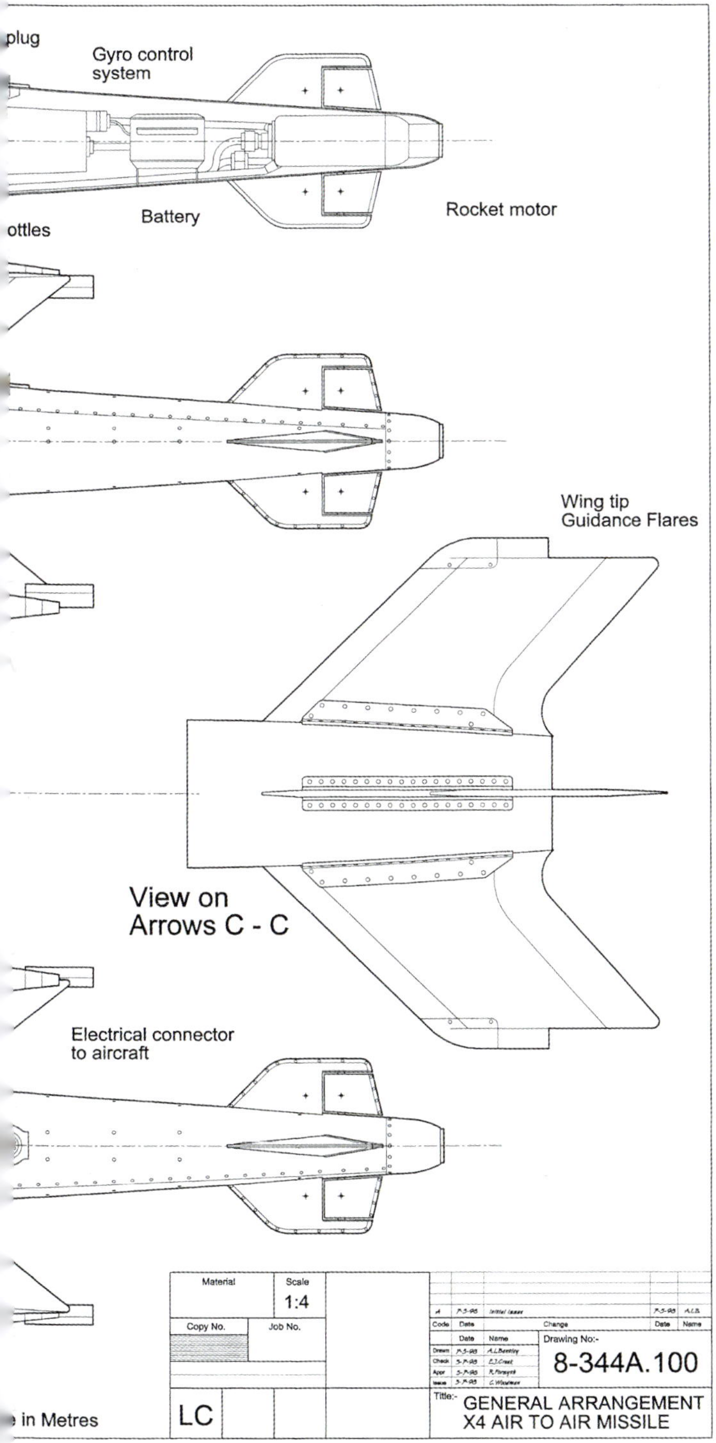

plug

Gyro control
system

Battery

ottles

Rocket motor

Wing tip
Guidance Flares

**View on
Arrows C - C**

Electrical connector
to aircraft

Material	Scale					
	1:4	A	P-5-95	Initial Issue	P-5-95	A.L.B.
Copy No.	Job No.	Code	Date	Change	Date	Name
		Drawn	P-5-95	A.L.Bentley	Drawing No:-	
		Check	5-P-95	E.J.Creed	**8-344A.100**	
		Appr	5-P-95	R.Forsyth		
		Issue	5-P-95	C.Woodward		
e in Metres	**LC**			Title:-	**GENERAL ARRANGEMENT X4 AIR TO AIR MISSILE**	

The pilot would aim at the target using a reflector sight and press the release button, which disengaged the gyro, fired the piercing detonators, armed the fuses, functioned the fin-tip tracer candles and released the missile all in the same instant.

On launching the X4 from the parent aircraft, a pyrotechnic train in the Rheinmetall *Kranich* fuse was initiated electrically, and seven seconds later it armed the proximity and impact elements. After a further 28 seconds, this train initiated the self-destruction element should the missile fail to find its target. Shortly after release, the X4 would reach a speed of some 250 m per second.

Propulsion came from a bi-fuel BMW 548 rocket motor developed by Dr. Ingenieur Helmut Zborowski that functioned from the reaction between *S-Stoff* (or *Salbei* – 'sage'), a nitric acid oxidant formed of 96 per cent nitric acid (HNO_3) and 4 per cent ferric chloride $FeCl_2$), and Tonka 250 rocket fuel, a composition of 57 per cent crude oxide monoxylidene with 43 per cent triethylamine.

The motor was formed of a double-compartment air bottle located within the helical fuel tanks that pressurised two aluminium fuel tanks delivering fuel to a combustion chamber and venturi. The motor was capable of providing an initial thrust of 125–145 kg, which fell progressively to 20–30 kg after 30 seconds.

ABOVE Test-firing of the Ruhrstahl X4 V59 from a Ju 88 over the waters of the Baltic during launch trials conducted at Karlshagen. (EN Archive)

© Jim Laurier

Although the X-4 was test-fired from the Fw 190, the plan was to deploy it primarily on the Me 262 jet fighter – each of the latter was to carry four missiles on underwing pylons. The weapon was controlled optically by the pilot using a *Knirps* joystick controller, sending course corrections via the wire link. Absolute precision was not required since it was anticipated that the missile would be fitted with a Ruhrstahl *Kranich* acoustic proximity fuse. French tests of an X-4 copy after the war found the performance of the missile extremely disappointing. Controlling a small missile from a fast-moving fighter using a simple joystick was a very inadequate form of guidance.

Stabilisation in flight was achieved by means of the four large swept wings fitted to the fuselage of the weapon, and the four small rear fins, the latter incorporating the solenoid-operated control surfaces through which two-dimensional directional control was achieved.

The first air-launched tests with the X4 were conducted in September 1944 at the Luftwaffe *Versuchsstelle* at Peenemünde-West using an Fw 190 as a carrier aircraft, and were considered partially successful, although weapons specialists always remained somewhat hesitant over the missile's deployment in numbers because of the volatility of its fuel system. Furthermore, it was thought that vulnerability of the parent aircraft to fighter attack would restrict the effective use of the missile in a combat situation, or at least restrict operations to missions against unescorted bomber formations.

Development is believed to have been officially terminated on 6 February 1945, although air tests continued through that month using Ju 88s. Test flights were also undertaken by an Me 262, with two X4s under the wings outboard of the jet nacelle, but they were not launched.

It was intended that all X4 components were to be crated and shipped directly from the factory, except the batteries, the S-*Stoff*, the detonators and the explosive charge. These items were to be delivered to units separately and were to be installed in the missile in the field when ready for use. Some 100–200 missiles are believed to have been completed, but many of the intended BMW 548 rocket motors were destroyed in air attacks on the company's works at Stargard.

After the war, in France, the *Arsenal de l'Aéronautique* manufactured a copy of the X4 for the *Armée de l'Air* that led to a run of missiles being used for tests, but these stopped in 1950 on grounds of fuel volatility.

Ruhrstahl X4 air-launched rocket	
Overall length	200 cm
Length of warhead	45 cm
Diameter of warhead at base	22 cm
All-up weight of missile before launch	60 kg
Weight of warhead	30 kg
Approximate fuel tank capacity: Salbei	4–5 litres
Tonka	2 litres
Maximum launching range	2,835 m
Glide ratio	1:5 to 1:6
Duration	30 seconds
Maximum acceleration	3 g
Maximum speed	805 km/h at 6,400 m

AIR-TO-AIR BOMBING

As early as September 1940, an engineer at the *E-Stelle* Rechlin had designed a high-explosive parachute fragmentation mine, the *Storkörper* SK 70/SK 106, which was intended to be dropped on enemy aircraft formations. In this device, a time release mechanism activated a small parachute that permitted, in turn, a slow descent of the mine towards its intended target.

According to Allied reports, at least one enterprising German fighter pilot serving in North Africa in February 1943 evolved a way of fixing four hinged containers to his Bf 109G-4, each one loaded with 34 standard infantry fragmentation hand grenades. Once packed into the container, the safety pins were removed and the detonator springs were held down only by the sides of the container and the tightness of the packing. One loaded container weighed just under 50 kg. The fighter would then be guided by a second fighter, which would report the height and speed of a targeted enemy bomber formation. The grenade-carrying Bf 109 would then fly at 1,000 m above the bombers and just so far ahead that vision of the leading bombers began to be observed off the wing, at which point the containers were dropped. The few flights that were made proved unsuccessful, primarily due to inaccuracy, but also due to the presence of escort fighters.

Throughout 1943, experiments using the Bf 109 and Fw 190 in this capacity had been carried out by elements of JGs 1 and 11 and E.Kdo 25 using SC 250 and SC 500 bombs in the defence of the Reich. The first attempts at air-to-air bombing were reported over northwest Europe by the USAAF in February 1943 when, during a raid to Saint-Nazaire on the 16th, the Eighth Air Force claimed that enemy aircraft dropped bombs on two crippled B-17s off the French coast.

In early March 1943 elements of Bf 109-equipped I./JG 1 based at Jever, in northern Germany, began experiments with dropping bombs onto USAAF bomber formations. The first batch of bombs arrived with the *Gruppe* on 8 March and 'training' began immediately. This consisted of the somewhat unlikely method of a Bf 109 attempting to drop bombs on sandbags that were towed endlessly across the sky by a Ju 88. It took ten days before the first practise hits were recorded.

Fw 190A-4 'White 2' of 1./JG 1, probably photographed at Deelen in the Netherlands in 1943, has been fitted with a 250 kg SC 250 bomb for air-to-air bombing. The aircraft had been assigned to I. *Gruppe* from IV./JG 1 during the reorganisation of the *Geschwader* in the spring, but it retains IV. *Gruppe's* markings and the *Geschwader* emblem. (EN Archive)

On 22 March, I./JG 1 was ordered to intercept a formation of bombers heading for Berlin. As had happened on previous occasions, there was insufficient time for the mechanics to fix bombs under the fuselages of the unit's fighters. However, the *Staffelkapitän*, Leutnant Heinz Knoke, passed control of his 2. *Staffel* to Feldwebel Hans-Gerd Wennekers and waited until a bomb was attached to his Bf 109. Slowly, he rolled along the runway. Suddenly a tyre burst and the aircraft became unbalanced. Knoke quickly fired a red flare and some 20 mechanics came running to his aid. Under Knoke's direction, they changed the tyre in record time. His aircraft finally took off, but it took Knoke 25 minutes to reach 9,000 m, the altitude necessary for an attack. He eventually caught up with the enemy formation as it headed for England after having set fire to the docks at Wilhelmshaven. Knoke placed himself above the leading bombers, but the defensive fire around him was intense. His left wing was hit and slightly torn, but the Messerschmitt still flew. At 1,000 m above the enemy aircraft, he released his bomb. It would take 15 seconds to detonate.

Slowly, the bomb fell towards a group of three bombers. Knoke counted the seconds until it exploded. The nearest bomber lost a wing and crashed into the sea, while the other two broke formation. Following his mission, Knoke's exploits became the subject of considerable interest. Reichsmarschall Göring even called to congratulate him on his initiative.

On 3 April, IV./JG 1, together with some 'bombed-up' Fw 190s, were scrambled from Deelen, but were unable to make contact with any enemy formation. On 17 April, the air-to-air bombing method was tried once again. This time, Knoke's bomb passed harmlessly through a bomber *Pulk* without inflicting any damage. The Eighth Air Force noted:

Armourers load an SC 250 bomb from a hydraulic trolley to the underside of an Fw 190A-4 of I./JG 1 based in the Netherlands in the spring of 1943. (EN Archive)

Inaccurate aerial bombing was reported. All in all, about 20 bombs were dropped; about half fell right through the formation. Believed to be 50 kg bombs, none of which burst closer than about 150 ft.

On 3 May, a number of Fw 190s from I./JG 1 went into battle equipped with bombs, but their efforts were without success. Conversely, on the 14th, during an American raid against the Kiel shipyards, things improved when three *Viermots* were apparently destroyed by bombs. Brig Gen Frederick L. Anderson, commander of the 4th BW, reported:

At 1313 hrs, I noted the first fighters in our vicinity; there were 23 airplanes off to our right front and they appeared to be P-47s making a sweep. However, they suddenly turned out to be Fw 190s. They turned sharply to the right, the whole 23 attacking in a string. I noticed three of them drop bombs on this attack; other members of our crew reported as many as six bombs. They dropped these bombs in head-on attacks, apparently with very short time fuses, in an effort to break up our formation. Two of the bombs came very close to the lead plane and exploded behind us.

The Eighth Air Force Bomber Command Narrative of Operations stated:

Three crews saw explosions within 50 yards of them, the shell or bomb disintegrating in a puff of black or green smoke from which streamers of black and green fluttered down, looking like strips of colored toilet paper. One aircraft ran through these streamers, some hit the nose of the ship, but doing no damage. Two aircraft reported aerial bombs with white parachutes; one aircraft saw three and the other one. None seen to explode. Aerial bombs were observed by two other aircraft, dropping down and exploding with blue, smoky bursts. No parachutes were attached. These bombs were seen bursting on the approximate position of the preceding formation. Two aircraft saw Flak bursting which began with the level of formation and continued above it. Pink explosion was followed by a puff of grey smoke. One crew saw a shell or bomb disintegrating into a number of shell fragments, each of which in turn exploded with a pink, smokeless burst.

Encouraged, the fighter-bombers continued with their tactics. On one occasion, on 28 July 1943, Unteroffizier Wilhelm Fest destroyed three bombers with a single hit. This was confirmed by the Eighth Air Force in its post-mission report:

Two crews reported bombs were dropped on formation by parachute. Loss of three B-17s are attributed to air-to-air bombing.

Despite this success, and regular attempts at bombing throughout May, June and July 1943, such methods were eventually abandoned because of the appearance of long-range Allied fighter escort.

However, also in July of that year, an engineer in the aerodynamics department of Focke-Wulf at Bad Eilsen proposed a method of dropping bombs onto 'flying targets' using an SC 500 bomb dropped from an He 111 carrier. Calculations were made, using as a theoretical target the 132 m² wing area of a B-17F. The bomb was intended to be dropped from a height of 100 m above the bomber, but as far as is known, the proposal never developed beyond the report stage.

A USAAF report dated August 1943 warned:

> Over and against the difficulties involved, however, must be set the potential rewards of success with air-to-air bombing. These find their simplest explanation in a plain statement of the difference in the size of the lethal burst between the best Flak and a 500 lb bomb. The largest lethal burst for Flak is 50 ft. The lethal burst of a 500 lb bomb is 300 ft. When it is considered that this burst could encompass as many as four of our Flying Fortresses in tight formation, it can be seen why air-to-air bombing, for all its difficulties, holds out such dazzling attraction. It is the considerable possibility, rather than the likelihood of any high proportion of direct hits by air-to-air bombing, which must compel us to regard it as a serious and increasing threat.

On a raid to Bremen on 13 November 1943, USAAF bomber crews reported seeing Fw 190s dropping bombs into their formation which 'lit up like flares'. On the 29th, E.Kdo 25 tried again, this time using a variation on the theme.

Gerät Liesel was a device developed by the *Forschungsanstalt Graf Zeppelin* based at Ruit near Stuttgart, using a 500 kg SD 500A *Splitterbombe* fitted with El.AZ (25) A, C or El.AZ (55) fuses suspended from at ETC 500 rack beneath an Fw 190. The bomb was to be rigged as a *Fallschirmbombe* (parachute bomb). The parachute was designed to fall at 700 km/h, but in triggering a brake device 100 m above the target, the speed would reduce to 500 km/h and a drop of 270 m could be achieved in 5.7 seconds. The tactical intention was that a *Schwarm* of Fw 190s fitted with *Liesel* was to fly in the same direction and height as an enemy bomber formation, then overtake it before turning back towards it, approaching from the front at an altitude some 100 m higher. In its fortnightly report for the second half of October 1943, Geyer stated on the *Kommando's* tactical intention:

> It is intended that when attacking from the front with a slight elevation, the bomb should be released about 800 m from the enemy formation. Attempts on target drones have had good results. The devices are available immediately in acceptable numbers.

In his following report, however, Geyer noted:

> Since use of the *Schlinge* and *Liesel* devices must be carried out with at least one *Schwarm* each, and the *Erprobungskommando* has only six Fw 190s with crews available for this purpose, in view of the importance of this operational trial, four additional fighter crews are requested, as well as the assignment of four Fw 190A-6s.

Work on the *Liesel* continued until the end of the year, but by then the *Kommando* still did not have its requested additional pilots or aircraft.

The British Air Ministry recorded in a report dated 15 January 1944:

> Very little success seems to have been achieved with air-to-air bombing. In recent reports of objects believed to be bombs falling through formations, the objects are frequently stated as having failed to explode. At the same time, there are many reports of Fw 190s and Me 109s being seen with belly tanks. In recent attacks some of the objects thought to have been bombs are recognised as jettisoned auxiliary fuel tanks. It is probable that this was the case with many earlier reports, and that air-to-air bombing has not in fact been used so frequently as was first thought. It has had little success.

The main tactical problem facing the Luftwaffe was that its piston-engined fighters needed too much time to attain the required height for bomb-dropping. This meant that although well within sight of the enemy, the conventional fighters waiting to exploit the weakness in the bombers' defensive capability caused by the bomb-dropping were forced, for their own safety, to hold back their attack until the bomb-dropping had taken place. Nor were German aircraft equipped with a suitable bombsight for such operations, and bombs were dropped according to the pilot's visual estimation.

It was also found that the Allied fighter escort tended to react immediately by despatching sufficient numbers of aircraft to a higher altitude to deal with the threat. Furthermore, even when dropped, it was discovered that the bomb types used had insufficient blast effect, and that their time fuses produced delays of varying and thus unreliable periods. However, such adversities did not mark the end of air-to-air bombing attempts.

Kommando Stamp

The idea of using the Me 262 jet interceptor in the air-to-air bombing role had originated with Oberst Berndt von Brauchitsch, Göring's chief adjutant. By deploying the Me 262 fitted with a jettisonable weapons canister containing semi-armour-piercing bombs equipped with time fuses designed to produce the maximum possible blast effect, von Brauchitsch believed that the jet's superior speed would overcome all the earlier disadvantages faced by the piston-engined fighters.

With such impetus, early in January 1945 four Me 262s were delivered to an experimental air-to-air bombing unit designated *Kommando Stamp*. This was led by Major Gerhard Stamp, a Knight's Cross-holder who had flown Ju 88s over the Mediterranean with I. *Lehrgeschwader* (LG) 1. He later transitioned to early *Wilde Sau* single-seat nightfighter operations, before taking command of I./JG 300, which provided high-altitude cover for the heavily armoured Fw 190-equipped *Sturmgruppen*.

In June 1944, Stamp had a chance meeting at Merzhausen airfield with a professor from the technical college at Braunschweig. The professor asked Stamp to test a barometric fuse that he had developed. Stamp promptly dropped the device from high altitude the

In January 1945, the bomber ace-turned fighter commander Major Gerhard Stamp (right) set up *Kommando Stamp* to test methods of air-to-air bombing using the Me 262. He is seen here in the spring of 1945 with Major Wolfgang Späte, who at the time was *Kommandeur* of I./JG 7. (EN Archive)

next time he was airborne over the airfield and results were positive. From that point, like von Brauchitsch, Stamp became a proponent for air-to-air bombing.

By 8 January 1945, *Kommando Stamp* was based at Lärz with four aircraft on strength and at least five pilots in addition to Stamp. As well as its pilots, the *Kommando* was assigned 40 military and civilian weapons specialists, including two professors from the technical colleges at Braunschweig and Brünn. The unit's headquarters was set up in a train located behind some aircraft hangars on the edge of the Müritzkanal that had a number of sleeping coaches, a dining car and another coach assigned for radio and technical equipment. As Stamp recalled:

> I admired both the elegant and powerful-looking design of the Me 262 and I was conscious that I was witnessing a new era in flight. Nosewheels and jet engines were completely new to me; the latter required a completely different pattern of take-off and landing procedures. The high cruising speed, previously not experienced, required a complete change in the 'feeling' of time and distance. Compared to other aircraft, it represented a completely new style and feel of flying – no noise and no vibration in the air. From an operational point of view, I was sure that the aircraft could, and would, do what I had proposed and what I was given the opportunity to do; to take 2 x 250 kg bombs, climb to 9,000 m, drop them into closed bomber formations, split them and create better attack conditions for our conventional fighter pilots.

Tactically, it was planned to use four Me 262s for an attack on an enemy *Pulk*, with each aircraft initially carrying one bomb, although some trials were later attempted with two in accordance with Stamp's intentions. The jets would fly echeloned back 10–15 degrees in a loose line astern formation with about 28 m between each aircraft, and were to fly next to the bomber formation and at the same altitude, assign targets, then make their approach from the front and some 915 m above the formation in order to avoid contact with any enemy fighter escort.

A coloured stripe was painted on the nose of each aircraft attached to the *Kommando* which slanted downward toward the front at an angle of 16 degrees below the horizontal. At the correct range from which he would commence his attack dive – about 2,740 m – it was intended that the pilot would use the stripe to line up the formation. The four-degree differential between the 20-degree attack angle and the 16-degree stripe was compensated for by the time it took to go into the dive at a speed of 750–800 km/h. At a distance of 550 m, the bombs would be released using a Type 89B two-second delay fuse, although trials were also conducted with barometric fuses – the *Baro 1* (which was hampered by the cold air flow around the aircraft), an acoustic fuse, the *Ameise* (which, due to its sensitivity, was susceptible to damage during transport) and an electric remote-controlled type known as the *Pollux* designed by Blaupunkt. The latter was to be used in conjunction with the FuG 16.

Having released their bombs, the Me 262s would then break away by 'split-essing' or climbing over the bomber formation and returning to base.

Various weapons arrangements were tested by the *Kommando* over the Müritz See, often using an Me 262 flown by Oberfeldwebel Hans 'Hanschen' Gross, to establish blast radius, dispersal during fall and fuse performance, although little data was gathered from these experiments since bombs often failed to explode and it was never really ascertained which was the most effective and practical arrangement. Ordnance types included the AB 500 container loaded with either 25 SD 15 Zt semi-armour-piercing bombs equipped with time fuses or 84 SD 2 'Butterfly bombs'.

Another variation was to load the AB 500 with 4,000 *Brandtaschen* (incendiary pellets) and enough explosive to scatter the pellets with sufficient velocity, similar to that of a massive shotgun blast, to penetrate the skin of a four-engined bomber. However, during testing, it was found that after the canisters opened, the bombs often hit each other and detonated too quickly, the resultant blast severely damaging the carrier aircraft. On one such occasion, the bombs detonated prematurely and Gross was forced to land in a meadow. Badly wounded, he was subsequently hospitalised for several weeks.

Leutnant Herbert Schlüter had served with I./JG 300 before joining *Kommando Stamp*, where he became involved in air-to-air bombing trials. (EN Archive)

Individual SC 250 (with a time fuse), SC 500, SD 250 and SD 500 bombs were also tested. Ultimately, however, it was decided to use the AB 500 filled with 370 kg of Triolin explosive in any eventual operations. To provide some degree of accuracy, in January 1945 Dr. Kortmann, a physicist at the Zeiss works in Jena, designed the *Gegner-Pfeil-Visier 1* (GPV 1), of which some 20 units were manufactured.

Leutnant Herbert Schlüter had been a fighter pilot with I./JG 300 flying Bf 109s on escort missions alongside Stamp. In September 1944, he was transferred to E.Kdo 262 at Lechfeld for training on the jet fighter, and then in early January 1945 he received further orders:

I was transferred to *Kommando Stamp* at Rechlin-Lärz. Purpose of the unit: to bomb bombers. What was innovative about our bombs was the barometric fuse. It consisted of a box that contained a barometer connected by a tube to the static outside pressure. The opening and closing of this tube could be operated with a button from the cockpit. The bomb consisted of a container made of sheet metal designed like a bomb with stabilisers as used by the Luftwaffe. The casing contained two identical bowl-shaped containers that were connected to the altitude stabiliser by means of a hinge pin. In front, both containers were closed. This container was used to spread small bomblets whereby the closed-off portion of the container was detonated by an adjustable fuse.

For the barometric bomb, only the casing was needed. A tube filled with explosives that was almost as long as the bomblet container was installed in the middle of the casing. Some 4,000 'incendiary packets' were placed around the tube. The packets contained round pieces of Magnesium which had been drilled open and filled with Thermit. Thermit is a mixture of aluminium oxide and pulverised iron. It is used to weld steel railway tracks. Thermit and Magnesium burn at a temperature of 2,000°C.

The use of W.Gr. 21 mortars persisted through to the closing weeks of the war. When the air-to-air bombing efforts of *Kommando Stamp* proved fruitless, the OKL disbanded the unit and it was integrated into the *Stabsstaffel* of JG 7, where its Me 262A-1as were used to test deployment of mortars and 55 mm R4M rockets. Two of the unit's aircraft are seen here at either Brandenburg-Briest or Parchim in the spring of 1945 undergoing fitment of launch tubes for 21 cm mortars. (EN Archive)

Values for the relative speeds of the Me 262 and the bomber formation, the relative altitude from which the bombs were to be dropped and the necessary ballistic figures for the type of bomb being used were adjusted in the sight before take-off. Although, as has been stated, the dive angle was set at 20 degrees, a lever to allow last minute manual override was fitted to the port-side of the cockpit. The pilot would then simply move the lever until its long axis was parallel to the horizontal plane in which the bomber formation was flying. He would then wait until the wingtips of a B-17 were framed by his reflex sight at a range of 550 m, at which point the bombs were released.

The pilots of Stamp's *Kommando* prepared for their task using a specially equipped Me 262 cockpit section assembled by Zeiss at Jena, into which had been fitted a GPV 1. A film projector then threw an image of an approaching bomber formation onto a white wall so that it appeared as if the cockpit were 'approaching' a target. Herbert Schlüter remembers the illusion as being perfect:

Zeiss had developed the new sight and we were to train with it. We met the men responsible for its development; there was Dr. Kortmann and Dr. Schneider, a mathematician and a few others who explained its workings. The training took place in an improvised cockpit with a ram sight, and ten metres in front of the mock-up there was a big screen. An original image of a frontal view of an American bomber formation was projected onto this screen; it was very realistic. At first the formation was small – just a dot, but it very rapidly became larger in front of our eyes. The Me 262 in a five- to seven-degree dive could quickly reach 940–960 km/h. The bombers flew at around 400 km/h. The simulator made it very clear to us that the closing speed was 1,350 km/h. We trained from morning till night with few breaks in between. After every 'attack', our co-workers from Zeiss told us how many 'hits' we had.

It took, on average, about five days for a pilot to become proficient at using the GPV 1. The *Kommando* returned to Lärz after about a week. As Herbert Schlüter recalled:

Finally, the day arrived. The *Kommandeur* had picked me to test the first bomb over Lake Müritz. I was to drop the bomb from an altitude of 8,000 m in the same manner as we had practised at Jena, but without the ram sight. I pushed the button to close the capillary tube between the barometric device and the static outside air pressure and climbed to 8,000 m. After a short dive, I dropped the bomb and went into a turn to observe the fireball. Nothing happened; the bomb did not explode. This was a big disappointment for all concerned. Dr. Schneider – the 'human calculator' – from Zeiss called his firm and told them about the failure.

Another try was planned for the next day. The barometric fuse was carefully examined but another failure was the order of the day. Another try had the same negative result. Then the fuses were tested with training bombs over the target area. Despite an intensive search, the problem could not be determined. So we looked for alternative possibilities. We tried without the use of a barometric fuse. Only much later was the cause of the problem found; the capillary consisted of a very thin rubber hose that could be closed by means of a solenoid and a push button from the cockpit. The rubber hose became permeable or porous in the cold and the fuse failed as a result. It would have been quite simple to have tested the device under operational conditions. A small chamber with a vacuum pump, an altimeter, some dry ice, a thermometer and a few aircraft instruments would have done the trick. The inventors of the barometric fuse – two professors from the university of Braunschweig and Brünn – were just too theoretical.

Engineers arrived from Blaupunkt to work on the radio-controlled *Pollux* fuses. Herbert Schlüter remembered:

The next series of experiments was conducted with radio-controlled fuses. Two engineers from the electronics firm Blaupunkt joined us and proceeded to install their devices into our aircraft and our bombs. Many bombs were then dropped over the test site, but the results were not good. Radio-controlled fuses proved to be unreliable and were discontinued.

Other tests were conducted with small fragmentation bombs that were dropped in large containers. The containers were blown open by means of a pre-set timing fuse and the small fragmentation bombs scattered like shotgun blast. At first, the experiments were conducted with 2 kg SD 2 bombs. The problem was that these bombs used in conjunction with the ram sight and the large container did not become live quick enough. The tail of the SD 2 bomb had a small propeller in the rear that began to rotate when dropped and, after a number of rotations, armed the bomb. In order to shorten this time, we manipulated the mechanical fuse. This was dangerous work, especially when carried out by amateurs. These experiments proved fruitless as well.

We tried even small fragmentation bombs. One bomb had a streamlined design and weighed about 2 kg. There was a tiny propeller in front on a short 2 cm tube into which were drilled several holes. The air passing through the holes determined the rotation of the propeller. After a certain number of rotations, the bomb was armed. But it still took too long to arm the bomb to make the system effective. So, in order to arm the bomb more rapidly, more holes were drilled into the tube. A mistake . . . as we were soon to find out!

A 250 kg bomb container was packed with these bombs. The bomb was to be tested over the target by a Staff engineer from the *E-Stelle* Rechlin. Gerd Stamp and his five pilots went to Rechlin to observe the tests. The Me 262 approached at only 400–500 m. The bomb container was dropped directly in front of us and we watched as it 'flew' just a few metres below and behind the aircraft. After a short time, the container opened and the small bombs dropped out like a shotgun blast. Then something happened which took our breath away. We saw that two of the bombs touched each other and exploded, which then set off a chain reaction of explosions. The result was a tremendous fireball – just a short distance behind the aircraft – maybe 100 m. We also could see that the aircraft must have been damaged, because it lost altitude and disappeared behind some trees. The pilot was lucky, since his aircraft was under partial control, and he crash-landed at about 400 km/h. He was badly injured, but it could have been worse.

We also tried out heavier bombs. The first time I dropped such a bomb, I made a surprising observation. Hans Gross and I had the task to each drop a 250 kg bomb with a time fuse from 4,000 m above Lake Müritz. We flew 40 m apart and had sight and radio contact. We were flying at high speed when I gave the order to drop the bombs. I felt a jolt as the bomb was released, and at the same time saw Hans' bomb drop. To my surprise, the bomb did not fall 'direction Earth', but only down to about three metres. It stayed in a horizontal position beneath the aircraft and continued along with the same velocity. I observed this for a few seconds. The timing device was running and we decided to peel away rapidly from the danger zone! Eyewitnesses on the ground later told us 'our performance' was impressive!

Trials continued throughout January 1945, but little practical success was achieved and the few experiments that were conducted were plagued with problems. A lack of fuel also hampered activity. In a report dated 3 February, Generalmajor Eckhard Christian, the Chief of the Luftwaffe, recommended to von Brauchitsch that *Kommando Stamp* be disbanded 'at once'. He wrote:

The essential objections are that the enemy will again immediately prevent bomb-dropping by using fighter escort at an even higher altitude in addition to the normal escort. By exploiting the necessarily long bombing run, enemy fighters using a 1,000 m height advantage have a clear chance of shooting down even jet aircraft. The present type of enemy formation offers the least favourable conditions for breaking up a formation.

By and large, the results of Major Stamp's experiment are still wide open. The procedures are still uncertain and have not been tested tactically. These considerations lead to the belief that, in spite of the improved technical position, no lasting success is to be expected. In the present situation, the necessary tests, personnel adjustments, material expenditure etc., do not appear justified.

In order to carry out the *Führer's* demands, the immediate need is to assemble all Me 262s for operation as quickly as possible. Six Me 262s were assigned to *Kommando Stamp*, and four have been allocated. It does not appear justifiable to divert aircraft of this type for an experiment which will not produce conclusive results in a short time period. The *Luftwaffenführungsstab* proposes that Major Stamp's project is dropped at once.

The Reichsmarschall's office gave its backing to Christian's report the same day and *Kommando Stamp's* aircraft and personnel were integrated into JG 7, becoming known as the *Stabsstaffel* JG 7. It continued operations, but the unit's role changed to that of developing the use of 21 cm W.Gr. 21 air-to-air mortars and 55 mm R4M rockets against Allied bombers.

LFA/RHEINMETALL-BORSIG *SONDERGERÄT* (SG) 116 *ZELLENDUSCHE*

The idea has often been expressed of having the firing mechanism of a weapon controlled by the target itself. Here we think of an alignment process such as takes place when aligning a rifle by hand. The rifleman sees the line of sight fluctuating around and across the target and endeavours to fire at the right moment. So as not to let this moment slip, he should be given a technical aid which fires automatically. We have to eliminate the human time-lag in the rifleman; that is the time which elapses from recognising the passing through the target by the line of sight until the rifleman's activation of the firing mechanism. A solution capable, technically, of being used for such a device has not yet been found. Such a device would be of special interest for firing from fighter aircraft, if the target demands employment of a large calibre. But for the fighter's usual kind of attack, the demands to be made on the detection device are very high: range over several hundred metres, with great accuracy in assessment (error, at the most, of 1/16°). Moreover, there would be difficulties in support and aiming off.

The method of attack, however, can be selected so that the conditions become simpler. In 1942, the author suggested attacking the target when flying close past it [Passierflug], by a shot fired perpendicular to the course of flight.

So wrote Dr. Ingenieur Paul Hackemann, former senior ballistics specialist at the Institute for Weapons Research at the LFA *Hermann Göring*, in a post-war report prepared by the London-based British Intelligence Objectives Sub-Committee.

In the summer of 1944, as part of its continuing efforts to devise weapons with which to shoot down Allied bombers, E.Kdo 25, which had been based at Parchim (an airfield roughly halfway between Hamburg and Berlin) since March, worked with several leading technical institutions, RLM test centres and arms manufacturers on inventive and technologically advanced forms of air-to-air armament. It was at Parchim that the *Kommando* carried out tests with the *Sondergerät* (SG – 'Special Apparatus') 116 *Zellendusche* ('Cell Shower'), a recoilless, single-shot, 30 mm weapon based around the barrel of an

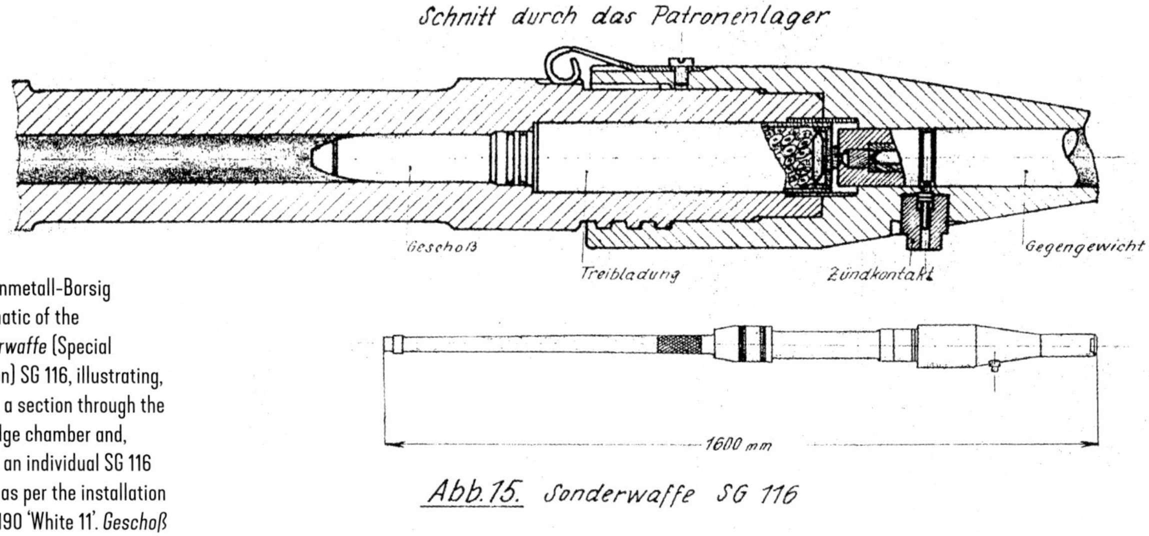

Schnitt durch das Patronenlager

Geschoß   Treibladung   Zündkontakt   Gegengewicht

1600 mm

Abb.15. Sonderwaffe SG 116

A Rheinmetall-Borsig schematic of the *Sonderwaffe* (Special Weapon) SG 116, illustrating, at top, a section through the cartridge chamber and, below, an individual SG 116 barrel as per the installation in Fw 190 'White 11'. *Geschoß* – Shell; *Treibladung* – Propellant; *Zündkontakt* – Ignition Contact; *Gegengewicht* – Counterweight. (Forsyth)

MK 103 cannon fitted to a breech block that was intended to be fired upwards as a fighter passed below a bomber.

Conventional Luftwaffe fighter attacks using machine guns and cannon against B-17s and B-24s were usually mounted from behind or from directly ahead of the bomber formation. In attacking from the rear, the fighter risked drawing the combined defensive firepower of multiple 0.50cal Browning machine guns mounted in the bombers' top turret, lower 'ball' turret, tail and waist gun positions as it passed through the formation to make its exit. In attacking from the front, although the intensity of the defensive guns was not so strong (even allowing for the twin-gun Bendix chin turrets on the later B-17G), the combined closing speed of the fighter and the bombers, as well as the narrower and much more challenging target profile, meant that the chances of shooting down a bomber were frequently beyond the capability of all but the most skilful *Experten*.

The tactical logic behind the SG 116 was that by approaching from the front of a target, the fighter could avoid the collective mass of a formation's defensive firepower by flying beneath it, while a bomber's underside presented a much larger, and closer, target to hit. Furthermore, the technology incorporated into the weapon meant that aiming depended more on the sighting apparatus rather than the human eye and the skill of deflection shooting needed with conventional guns. Horst Geyer, the former commander of E.Kdo 25 recalled to the author:

> The intention with the SG 116 was to make a frontal approach, fly under an American bomber and release the shot, which was fired by means of an explosive charge built into the base of the tube designed to be mounted into the side of an Fw 190 fuselage. The intention was good, and the aiming technology impressive. We believed that with such a weapon we could inflict fatal hits in the B-17's and B-24's wing fuel tanks.

Fw 190A-5 fitted with a 30 mm
SG 116 *Zellendusche* upward-firing
cannon installation.

Although the concept of the SG 116 was devised by the LFA *Hermann Göring* (see Chapter One), the weapon was built by Rheinmetall-Borsig. It was, essentially, a reversed development of the 7.7 cm SG 113a airborne recoilless anti-tank weapon, which comprised a vertically mounted barrel, loaded with a 45 mm armour-piercing shell, fitted into the fuselage of an Fw 190. A radar installed in the aircraft would detect the echo impulse given off by a tank moving on the ground, triggering the weapon. The shell would be fired downwards at the tank, with success at hitting the target assured at a range of 200 m. The weapon had a high muzzle velocity because of the mass of its counterweight.

Three 1.6 m-long, obliquely
mounted 30 mm MK 103
cannon barrels built into the
fuselage of Fw 190 'White 11'
of E.Kdo 25 as the
Sondergeräte 116
Zellendusche installation in
mid-1944. (Forsyth)

The LFA raised interest in the weapon with the RLM and a specification order was issued to Rheinmetall to develop a similar weapon of 3 cm calibre for anti-bomber work. It was to fire a shell with tracer at a velocity of 860–900 m per second. Tactically, it was at first envisaged that an Fw 190 would take up a position directly astern of a bomber (apparently regardless of defensive fire from the tail, ball turret and waist gunners), attempting at the last moment to fly immediately below the aircraft and within 50 m of it. When it was suggested by Allied intelligence officers to a former pilot of E.Kdo 25 that this represented a somewhat dangerous tactic, he replied that the SG 116 was always viewed as being 'in the nature of an inventor's experiment'.

The resulting weapon was developed in a very short time using stocks of MK 103 cannon barrels (see Chapter Two). The SG 116 comprised a rifled MK 103 barrel without a finished chamber, and the breech block with the ignition system. The breech block was connected with the barrel by means of threads and secured against rotation by a spring. The complete round consisted of the 3 cm shell, a cardboard container and the propelling charge, and the counterweight with the firing mechanism.

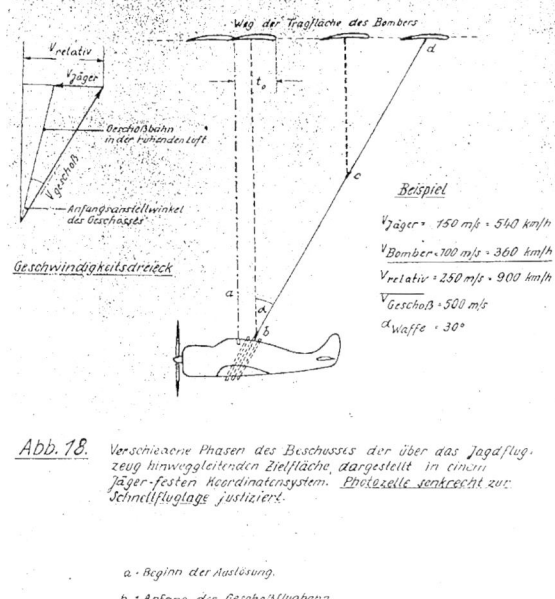

Abb. 18. Verschiedene Phasen des Beschusses der über das Jagdflug-
zeug hinweggleitenden Zielfläche, dargestellt in einem
Jäger-festen Koordinatensystem. Photozelle senkrecht zur
Schnellfluglage justiziert.

a = Beginn der Auslösung.

b = Anfang der Geschoßflugbahn.

c = Einem beliebigen Punkt der Geschoßflugbahn zugeordnetes Bild.

d = Augenblick des Auftreffens des Geschosses
auf der Zielscheibe.

t₀ = Relativweg während des Verzuges von Auslösegerät
und Schußentwicklung.

ABOVE Rheinmetall-Borsig diagram to show the various phases of firing
from below at the wing of a bomber using the SG 116 mounted as fixed
armament in a fighter with a photoelectric cell adjusted to a
perpendicular angle. The key, in German, is as follows: a – Commencement
of triggering process through photoelectric cell; b – Commencement
of shell trajectory; c – Image assigned at any point on the trajectory;
d – Moment the shell hits the target; t_o – Relative path during delay
between activation of trigger device and firing process. (Forsyth)

Three barrels were to be fitted on the left fuselage of an
Fw 190, the forward barrel being just aft of the point where the
rear of the cockpit joined the fuselage when closed. The distance
between each barrel was about 15 cm. The barrels pointed aft,
but were slightly displaced from the parallel, the foremost being
set at an angle of 74 degrees, the next at 73 degrees and the
third at 72.5 degrees to the horizontal axis of the aircraft. They
projected about 50 cm above and 25 cm below the fuselage.
In most cases where this arrangement was installed, the
Focke-Wulfs' two outer-wing 30 mm cannon were removed.

After being activated by the automatic firing mechanism
described below, the propelling charge was ignited, causing the
shell and the counterweight to move. By regulating the travel
with the differential in weight, recoillessness was obtained as
the shell and the counterweight left the weapon simultaneously.

The weapon was to be activated and fired automatically using
a photoelectric cell known as a *Magisches Auge* ('Magic Eye'),
developed under the supervision of Dr. Paul Hackemann (see
also Chapter Four) and Dr. R. Schwetzke of the *Institut für
Waffenforschung* (Institute for Weapons Research), a part of the
LFA, and later manufactured by AEG. The device was built into
the fuselage of the Fw 190 immediately forward of the barrels
and comprised four reduction lenses placed one below the other,
with a photoelectric cell fitted to the lowest lens. This was
connected to a solenoid on which a contact arm functioned.
When the photoelectric cell was activated by an image in the lens,
the solenoid was energised and the circuit to the barrel firing gear
completed. The diameter of the external part of the 'Magic Eye'
was 8 cm. The maximum range from the target aircraft at which
the 'Magic Eye' was sufficiently sensitive to operate the firing gear
was around 55 m.

As a side project, development work in the field of infrared tracking was undertaken by
Diplom-Ingenieur Schmude at the Ernst Ohrlich Institut in Danzig under the codename
Butterblume while another study was undertaken at Tarnewitz in association with the AEG
and Carl Zeiss concerns to develop an infrared-based triggering system for night or bad
weather operations. In this regard, Hackemann noted:

There was a need to find a method of firing over or under an enemy aircraft at night
or in fog. We intended to develop an additional apparatus for the photoelectric cell
consisting of an infrared radiator, so that the infrared rays irradiated the opponent.
The reflected rays would then return, and thus give an effect which triggered
automatic firing.

Such work ceased with the end of the war.

E.Kdo 25 commenced trials with the SG 116 fitted into an Fw 190F-8 at Parchim on 1 July 1944. As Horst Geyer recorded:

An optical automatic release system had also been developed. On one occasion, on 1 July 1944, General Galland, the commander of our fighter forces, and some of his staff, made an inspection trip to Parchim and we demonstrated the weapon using an Fw 190 fitted with three such tubes mounted immediately behind the cockpit and each loaded with a 3 cm M-shell. We put an NCO pilot into the specially rigged fighter and arranged for an Fw 58 *Weihe* to fly simultaneously overhead, about 200 m above the Fw 190. The *Weihe* was to tow a target drogue. The Fw 190 flew in very low, at about 100 m, so that we on the ground could observe the weapon being used to its full effect.

In the interests of safety, and because the Fw 190 was not a large aircraft, we developed a firing system designed to allow the pilot to fire only one tube at a time, thus minimising and avoiding the risk of any blast damage from all the tubes firing at once. Unfortunately, however, the NCO made a mistake and fired all the tubes simultaneously. There was a loud explosion in the air with a huge cloud of smoke. Then, emerging from the smoke came the *Weihe*, flying gracefully on, undamaged. However, the Fw 190 was destroyed in the process. Fortunately, the pilot somehow got out, but like the *Weihe*, the drogue was also untouched. The pilot of the Fw 190 landed by parachute and limped up to Galland and I. He pulled off his flying helmet and his first words were 'Permission to have a cognac, *Herr General.*'

Later that month Geyer departed E.Kdo 25 and his position as *Kommandoführer* was taken by Major Georg Christl, a Knight's Cross-holder who had previously led Bf 110-equipped III./ZG 26. By the beginning of August, E.Kdo 25 had been redesignated *Jagdgruppe* 10 (J.Gr.10), although it retained the structure of the *Kommando's Stab* and three *Staffeln*.

By 22 August 1944, J.Gr.10 had commenced fitting the photoelectric cell, which by that time was under manufacture as the *Wurzen* at the HASAG in Leipzig, into 19 of its Fw 190s at Tarnewitz. Trials were also conducted using an He 177 of the *Kampfstaffel* as a 'target', but these proved unsuccessful and, for the time being, further development was dropped while the various parties involved in the weapon's manufacture went back to the drawing board. By 18 September the decision had been taken not to use the SG 116 operationally until greater accuracy could be assured, although fitment of the optical devices was nearing completion.

On 26 September 1944, the commander of the RLM's *Technisches Amt* calculated that in a pass below a bomber at 50 m, a fighter equipped with five SG 116 barrels could look to achieve three hits on its target. He further calculated that if 12 barrels were fitted into an Me 262 jet interceptor, a type which had recently made its operational debut in the West with E.Kdo 262 (to which E.Kdo 25's first commander, Horst Geyer, had been posted to lead), then a pass at 50 m would ensure 12 hits and at 100 m, six hits.

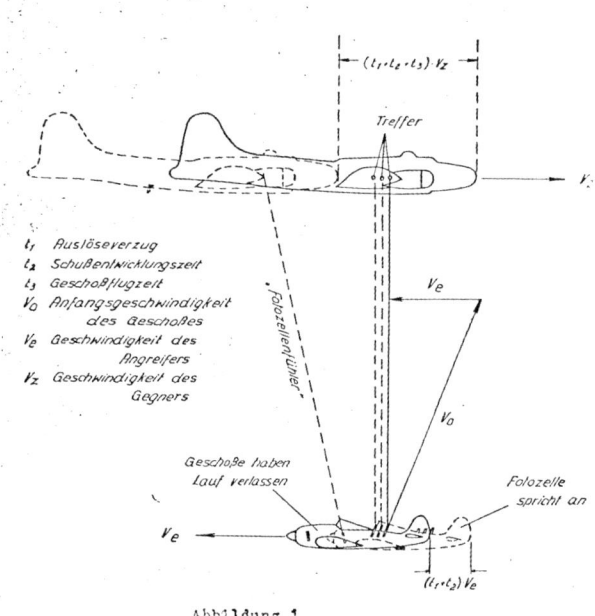

Optimale Bewaffnung und Treffwahrscheinlichkeit
bei Senkrechtschuß mit automatischer Schußauslösung.

Die Vielzahl der laufenden Bewaffnungsprojekte für Senk-
rechtschiessen mit automatischer Schußauslösung gibt Anlaß
zur Untersuchung der günstigsten Bewaffnung in Bezug auf
Kaliber, Geschoßgeschwindigkeit, Auslösegerät und Justierung
aufgrund der z.Zt. vorhandenen Unterlagen. Das Schießver-
fahren veranschaulicht nachstehende Abbildung 1.

l_1 Auslöseverzug
l_2 Schußentwicklungszeit
l_3 Geschoßflugzeit
V_o Anfangsgeschwindigkeit
     des Geschosses
V_e Geschwindigkeit des
     Angreifers
V_z Geschwindigkeit des
     Gegners

Abbildung 1

ABOVE This document, issued by the *E-Stelle* Tarnewitz on 30 December 1944, details the ballistics calculations relative to the SG 116. (Forsyth)

In his monthly work report for September, Christl admitted that the automatic triggering device in the *Zellendusche* was not proving reliable. However, in ground tests, it was found that when fired at a wing salvaged from a crashed B-17 from a range of 60 m, the equipment and detonator functioned correctly, at a velocity of 845 m per second.

In the air, tests using the SG 116 were conducted against an Fw 58 and an He 177 that had been rigged with specially constructed automatic steering devices. After a pre-determined period of time, the pilots of the aircraft bailed out and the machines flew on unmanned, at which point the Fw 190s made their attacks at altitudes of 1,000 m against the Focke-Wulf and at 6,000 m against the Heinkel. It was noted that problems arose at heights in excess of 6,000 m when the automatic triggering device became too sensitive, resulting in premature and random firings. The *Weihe* was hit and damaged, but the Heinkel flew on undamaged and is believed to have eventually crashed into the Baltic as planned. Whatever the case, the results in the air varied wildly from the optimistic results calculated by the *Technisches Amt*.

By the end of the first week of October, 18 of J.Gr.10's *Wurzen*-fitted Fw 190s had returned to the *Gruppe's* base at Parchim from the *Erprobungsstelle* at Tarnewitz, where the circuitry work was completed. One machine remained at Tarnewitz, along with one of the *E-Stelle's* Focke-Wulfs, which were to take part in further firing trials against another He 177. These resulted in only one Fw 190, equipped with five barrels, firing two shots and achieving two hits, but with three barrels suffering misfires. The other Focke-Wulf failed to get off the ground due to technical faults.

Tests continued through until the end of December 1944, and saw a series of revisions and enhancements to the weapon and its firing system. It was planned to mount the weapon in three different ways in the Fw 190 for operations: firstly, four barrels in a rhomboid mounting in the fuselage; secondly, six barrels in two triangular blocks; and thirdly, three barrels in a line, with a 'fan' spread with each barrel angled at two degrees. However, nothing more materialised in terms of operationally ready equipment.

By that stage of the war, in reality, the Allied air forces had control of the skies over what remained of the Western Front, and the USAAF's bomber groups and their fighter escorts were operating with increasing impunity over the Reich.

The general specifications for the SG 116 were as follows:

Weapon	
Calibre	30 mm
Muzzle velocity	860 m per second
Muzzle velocity of counterweight	200 m per second
Weight of projectile	0.315 kg
Cyclic rate	Single-loader
Weight of weapon (barrel and breech)	28 kg
Length of weapon (empty)	1,600 mm
Length of weapon (loaded)	1,620 mm
Assembled weight of three weapons for use	96 kg
Maximum diameter	105 mm
Ammunition	
Weight of projectile	315 g
Length of projectile	140 mm
Weight of explosive	72 g
Weight of propellant charge	125 g
Weight of counterweight	1.35 kg
Weight of complete round	1.80 kg
Length of round	625 mm
Firing system	Electrical
Type	HE
V⁰ of shot	845 m per second
V⁰ of counterweight	170 m per second

RHEINMETALL-BORSIG *SONDERGERÄTE* SG 117, SG 118 *ROHRBLOCK* AND SG 119 *ROHRBATTERIE*

In recognising the operational value of a greater density of fire from fixed armament – what was termed *Schrottschussverfahren* (volley or grapeshot method) – as well as the opportunities offered by *Magisches Auge* photoelectric cell triggering technology (see SG 116), so in the autumn of 1944 German technicians at Unterlüss identified the need for a range of multi-barrelled aircraft weapons. In their report, Rheinmetall specialists Flugbaumeister Diplom-Ingenieur P. Riecker (in charge of air-testing) and his colleague Dr. Kokott stated:

The idea of relieving the pilot as much as possible of any additional mental strain, especially in regard to the technical aspects of gunnery, which, in any case, requires certain training, and of eliminating the usual errors in point-fixing of aim and lead, led

TOP LEFT AND RIGHT Front and rear views of the 30 mm seven-shot Rheinmetall-Borsig *Sondergeräte* SG 117 salvo-firing 'barrel block', intended for vertical or horizontal installation in fighter aircraft for deployment against bombers. This weapon had been set up on a test bench by Rheinmetall-Borsig at Unterlüss. (Forsyth)

ABOVE RIGHT Rheinmetall-Borsig sketch showing 'The Principle of the Barrel Block (Bundled arrangement of 7 MK 108 Barrels)'. (Forsyth)

ABOVE This example of an SG 117 was damaged during firing tests. Note the torn barrel and barrel binding. (Forsyth)

to a special form of armament in combination with an automatic firing release while in *Passierflug* [passing flight]. The operation was to be effected by passing below the enemy as closely as possible – within about 100 m. The weapons were mounted vertically to the direction of flight and were fired as multi-barrel, concentrated fire [*Schrottschuss*], or released in a corresponding succession, timed according to the size of the lead errors to be covered.

To achieve this in the quickest and simplest way, and in order to use existing ammunition, Rheinmetall decided to base such a weapon on its 30 mm MK 108 cannon shell. According to Kokott:

> The weapons were to have an extremely high rate of fire of up to 20,000 rounds per minute. The intended tactical operation was the surprise attack from out of the sector where defence was weakest – that is generally from below, where the fighter brought its entire firepower to bear in *Schrottschussverfahren* from the shortest range. The rounds were timed to follow one another so that a barrage was formed around the enemy through which he had to fly. In order to obtain a barrage of sufficient concentration, rates of fire of more than 10,000 rounds per minute were needed. The prospects of such an attack were bound to be exceptionally favourable.

Fw 190A-5 fitted with twin 30 mm SG 117 *Rohrblock* upward-firing cannon batteries.

What emerged as the SG 117 *Rohrblock* (tube or barrel block) comprised a battery of tubes, each containing one 30 mm cannon round, intended for vertical instalment, initially at least, into an Fw 190. The barrels were clustered cylindrically, held together by a screwed-on 5 mm metal brace, running to a breech block and fired by means of an electric connection. The 30 mm cartridges were loaded into the tubes on the ground and left the tubes sequentially when fired. This design had the benefit of requiring minimal alteration to the aircraft, as well as having minimal recoil. The tubes were 5 mm thick so as to require the minimum space and weight. The spaces between the tubes, however, were filled with a concrete-based compound that helped in absorbing the stress caused from firing.

A wooden model of a B-24 Liberator is used to demonstrate the principle behind the *Schrottschussverfahren* (volley or grapeshot method) of firing as used by the SG 117, SG 118 and SG 119 special weapons. Such an intense barrage of fire would destroy a bomber. (Forsyth)

At the base of the tube, directly beneath a loaded projectile, was a layer of a compressed fabric-based compound, then a spring, through which ran the firing pin and connecting wire. The ignition wires were grouped behind a sheet metal disc and led to the triggering device via a sensor that guaranteed ignition for 20,000 shots and was adjustable.

In action, as the first round was fired, the entire *Rohrblock* recoiled along a guide rod that contained seven contact points, each of which closed and fired the tubes in succession as the unit recoiled downwards.

The following development, the SG 118, was based on the SG 117 but used three seven-tube *Rohrblocks* assembled rigidly around a central shaft. The shaft was attached to a coil spring, which rotated each block into firing position. This created a 21-shot weapon, but it was intended as a horizontal, rack-mounted installation for suspension

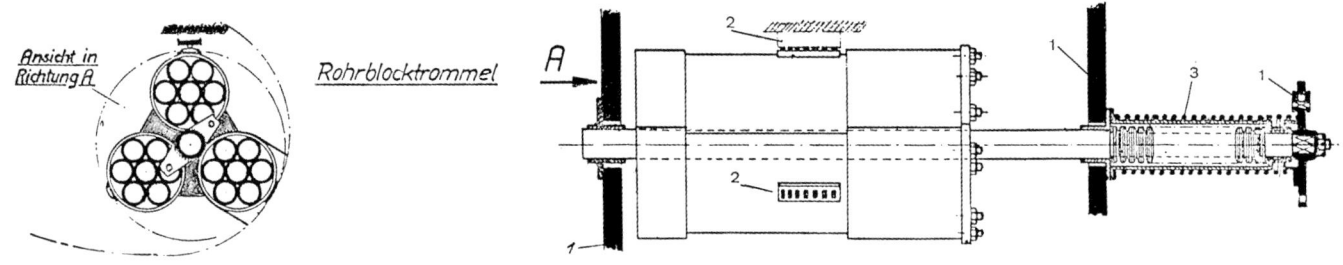

Ansicht in Richtung A

Rohrblocktrommel

Rheinmetall drawings of the 21-barrel SG 118. (Forsyth)

from the fuselage of a high-speed aircraft or from its underwing surfaces. In this configuration, an eighth contact was provided, which triggered the first shot of the following block.

The central shaft was to be fitted to the aircraft by means of two flanged bearings, with the shaft being able to slide within the bearings. In order to absorb the projected recoil of some 2,000 kg, a buffer spring – composed of plate springs – was to be fitted on the shafts. A recoil of approximately 22 mm was anticipated. The recoil shock would be transmitted to the springs by means of a disc that was pushed along the shaft by a key. The other end of the springs would butt against a tubular cartridge by means of another disc. The tubular cartridge was to be connected by means of a flange with the rear flange bearing. At the end, the shaft would have carried a sliding sleeve, a locking disc and a castellated nut. This nut would have made it possible to modify the tension of the plate-shaped springs, but care would have had to be taken not to allow any axial play on the shaft. The recoil action of the drum pressed the rear disc rearward, through the sliding sleeve, the locking disc and castellated nut. The plate-shaped springs would then be compressed forward and partly absorb the counter-recoil energy.

A strong, pre-loaded torsion spring applied to the tubular sleeve connecting on one side to the flange of the tubular sleeve and on the other side to the locking disc would have turned the drum in one direction. The free rotation would be controlled by a latch in the recess of the key way of the locking disc. The latch, fixed to the sheet metal segments and the hub would have taken part in the back-movement of the drum. The hub was to slide on the pin. The pin would be held by the bearing fastened to the aircraft. After the first barrel block had been fired, the pilot was to release the latch by means of a cable and the tension of the spring would make the drum revolve by one section, until the latch reached the recess. After firing the second barrel block, the pilot was to release the latch and the spring would turn the drum by one section until the latch reached the recess. In this position the third block could be fired.

Loading shells on the ground would have begun with the third block, turning one section backwards by hand, until the latch was brought in front of the surface. After the second block had been loaded, the drum would again be turned by hand until the latch was in front of the surface, and then the first block was loaded.

The firing of each block was to have been effected by means of an electric cylindrical switch. The seven rounds of one barrel block followed each other at a distance of six metres. The seven firing cables of the first barrel block led to a contact case on the second barrel block. The contact case was provided with sliding elements that pressed the contact springs of the contact case. The contact case was to be fixed to the aircraft. From this contact box, the firing cables were to be conducted to the cylindrical switch. The firing cables of the second barrel block were to be linked to the contact case on the third barrel block, and then to the case on the first barrel block.

For loading, the cover of the third barrel block was to be removed and the cartridges with the contact cases were inserted into the barrels. The cover was screwed on the barrel block again after the firing cables were fed through the corresponding holes in the cover. The firing cables would then have had to be connected numerically to the contact case on the second barrel block. The loading of the second and first barrel blocks was carried out in a similar way.

The pulling cables of the left and right drums would have been united into one cable, which was brought through the different block pulleys near the pilot and ended there in a three-way lever. There would have been a cover in the fuselage wall for the purpose of servicing the installation. The drum would have been sealed internally by a coating between the ribs.

According to Riecker and Kokott:

In a frontal attack by a fighter equipped with standard fixed armament, the personal danger to the pilot is very much reduced. However, to the same extent, if not more, his prospects of a hit are diminished, as the duration of the attack is shortened and the launching of an attack is disproportionately more difficult than from the rear, owing to the absence of automatic directing into the enemy's line of flight. The full advantages of a frontal attack are offered by *Passierflug* through the factor of surprise, as well as by the simplified task for the pilot of such a method of approach. Except for the pressing of a *Bereitschaftsknopfes* [action button] on the control column, the pilot has only the task of flying beneath the enemy at a distance up to 100 m, and of endeavouring to bring the shadow of the enemy aircraft onto his optical release. Aiming, lead and distance observation for the opening of fire, and correction of the cone of fire by means of sight or tracer trajectory, are not necessary.

From late 1944, an aircraft considered as a viable carrier for the SG 117 and the SG 118 was the He 162 *Volksjäger*. In a report prepared for the Allies in July 1945, Dr. Ingenieur Josef Schoetz, a ballistics specialist at Rheinmetall-Borsig, explained that:

Combat tactics deployed by fighters against bombers consisted mainly of frontal or rear attacks. However, it became evident during 1943–44 that these tactics failed to guarantee successful destruction of the bomber.

Representation of an He 162A-2 of 1./JG 1 fitted with 30 mm SG 118 internal horizontal *Rohrblock* cannon batteries and underwing pods.

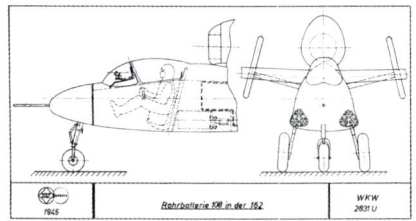

RIGHT The diminutive He 162 *Volksjäger* was seen as the carrier aircraft for the SG 117 and SG 118 systems. This example is He 162A-2 Wk-Nr 120074 'Yellow 11' of I./JG 1 seen at Leck in the days following the German surrender. Posing alongside the fighter is Oberleutnant Karl-Emil Demuth, whose 16 victories have been marked on the aircraft's port tail fin – none of these were claimed in an He 162. The aircraft also carries the emblem of I./JG 1 and a small '20' next to the main tactical number. Note also the dual-coloured protection plate inserted into the engine intake bearing the number '24'. (EN Archive)

ABOVE A Rheinmetall-Borsig drawing from 1945 showing the intended installation of two SG 118s in specially created troughs on either side of the lower fuselage. This would have provided a salvo or volley shot using a total of 42 barrels. Additionally, a single SG 117 battery would have been suspended in a pod beneath each wing, creating, potentially, 56 shots and effectively turning the He 162 into a *Zerstörer*. (Forsyth)

In the direct frontal and rear attack the bomber offered a small target which required accurate approach and accurate fire. This situation was made more challenging by the defensive fire from the bomber formation, which was often intense enough to destroy the fighter. The development of high-speed, jet-propelled fighters of approximately 800 km/h combined a high velocity with the minimum size of aircraft and a flight duration of only a few minutes – for example the He 162.

Schoetz's point was that for the He 162 fighter (which was envisaged to be produced in great numbers, but dogged by limited combat endurance) to be tactically effective against *Pulks* of heavy bombers

Ba 349 *Natter* fitted with 30 mm
32-barrel *Große Rohrbatterie 108*
installation with M-shells.

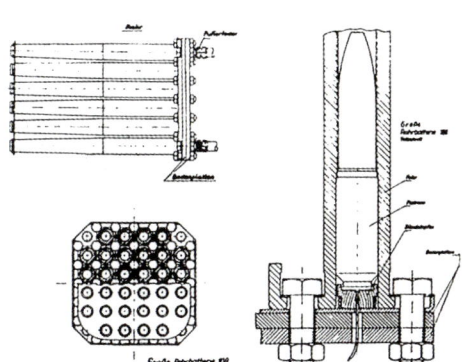

ABOVE LEFT Variation on a theme: schematic of the proposed *Große Rohrbatterie 108* which consisted of 32 barrels, each loaded with one 30 mm MK 108 shell, here shown installed in the Ba 349 *Natter* VTO interceptor. From upper left clockwise: side view of the weapon mounted to an armoured bulkhead; front view of 'honeycomb' installation in nose; rear end of a single barrel, with shell and ignition wiring in situ; close-up cross section of weapon, showing forward supporting plate partly cut away; close-up side view of weapon showing the barrels, buffer springs and base plate. (Gooden)

ABOVE The *Große Rohrbatterie 108* on a test stand. Note the multiple electrical cables for the sequential ignition of the 32 shells. (EN Archive)

LEFT A honeycomb installation of MK 108 barrels fitted into the nose of a Ba 349 *Natter* during firing trials at Heuberg in early 1945. (EN Archive)

with massed defensive firepower, it would need to be able to deliver the maximum weight of firepower in the quickest possible time. With rigidly fixed machine gun or cannon armament, such as the standard MG 151 or MK 108 installation in the He 162, a pilot was required to fly directly at the target and remain 'locked on' to it in his sights for several seconds in order to ensure a sufficient number of strikes with a comfortable degree of probability. In order to reduce this period, during which the pilot and fighter would be exposed to defensive fire, the Germans endeavoured to develop weapons offering a 'spray fire', 'fan fire' or *Schrottschuss* effect, resembling a wide 'blast' or arc of fire, rather than a single stream.

Thus it was proposed to install two SG 118s in the fuselage where the usual MK 108s or MG 151s were housed, meaning that a pilot would be able to fire a salvo of 2 x 7 rounds at a bomber from the fuselage position. Additionally, a single SG 117, adapted as a horizontal SG 118 with a mounting rail, could be slung beneath each outer wing, offering a further 14 rounds in total. The pilot of the He 162 would fire the weapons by means of a trigger-button on his control column, but a selector switch would allow him to fire salvos either from two batteries in the fuselage-mounted revolving drum or the two wing-mounted blocks, or to fire from the drums and wing blocks simultaneously.

The underwing SG 117s (as SG 118s) were to be fitted with an aerodynamic, faired cover, which could be jettisoned by igniting combustible bands just before opening fire.

In order to achieve an even greater density of fire, seven tubes of the SG 117 type – a weapon which was deemed to be successful in principle – were combined to form a battery of 49 tubes known as the SG 119. With a weight of 200 kg, in contrast to the SG 117, the SG 119 was not low recoil, although it was proposed that recoil would be absorbed by a cladding consisting of, bizarrely, peppercorns. The battery was mounted to move downwards against a buffer spring assembly, with a maximum recoil force of approximately 6,000 kg. It could only be installed in an aircraft fuselage, and the recoil was absorbed by the aircraft. According to Kokott:

> The separate barrels of the SG 119 could be so adjusted that the most appropriate distribution of the rounds was obtained against the target. The most favourable dispersion pattern could be determined in advance according to the tactical approach and the errors in lead which had to be overcome.

A further development showed that the recoil of the pipe blocks could also be achieved by nozzle action, so that it was no longer necessary to shoot out the entire pipe block. Tests were carried out with a nozzle-equipped tube and, in order to save weight and to keep the diameter as small as possible, a sleeveless cartridge made of nitrous cardboard was also used. The first attempts showed good values and complete rebound. However, they could not be completed due to the worsening war situation.

SG 500 *JÄGERFAUST* (FIGHTER FIST)

Results from tests with the 30 mm SG 116 *Zellendusche* encouraged German engineers to consider a similar, vertically firing volley weapon incorporating the same optical cell triggering method but of a heavier calibre, ideally 50 mm. The problem was that a 50 mm gun with a muzzle velocity having the same impact as the SG 116 would be too heavy, and therefore the barrel and counterweight would have to be redesigned. Instead of a fan-like arrangement for simultaneous fire, a high rate of consecutive 'fan fire' from several barrels would cover the same field along the same principle as utilised in the SG 117.

Development of this weapon, which became known as the SG 500 *Jägerfaust* (Fighter Fist), was undertaken by HASAG of Leipzig, the firm which had also developed the photoelectric cell for use in the SG 116.

Because of the 50 mm calibre, muzzle velocity of the SG 500 was limited to 400 m per second, as a result of which range to the target was reduced to 50 m to improve efficiency.

The *Jägerfaust* consisted of a rifled tube which was closed at the bottom, thus forming a cartridge case. The propellant charge and the primer were housed in a paper envelope. The projectile had a pre-engraved band and was fed from the muzzle into the rifled section of the tube. SG 500 tubes were produced from round bars, using the same machines which manufactured the 50 mm cartridges. In the final finish, there was a constant rifling of eight degrees in a tube without machining. However, since a tube was for one-time use only, there were no strict requirements for strength or exact rifling. An ignition cable was inserted through a hole in the base of the tube and the end clamped to an ignition circuit.

The propellant charge was similar to that of HASAG's *Panzerfaust* recoilless anti-tank weapon. It was enclosed in a cardboard cover, which also protected the fuse from damage. The fuse, which comprised a tricinate pill with a 500 g charge, was identical to that used on other *Sondergeräte* weapons and reduced the ignition period to a minimum.

The projectile was an extremely thin-walled M-shell designed especially for low acceleration. Any issues that might have resulted from wear and tear or conditions arising from external ballistics were not considered important since the distance of flight would have been a maximum of 100 m.

As a further development, a projectile stabilised with steel fins was planned, although there was a risk that the adjacent tubes could influence the path of the projectile and possibly damage the fins with their gas pressure. The fin-stabilised projectiles would have had considerably larger pressure than those stabilised by rifling. Due to safety considerations and production costs, it was decided to concentrate on rifled projectiles only.

Under the jurisdiction of the *E-Stelle* Tarnewitz, an SG 500 was installed experimentally in a five-tube arrangement in each wing of an Fw 190. The tubes were located in one plane parallel to the longitudinal axis of the aircraft, firing fanwise vertically upwards so that the dispersion over a target at 100 m did not exceed 16 m. To facilitate the path of the projectiles, a thin-walled aluminium tube was installed in each wing and the weapon was located between shearing pins. The loose ignition cables were fastened to clamping screws.

ABOVE Two Me 163B rocket interceptors of II./JG 400 photographed at Husum in April 1945. Had the proposed SG 500 *Jägerfaust* installation been fitted, the tubes would have been visible on the wing uppersurfaces of 'Yellow 2', the aircraft at left in this photograph. (EN Archive)

RIGHT Diagram illustrating the sequential firing process of the SG 500 from an Me 163 passing beneath a B-17, and the fall of expended cartridges. (Forsyth)

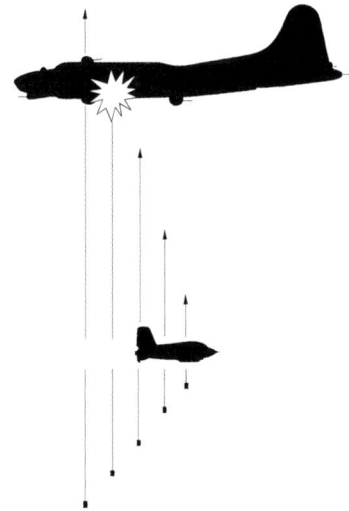

The firing of the five tubes from one wing was effected simultaneously by means of an ignition impulse actuated by the automatic photoelectric cell. If the striking effect was not considered sufficient, provision was made to fire all ten rounds simultaneously, thus excluding the possibility of a second attack. It was also possible to fire the various barrels consecutively, as with the SG 119, by means of a timing device that determined the rate of fire. In such an instance, the tubes would have been arranged in parallel rather than in a fan-like formation.

After initial testing with single tubes in each wing of an Fw 190F-8 was conducted successfully at Tarnewitz, towards the end of October 1944 Oberst Gordon Gollob, a veteran fighter ace who was overseeing development work on the Me 163, Hauptmann Rudolf Opitz, the acting *Gruppenkommandeur* of the Me 163-equipped I./JG 400 and Hauptmann Anton Thaler, commander of the rocket-interceptor evaluation unit E.Kdo 16, based at Brandis, visited HASAG in Leipzig to inspect work on the SG 500 and watch a demonstration. Gollob and Opitz felt the weapon had potential and, subsequently, a contract for production was issued. For installation in the Me 163 the weapon would be adapted to fire 2 cm calibre rounds.

In mid-November 1944 tests were conducted using Me 163 BV 45 Wk-Nr 16310054 C1+05 of E.Kdo 16 at Brandis. Technicians from HASAG worked closely with personnel of the *Kommando*, and preparation of a prototype weapon installation for operational trials proceeded without any problems over the course of a day, although the fitment of the

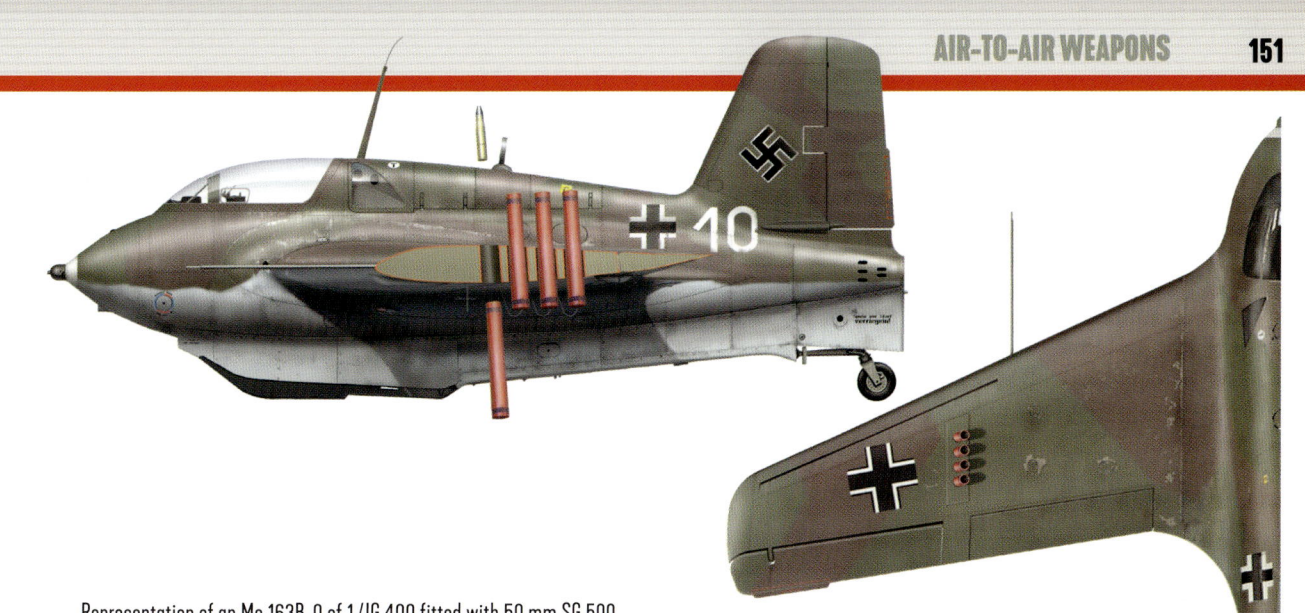

Representation of an Me 163B-0 of 1./JG 400 fitted with 50 mm SG 500 *Jägerfaust* vertical, four-barrel, single-shot installations.

photoelectric cell took longer. This was eventually installed close to a radio aerial near the Me 163's ammunition bay. The SG 500 was first fired against a scrap wing, with no damage being sustained by C1+05, before Leutnant August Hachtel of the *Kommando* took off on 13 November with half-full fuel tanks and the aircraft's standard MK 108 cannon removed. Hauptmann Thaler reported:

> The pilot was completely satisfied with his flight: the weapon installation had no effect on the length of the take-off run, tight turns at 700 km/h did not affect the aerodynamic or mechanical performance of the aircraft and the weapon installation was not deformed in any way. After Hachtel had performed a series of tight turns, he flew straight across the airfield and fired the weapon. The noise was somewhat louder than that heard when firing the MK 108 cannon. The aircraft did not deviate from its course and the pilot felt no vibration as the weapon was fired. The barrels released themselves as expected and the pilot made a good landing.

Hachtel then performed a second flight in front of Generalleutnant Galland, the *General der Jagdflieger*, who was visiting Brandis. Things went well, and Thaler noted:

> The fitting of the vertical 'shot gun' installation in the wings of the Me 163 is very simple and cheap, and does not degrade the aircraft's performance in any way. The only thing remaining to be proved is that the optical release system functions without problem.

One area where the pilots of E.Kdo 16 and HASAG differed was the optimum range at which the SG 500 was deemed to be effective. HASAG had calculated that the weapon would be most effective when an Me 163 was flown 150-100 m below an enemy aircraft, in which case the pilot would have a 35 per cent chance of striking the enemy aircraft, but in the opinion of the *Kommando* that distance was too great. The chance of a hit would increase significantly if the Me 163 flew 50 m below.

Hachtel made two further flights, this time as mock attacks against a Bf 110 flown by Thaler. The first flight, at 80–100 m below the Bf 110, was accomplished without difficulty, but Hachtel considered the distance between the two aircraft to be too great. During the second flight, Hachtel's Me 163 was fitted with a photoelectric cell, and he flew 30–40 m below the Bf 110. Unfortunately, film taken during the flight did not show anything as the Me 163's speed was too high. In a second flight, the weapon was loaded with 2 cm tracer shells. Their trajectories were filmed with a gun camera fitted to the aircraft so that their flight could be measured exactly.

The *Kommando* also considered a number of methods to prove that the optical release system functioned efficiently, and proposed a mock attack against an He 177 bomber based at Brandis using practice ammunition to assess shell dispersion and to determine where the target was hit. However, no further flight-testing could be conducted because J2 fuel was not available.

Thaler was of the opinion that should the weapon's accuracy prove as good as the theoretical estimates, it would not be necessary to fire all of the shells simultaneously, and he suggested to HASAG that only two or three shells should be fired at the same time, so that two or three bombers could be engaged as an Me 163 made a *Passierflug* below. HASAG concurred. Thaler also suggested that the shells should be capable of being fired by the Me 163 when flying at different speeds. The required switch necessary was already available, and had been installed in the aircraft, and it allowed the weapon to be fired at any time during three different speed ranges: either when the Me 163 attacked from astern; at one speed range when in a head-on attack, so that it became irrelevant whether the pilot was flying at 700–800 km/h or 600–700 km/h with thrust, or at 500–600 km/h with the engine shut down; or in making a frontal attack. Whatever the case, the switch would prevent the weapon being fired until the aircraft was being flown within the relevant speed range.

It was not until 10 April 1945 that the SG 500 was used in combat. During the early evening, in clear weather, 230 aircraft of RAF Bomber Command attacked the railway yards at Engelsdorf and Mockau in the outer districts of HASAG's base city of Leipzig. At least one Me 163 of 2./JG 400, flown by Leutnant Friedrich Kelb, was scrambled from Brandis to intercept the bombers. According to a subsequent German report:

At about 18.00 hrs a strong formation of enemy bombers appeared over the city. An Me 163B with 'vertical armament' went into action. The aircraft came unheeded below the enemy machines. The armament was immediately discharged. A Boeing B-17 [sic] at once fell out of the sky like a burning torch, none of the crew being able to make their escape. Two other aircraft were so badly damaged that they also crashed after flying over our airfield.

In fact, Kelb had fired his SG 500 at the Lancaster flown by Sqn Ldr C. H. Mussels of the Royal Canadian Air Force's No. 405 Sqn as it completed its first pass over the target. The projectiles from the *Sondergerät* shot away the Lancaster's rear gun turret and starboard rudder. Watching from the ground at Brandis with the aid of a long-range Flak telescope was Feldwebel Hans Hoever, who worked with the radar section of 2./JG 400. He recalled observing Kelb's flight:

> I saw him head towards the lead aircraft of the bomber formation which was flying at a height of about 8,000 m. I thought he wanted to ram it, but just at the moment as he passed about 100 m below the aircraft, the bomber exploded in a cloud of smoke and flames. I had never before seen a bomber so easily destroyed as that attacked by Leutnant Kelb.

The Lancaster began to dive out of control, but somehow Mussels managed to level out. Escorted by P-51s, he flew the crippled bomber as far as the British coast, after which he ordered his crew to bail out. Mussels remained with the aircraft and was able land it at the nearest airfield. He was awarded the Distinguished Service Order for his efforts to add to his Distinguished Flying Cross.

Despite his fuel being spent, Kelb evaded the Mustang escort and land back at Brandis. It was the first and only time that the SG 500 was known to have been used in combat. The HASAG works were destroyed in the bombing.

The general specifications for the SG 500 were as follows:

Leutnant Friedrich Kelb of 2./JG 400 enters the cockpit of Me 163B Wk-Nr 190579 at Brandis. On 10 April 1945 he became the only pilot to fire the SG 500 operationally when he attacked a Lancaster of No. 405 Sqn RCAF. (JG 400 Archive)

Calibre	50 mm
Projectile	M-shell
Weight of projectile	1 kg
Explosive charge	400 g
Muzzle velocity	400 m/sec
Total weight	7 kg
Outer diameter of the tube at the muzzle	65 mm
Outer diameter of the tube at the cylindrical section	73 mm
Length of the weapon	520 mm
Gas pressure at muzzle	900 kg/cm^2
Range	100 m

Discovery of only known surviving SG 500 *Jägerfaust*

In the summer of 2011, the barrel of an SG 500 was discovered during an explosive ordnance clearance in the Leipzig area. This was confirmed by checking the weight and dimensions of the artefact with a report on the weapon prepared by Oberingenieur. (DPhil) Walter Grasse of Rheinmetall-Borsig AG at Unterlüss in September 1946.

An interesting detail is that at the lower end of the barrel there are three spring catches. These were most likely intended to give a tight fit and attachment between the barrel and the outer aluminium tube. Each spring element is 6.5 cm in length and 9 mm wide. Inside the barrel there are 24 spiral grooves. The outside of the barrel is slightly conical which differs from the known drawings.

ABOVE The SG 500 barrel photographed after sandblasting. (Hans-Hermann Cammann/JG 400 archive)

ABOVE Three spring catches were located 5 cm from the lower barrel covered end. (Hans-Hermann Cammann/JG 400 archive)

ABOVE The lower, covered end of the barrel has a 4 mm drilled hole for the electric ignition wire. (Hans-Hermann Cammann/JG 400 archive)

ABOVE Internally, the barrel of the SG 500 had 24 spiral grooves. (Hans-Hermann Cammann/JG 400 archive)

LEFT During the ordnance clearance near Leipzig, a 5 cm Mine shell for the SG 500 was also discovered which had to be blown up. (Hans-Hermann Cammann/JG 400 archive)

SONDERGERÄTE 2 cm *BÜRSTE*, *HARFE* AND *SCHLITTER* AND 3 cm *BOMBERSÄGE SALVENFEUERWAFFEN* (VOLLEY WEAPONS)

During 1944, aside from the already described *Sondergeräte* series of air-to-air weapons, several other systems based on photoelectric cell-firing technology were proposed for combat against enemy bomber formations. Some took their influence from a proposal drawn up for a vertically mounted weapon incorporating eight MG 151/20 barrels by Stabsingenieur Poppendieck of *Abt*. E4 at the *E-Stelle* Werneuchen, where he had also worked on the *Spanner II* infrared targeting periscope.

One such weapon was the 2 cm SG *Bürste* (Brush), a rigid, vertically mounted, spin-stabilised, volley-shot design for firing upwards, believed to have been the brainchild of Luftwaffe officer Oberleutnant Schlitter. The name *Bürste* reflected the weapon's 189 barrels, each just 200 mm in length, giving the appearance of the dense hairs of a brush. The *Bürste* was intended to fire 15 mm incendiary shells to achieve the equivalent rate of fire, in three volleys, of 20,000 rounds from an MG 151 at a velocity of 200 m per second. The smooth barrels were drilled to 25 mm calibre using a sabot. However, the ballistic calculations relative to the smaller 2 cm calibre were considered ineffective and led to cancellation of the project.

Another design based on a 25 mm barrel with 2 cm MG 151 ammunition was the SG *Harfe* (Harp), which comprised one or two rows (one row either side) of 20 smooth tubes mounted to the fuselage side of an Fw 190 perpendicular to the direction of flight, giving the appearance of a harp. The forward tube of each line was placed just aft of the rearmost point of the Focke-Wulf's closed canopy glazing. The spacing between each tube was about 1 cm. Inspiration for the design came from an experimental 12 cm perpendicular, fuselage-mounted recoilless tube intended to fire an 8.8 cm Flak incendiary shell either upwards or downwards that would disperse 300 incendiary pellets, but it was not considered stable.

As with the *Bürste*, the projectiles for the *Harfe* were to be fired from barrels bored by a sabot. Conceived by HASAG in Leipzig, the tubes were recoilless and electrically ignited, with powder gas being burned and discharged from an outlet in the base of each tube. The tubes were set at an angle of 74 degrees to the horizontal axis of the aircraft. After firing, the tubes were jettisoned so as to avoid any aerodynamic interference to the fighter.

E.Kdo 25 first trialled the weapon in August 1944 when it fired 30 rounds of 2 cm incendiary shells simultaneously without any problems. An aerodynamically improved projectile was introduced that featured three flexible steel fins that folded up when loaded into the upper end of the tube and sprang into position when fired. Air measurements in trials showed a satisfactory hit pattern and sufficient stability. Tests continued the following month, and the *Kommando* reported:

> Modified ammunition was delivered by HASAG for the *Harfe* device which was also given to the *E-Stelle* Rechlin in order to check the projectile trajectory. The new projectile has extendable stabilising fins. Ground-firing tests and air-to-ground firing have given perfect results. No ricochets have been found so far. The air firing was conducted from heights of 100 to 250 m to the ground at a horizontal speed of 420 km/h. The longitudinal dispersal of the shots from the first to the last barrel was measured at approximately 14 m at an altitude of 100 m and approximately 29 m at a height of 250 m.
>
> Ground-firing tests at a firing range from 100 m to the target were fired simultaneously from ten barrels, with an angle setting for the barrels of 0 degrees and with barrels arranged one above the other over a radius of 70 cm. There was a spread of 75 cm across the target.

The projectile trajectory is to be measured at the *E-Stelle* Tarnewitz in the next few days, as soon as the tracer ammunition has been delivered for measuring.

However, as with the *Bürste*, the effect of the 2 cm M-shells was deemed insufficient, so development of the *Harfe* was stopped in favour of larger calibre armament.

The general specifications for the SG *Harfe* were as follows:

Calibre of barrels	25 mm
Calibre of ammunition	20 mm
Length of the weapon	1,250 mm
Length of barrel section at 25 mm cal.	720 mm
Outer diameter of barrel	33 mm
Assembled weight	34 kg
Velocity	325 m per second
Projectile weight	0.255 kg
Charge weight	0.135 kg
Explosive	0.019 kg
Propellant charge	0.120 kg
Length of projectile	640 mm
Length of charge	110 mm
– of which tail unit	27 mm
Length of propellant section	530 mm

Oberleutnant Schlitter, also proposed a 2 cm system which became known, briefly and eponymously, as the *Schlitter* or *Schlitter-Projeckt*. In this design he proposed an obliquely mounted, multi-barrel weapon that would fire projectiles upwards or downwards following an optically operated triggering. After due consideration, and some revision to the design by the LFA *Hermann Göring*, initial construction was taken over by the Reichswerke *Hermann Göring* at Salzgitter. The weapon emerged as the *Handfeger* (Hand Brush) and took the form of no fewer than 60 2 cm tubes loaded with standard 20 mm MG 151 shells. It was intended as a single-shot weapon that had to be primed before take-off of the carrier aircraft, and it could not be reloaded during flight. The projectiles were fired at the equivalent rate of 20,000 rounds per minute, with a velocity of 180 m per second. Recoil was kept within a limit of five to six tons by means of a sprung baseplate.

Ultimately, the weapon was not released for production for the same reasons as the *Harfe*.

Meanwhile, as a conceptual progression of the *Handfeger*, 60 3 cm tubes were assembled on a spring-loaded platform mounted behind the pilot based on a recoilless system and discharging balancing gas downwards to become a basic prototype of the SG *Bombersäge* ('Bomber Saw'), but the adverse course of the war prevented any further practical development.

A table was drawn up on 30 December 1944 by *Abt*. IIE at the *E-Stelle* Tarnewitz that compared the number of hits expected on a B-17 by a fighter equipped with the listed weapon during a pass below at 50 m distance. This indicated that while the *Bürste*, *Harfe* and *Schlitter* weapons were heavier, greater impact could be expected from them:

Weapon	Calibre	Barrels	V⁰m/sec	Operational Wt.	Anticipated Hits to:			
					Fuselage	Inner Wing	Engine	Outer wing
Bürste	15 mm	189	200	382 kg	40	16.6	20.3	12.2
Harfe	20 mm	15–20	325	37.8 kg	26.2	16.6	20.3	12.2
Schlitter	20 mm	60	385	107.2 kg	23.6	16.6	20.3	12.2
SG 116	30 mm	3	845	102.2 kg	4.9	4.5	4.9	3.5
SG 117	30 mm	7	480	38 kg	6.7	5.1	5.9	3.8
SG 500	50 mm	5	400	45.6 kg	2.8	2.2	2.3	1.8

CHAPTER FOUR

AIR-TO-GROUND WEAPONS

LFA SG 113A RIGID, VERTICALLY MOUNTED ANTI-TANK WEAPON

Heading one of the three divisions of the *Institut für Waffenforschung* (Institute for Weapons Research) of the LFA *Hermann Göring* in 1944 was Dr. Ingenieur Paul Hackemann. A native of Bocholt, the 38-year-old Hackemann had studied mechanical engineering at technical college in Aachen and entered the field of ballistics in 1931, joining Rheinmetall at Düsseldorf as a scientist in the ballistics department in 1933. Two years later, he was appointed to head a department at the *Deutsche Versuchsanstalt für Luftfahrt* (DVL – German Aviation Research Institute) in Berlin and

RIGHT AND FOLLOWING PAGE The large, funnel-like dorsal and ventral casings of the downward-firing SG 113A six-barrelled 7.7 cm anti-tank weapon are seen fitted to Hs 129B-0 Wk-Nr 0016 at the LFA *Hermann Göring* at Völkenrode in 1944. Tests produced mixed results. (EN Archive)

ABOVE View from above, showing the cluster of six barrels of the SG 113A. The weapon's 45 mm projectiles were fired without upwards recoil, the barrels being locked down by iron counterweights five times the weight of the projectiles, and they flew away upwards when the weapon was fired. (EN Archive)

qualified as a civil pilot, gaining his B.1 qualification the following year. He then pursued an acclaimed career in the field of aeronautical ballistics at the LFA in Braunschweig as the *Abteilungsleiter* for the *Institut für Kinematik*.

In 1942, Hackemann suggested attacking an aerial target as an aircraft flew past it at close range or in 'passing flight', using a shot or shots fired at an angle perpendicular to the course of flight. His suggestion was first applied to attacking tanks, whereby an aircraft would fly over a tank at low altitude and fire downwards from between five and 20 metres. Air trials ensued at Völkenrode using a twin-engined Fw 58 and Fw 189, in which the pilot approached the mock targets in either a shallow dive or in low-level, horizontal flight without any aids. These trials demonstrated that such a method could be carried out with sufficient accuracy. Hackemann wrote of his proposal:

The pilot need only watch the side and the time. If he does not succeed in flying vertically over the target, he can always manage to put the vertical axis through the target by banking the aircraft. The accuracy of alignment required here is low because of the short distance from the target. The alignment process is thus much simpler than in rigid forward firing.

However, the challenge remained to find a method to bring about firing at the moment the aircraft passed directly over the target. Initial thought was directed towards a reflection process with closely focused ultra-short waves, but it was not possible to produce such a system in a short time and without prohibitive cost. Thus, as an interim measure, it was decided to use the recently introduced Siemens FuG 101 precision radio altimeter for twin-engined aircraft that had a measuring range of 150–170 m and accuracy to within two metres. However, this device would also pick up other entities such as trees and houses on its indicator.

The device was installed into an Fw 58 along with an oscillograph and a photoelectric cell which was directed downwards. As a target, the test used a hangar at the edge of a wood, in front of which had been erected a light barrier. As the Focke-Wulf flew over the barrier, a distinct mark was produced on the oscillogram. 'The first test results showed that the altimeter had obviously registered the front of the hangar,' noted Hackemann, but after flying at low-level over smaller buildings and huts, the sizes of which were below the minimum recognisable on the FuG 101, it was realised that the system would not be effective if deployed against tanks.

Eventually, an electrostatic release system, introduced by Dr. Robert Schwetzke, a colleague of Hackemann's at the *Institut für Kinematik* in Braunschweig who specialised in electrical engineering, was linked to an MG 131 machine gun. These, along with an antenna to detect distortions, were fitted into Fw 189 DU+UW, and trials carried out against 'a sheet iron box corresponding to the average dimensions of a tank'. The tests proved generally successful, with a strike rate in the region of 80 per cent, although it was found that the carrier aircraft could not exceed an altitude of ten metres above a target; if it fired at a height any higher, ground interference risked firing activation at the wrong moment. There were also two defects: firstly, triggering could be activated by any elevation that was steep enough above ground level such as a tree, a small rise or a wall, and secondly, the resulting strength of an electric field which fluctuated considerably meant that a pilot would have to adjust the sensitivity of the equipment just before making an attack, thus distracting him. The great iron mass of a tank caused a distortion in the magnetic field of the earth.

A solution to this was found in a magnetic probe developed by the *Forschungsanstalt Graf Zeppelin*

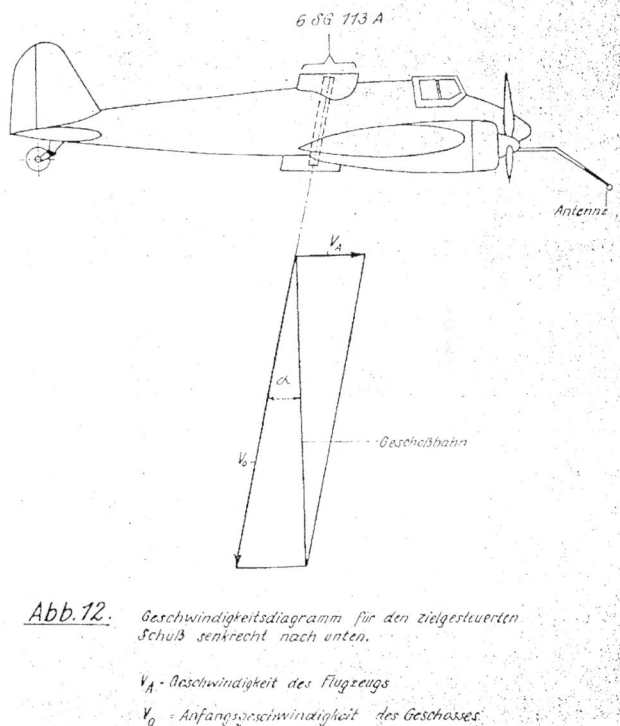

ABOVE 'A Speed Diagram for a Targeted Shot Fired Vertically Downwards' as produced by Rheinmetall-Borsig, showing the path of an SG 113A projectile fired towards the ground from an Hs 129. V_A – Speed of aircraft; V_0 – Initial speed/velocity of projectile. (Forsyth)

BELOW A film still captures an Hs 129 flown by Unteroffizier Walter Raufelder of 10.(Pz.)/SG 9 during tests with the SG 113A at the LFA *Hermann Göring* at Völkenrode on 18 January 1945. The Henschel passes over a captured T-34 at the moment the SG 113A fires. The projectile hit the tank but it ricocheted off. (EN Archive)

near Stuttgart, and with which it was possible to measure a magnetic field statically in strength and direction, with results independent of the speed of a carrying aircraft. Such a device was installed into an Fw 190, and results showed a 70 per cent hit rate that was deemed to be operationally acceptable.

Attention then turned to the design and build of a suitable weapon that could function alongside the automatic firing system. The weapon became known as the *Sondergeräte* (SG) 113A, and less formally as the *Förstersonde* ('forester's probe' or 'tube/barrel'). Hackemann described it thus:

> The final model consisted of a smooth-bore barrel, 7.5 cm in diameter and 160 cm long without breechblock. A 4.5 cm armour-piercing shell formed as a flanged projectile was fired, without recoil, through this barrel. The barrel was locked down by a loose mass of iron five times the weight of the projectile, which flew away upwards on shooting. The distances travelled by the projectile and mass of breechblock inside the barrel until the relative muzzle is opened are in inverse proportion of 5:1.
>
> Construction was in the hands of Rheinmetall-Borsig. With a barrel weight of 50 kg, the firm succeeded in firing the projectile at about 700 m per second. The projectile required further development work until it was possible to penetrate 45 mm armour plate of 120 kg/mm² tensile strength at 60 degrees. The effect produced appeared to be sufficient, as the armour-plating of most enemy tanks was thinner on top. Although the projectile was not stabilised, it flew faultlessly over the required distance of ten metres, despite lateral airstream. Penetrations into pasteboards were always elongated to an extent corresponding to the oblique position of the projectile when fired from the aircraft.

The target detection antenna fitted to the nose underside of Hs 129B-0 Wk-Nr 0016 for trials with the SG 113A at Völkenrode. (EN Archive)

In these firing range tests conducted at Unterlüss, muzzle velocity of 650 m per second was obtained, while velocity of the installed counterweight was 125 m per second and gas pressure was around 1,700 kg/cm². From what Hackemann states, and from known velocity figures, this would have been sufficient to penetrate the top armour of Soviet T-34 and American Sherman tanks.

In 1944 three Hs 129 ground-attack aircraft, Wk-Nrs 0016, 0249 and 140499, were equipped identically at the Deutsche Lufthansa Institute at Berlin-Staaken for testing with the SG 113A, together with an improved release mechanism developed by Opta-Radio. The Hs 129 was selected because it offered good visibility downwards and forwards, which were very advantageous factors when flying *over* a target.

Six barrel tubes were mounted behind the cockpit, close to the centre of gravity, and inclined eight degrees rearward in order to compensate for the projectiles' line of flight. Externally, a detector antenna was fitted below the nose, extending two metres forward to avoid interference from the propellers. The pilot would operate two switches – a push-button safety device to prevent accidental firing of the weapons, and a selector to allow firing singly or as a salvo.

In order to determine exactly how low an aircraft could fly over a target before it incurred damage from blast, an Hs 129 was suspended at various heights between two masts and explosive charges detonated beneath it. These revealed that damage would not be an issue as long as the aircraft did not descend below four metres.

The first type of projectile to be tested was a 3 kg 7.5 cm hollow-charge, armour-piercing shell carrying 500 g of explosive, but one such round failed to destroy an armoured vehicle. As another option, a sub-calibre, armour-piercing, high-velocity 77/45-type projectile was tested. This used a 77 mm-diameter aluminium sabot and a 45 mm round, which detonated only after penetrating armour, and is the one described earlier by Hackemann that was used in ground tests at Unterlüss.

The complete round consisted of the projectile and a counterweight that were connected to one another with a notched rod, which assured that the round was held together. The propelling charge was assembled in a cardboard cartridge around the notch staff placed in between the projectile and the counterweight. Ignition of the charge was accomplished by a primer located in the counterweight. Firing of the primer and the propelling charge was obtained by contact with the shot release mechanism. Upon build-up of sufficient gas

Three views of two Fw 190F-8s at the *E-Stelle* Tarnewitz fitted with wing-mounted SG 113A fairings each containing two 7.7 cm barrels. Note the bulges on the wing uppersurface which were the upper fairings for the photoelectric detectors. (Forsyth)

pressure to break the notched rod, the projectile and the counterweight started moving together. By regulation of the travel, recoillessness was obtained by the projectile and the counterweight leaving the barrel simultaneously.

On 9 October 1944, Unteroffizier Walter Raufelder, an experienced Hs 129 pilot serving with 10.(Pz.)/SG 9, collected Wk-Nr 0249 from Berlin-Staaken and ferried it to Völkenrode, where it was joined by Wk-Nr 0325 a short time later. Between 13 October and 18 January 1945, Raufelder undertook 17 test flights associated with the SG 113A in both the available Hs 129s and in the earlier mentioned Fw 189. These tests produced mixed results. It was found that one projectile hit the target tank in a thick-welded join and ricocheted off, while another grazed the turret and tore away a large piece of 15 mm armour from the engine compartment. The large number of metal splinters found in the compartment were deemed to have destroyed the engine. Similarly, a third projectile, having penetrated 20 mm armour, blew apart and left splinters in the turret. But there were also misfires, and in one of three test-flights against a captured T-34 on 18 January 1945, the projectile fired late and impacted with the ground 30 m away from the tank. According to Hackemann:

> A tank can be attacked by an aircraft equipped in such a manner both in the usual shallow dive approach and in low-level, horizontal flight. Several Luftwaffe 'ground-strafers' agree that the latter method of attacking gives better protection against anti-aircraft fire. In most of the test-flights carried out, the ability to go into action even under very poor weather conditions, especially with low-lying clouds, proved especially advantageous. Successful test-flights were carried out often under conditions making a shallow dive attack impossible.

Despite any sense of optimism on Hackemann's part, the SG 113A installed in the Hs 129 for anti-tank purposes met with a cautious response from both the Luftwaffe and engineers at Tarnewitz. Indeed, Unteroffizier Raufelder felt the weapon presented considerable danger to the aircraft and pilot, who risked being blown to pieces as a tank exploded so close below. Undaunted, Hackemann speculated that, 'Only employment at the front would have produced a final judgement on this method of attack.'

Aside from trialling the SG 113A on the Hs 129, in late 1944 tests were also conducted using the Fw 190. In this instance, two barrels were fitted into vertically mounted, aerodynamic fairings. The complete unit was then installed through a cut-away section close to the root of each wing which extended above and below the upper- and undersurfaces. The photoelectric cell was also housed in the wings outboard of the SG 113A. The weapon is known to have been fitted to Fw 190 V75 Wk-Nr 582071 and Fw 190F-8s Wk-Nrs 586586 and 933425, and firing tests were carried out against a captured Russian tank at Tarnewitz. As with the Hs 129, results were mixed, accuracy questionable and a shortage of fuel and the required ammunition brought a halt to further development in February 1945.

The general specifications for the SG 113A were as follows:

General Data	
Type of weapon	Recoilless, single-loader
Calibre	77/45 mm
Muzzle velocity of projectile	650 m/sec
Muzzle velocity of counter-weight	125 m/sec
Weight of total round	1.9 kg
Cyclic rate	Single round
Weight of weapon	48 kg
Length of weapon	1,600 mm
Ammunition	
Type of projectile	Armour piercing
Total weight of round	1.9 kg
Length of projectile	140 mm
Weight of explosive charge	18 g
Weight of propellant	360 g
Weight of case	12 kg
Length of case	530 mm
Type of firing	Electrical
Tube	
Calibre	77 mm
Length of tube	1,600 mm
Weight of tube	48 kg
Type of tube	Smooth
Maximum gas pressure	1,700 kg/cm^2
Designed gas pressure	2,200 kg/cm^2

AIR-LAUNCHED FLYING BOMBS

The concept of launching Fieseler Fi 103 or *Flakzielgerät* (FZG) 76 flying bombs from carrier aircraft against targets in Britain was first considered in April 1944. The advantages to this method of deployment over fixed launch sites, which were vulnerable to air attack, were seen as offering an extension of the flying bomb's endurance, an expansion of its range of targets and that launching aircraft would be able to attack from any direction without having to be in close proximity to the target.

Air-launching trials had been carried out over the Baltic using an He 111H-16 fitted with a PVC 1006 rack and protective metal sheets to the tail assembly and areas of the starboard wing undersides, together with the necessary switchbox for the flying bomb. To counter this additional weight, as much defensive armament as possible was removed.

He 111H-16 CK+UE releases an Fi 103 flying bomb during a test at the *E-Stelle* Karlshagen in September 1943. For this test the bomb has been suspended from the port wing, whereas when the weapon was launched operationally by III./KG 3 and the *Gruppen* of KG 53 during 1944, bombs were suspended from the starboard wing. (EN Archive)

In late June 1944 the personnel of Ju 88-equipped III./KG 3 were assembled at Grieslienen. Here, the unit was reequipped with He 111s and fresh crews, several of whom had torpedo-bomber experience and who were hurriedly retrained to fly the Heinkel. As part of this training they were also instructed in flying He 111s with a single torpedo-shaped cement 'bomb' carried under the starboard mid-wing section. The bombs were fitted with wooden wing stubs to simulate the design and weight of a FZG 76 flying bomb, hundreds of which had been fired at the British Isles from launch sites in France from mid-June 1944. Powered by an Argus pulse-jet engine and weighing 2,150 kg, the FZG 76 carried an 850 kg warhead of Amatol high-explosive and had an operational range of around 250 km.

Towards the conclusion of III./KG 3's initial training process a detachment led by the *Gruppenkommandeur*, Major Martin Vetter, travelled to Karlshagen on the Baltic coast. Here, crews made a limited number of flights in He 111s fitted with flying bombs, and by completion, 54 trial bombs had been launched. Vetter and his crews then returned to Grieslienen, where more He 111s had been delivered fully adapted as flying-bomb carriers.

From April 1944, the RLM commenced a programme of adapting He 111H-16s and a few H-20s at the Lufthansa workshops at Oschatz and Stuttgart-Böblingen to carry flying bombs. Internal bomb racks were removed and a purpose-made carrier formed of metal sheeting was fitted to the area surrounding the mid-wing fuel tank inboard of the engine nacelle on the starboard wing. A single FZG 76 was then suspended from two inclining lugs. To maintain lateral stability, a spring-loaded 5 cm diameter buffer tube was attached close to the trailing edge of the wing and partly within it, and protruding for about 50 cm outside the wing undersurface. The tube contained a coiled spring that was compressed by a buffer rod which was pushed into the tube and carried a flat buffer head about 20–25 cm in diameter. When the flying bomb was in position, the buffer head pressed from above onto the wing of the flying bomb. The buffer would hang loosely from aircraft returning from operations. So adapted, such He 111s were referred to in some quarters as 'H-22s', but to all intents and purposes, the aircrews and units reported aircraft under their original sub-variants (i.e., H-6s, H-11s, H-16s and H-20s).

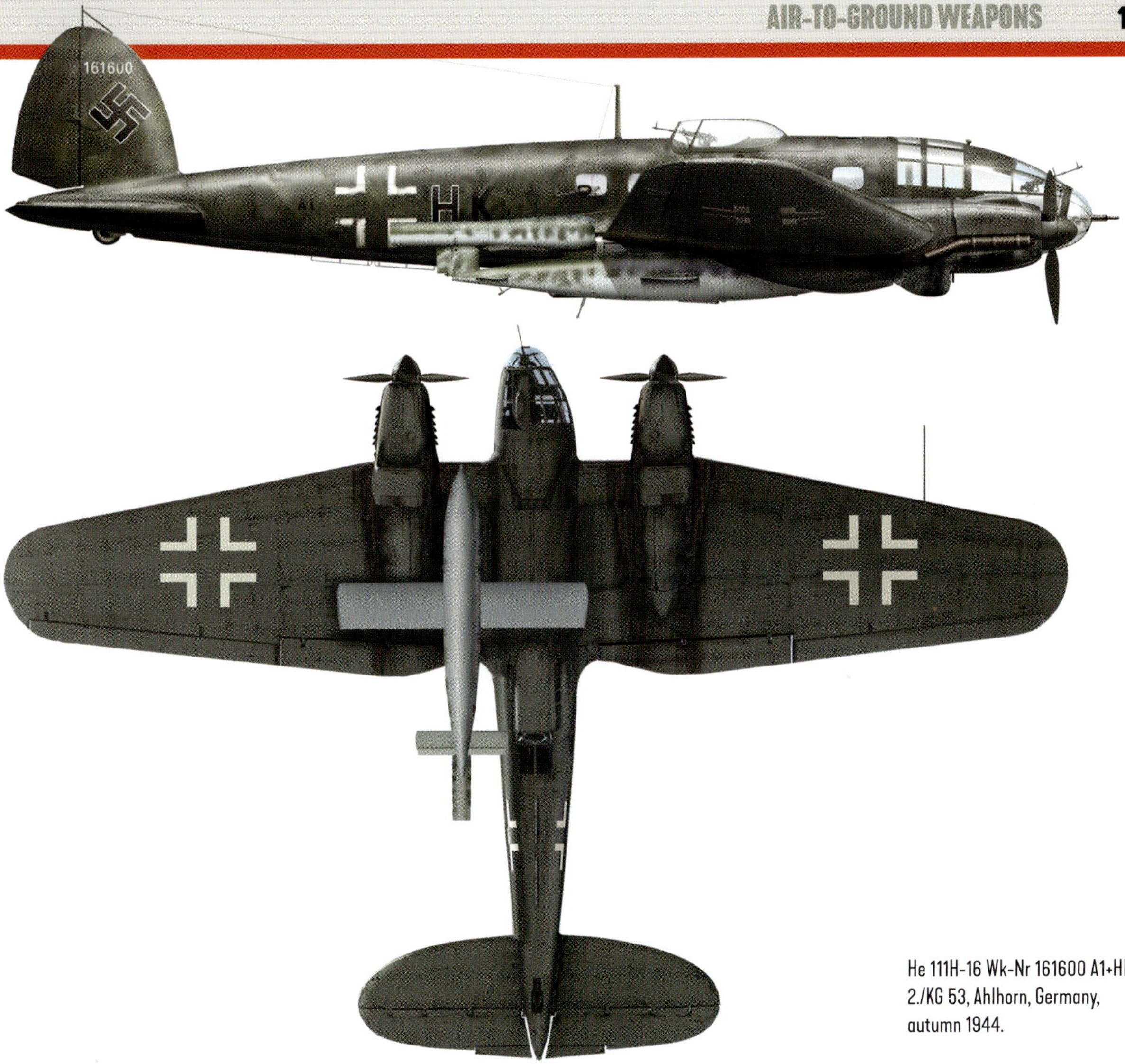

He 111H-16 Wk-Nr 161600 A1+HK of
2./KG 53, Ahlhorn, Germany,
autumn 1944.

He 111H-16 Wk-Nr 161600
A1+HK of 2./KG 53 carrying an
Fi 103 flying bomb beneath
its starboard wing at Ahlhorn
in the autumn of 1944. Note
the igniting coil fitted to the
top of the Fi 103's pulse jet
engine which runs from just
below the beam window. The
dorsal turret is not armed.
(EN Archive)

In July, III./KG 3 had relocated to Venlo, in the Netherlands, from where the *Gruppe* commenced offensive operations against Britain, air-launching FZG 76s, or 'V1s' as they had become known (for *Vergeltungswaffe* 1 – 'Vengeance Weapon 1'). Early missions were flown against London and Southampton, from where much materiel was being shipped to the Allied armies in Normandy. Sorties were mounted at dusk or in darkness, with the number of aircraft taking part varying depending on serviceability and the availability of flying bombs, which were delivered to Venlo by road and rail and unloaded there by local civilian labour. Over 15 nights throughout July, groups of between ten and 20 He 111s of III./KG 3 fired 277 V1s, mainly against London, and on three occasions against Southampton.

On 17 August 1944, Oberstleutnant Fritz Pockrandt's KG 53 was transferred from East Prussia to eastern France, and in September it also started preparing for air-launched operations using adapted He 111H-16s and H-20s. Returning from France, I./KG 53 was disbanded at Neubrandenberg and reformed almost immediately with a nucleus of crews from III./KG 3 and KG 55. The *Gruppe's Staffeln* moved to Varelbusch, Ahlhorn and Vechta, while II./KG 53 under Major Herbert Wittmann trained up at Reppen, before moving to its operational bases at Bad Zwischenahn, Jever and Wittmund by October. At the same time, Major Allmendinger's III./KG 53 moved to Leck, Schleswig and Eggebeck. Crews, usually in cadres of eight or nine at a time, continued to be sent for ten-day training courses where pilots flew with a practice flying bomb without a warhead, while flight engineers were given training in the V1 launch mechanism.

To prepare for an air-launched V1 mission, flying bombs were towed from their storage area to a servicing shed or area, where they were fuelled. The He 111s were moved in pairs to their refuelling point and then taxied to the V1 area, where the flying bombs would be hoisted onto their carriers by specially trained armourers. After loading had been completed, the He 111s were moved back to their dispersal.

Once airborne, to launch the V1, the pilot and observer used a machine known as the *Zahlwerk*, which was similar in function to a tachometer. Before a flight, the observer pre-set the device using a number (usually between 2,000 and 3,000), and later, when a certain point on the aircraft's course had been reached, the observer flicked a switch on the *Zahlwerk*, which was housed in a box located slightly above and in front of the pilot. This activated the pre-set number to run down, with the countdown being replicated on another element of the *Zahlwerk* system in the flight engineer's position. At '100', the observer issued a warning over the intercom for the engineer to prepare for launch, and at '25', the engineer pressed a black button on the *Zahlwerk* (the *Anstellknopf*) that started the pulse-jet engine on the FZG 76 (a red button – the *Abstellknopf* – could terminate launch if necessary). When the countdown reached zero (about ten to 15 minutes), the engineer released the bomb.

The 'standard' tactic of the air-launchers was to leave their airfields in darkness in small formations and to fly a straight course across the Netherlands and the Dutch coast around Den Helder, Alkmar, Bergen aan Zee, Zandvoort or Ijmuiden. Laden with their heavy external loads, they would then slowly cross over the North Sea at extremely low level towards the east coast of England in order to avoid being picked up by British radar. At about 100–150 km off the British coast (usually between Great Yarmouth and Orford Ness) the Heinkels would climb to about 450 m and release their FZG 76s in the aforementioned manner, after which they would turn back across the North Sea using cloud cover whenever possible or low-level flying to make good their escape back to their bases along the same course.

The problem was that the exhaust flame of a V1, once launched, often gave away the Heinkel's position to patrolling RAF Tempest V, Mosquito or Beaufighter nightfighters, which were then able to acquire and attack the intruders. Furthermore, flying at low-level for three to five hours in darkness so close to the sea introduced the real risk of unintentional contact with the water. Crews thus wore bulky immersion suits and life vests that would have added discomfort to what were already very hazardous and draining operations.

Throughout October, III./KG 3 mounted regular attacks aimed at London, although most of the flying bombs it launched fell well clear of the capital. Launchings were also plagued by technical problems – on the night of 12/13 October, for example, of 17 FZG 76s released, only 14 went off under control. Two failed and one had to be emergency released prematurely owing to its carrier aircraft suffering engine damage from a nightfighter attack. Additionally, as autumn set in, the weather deteriorated, bringing rain, cloud and wind. The British anti-aircraft artillery, radar and nightfighter defences were also bolstered, although the crews from the Mosquito squadrons of the Air Defence of Great Britain found 'Heinkel-Hunting' to be some of the toughest flying they had ever attempted.

On 21 October, III./KG 3 was effectively absorbed into I./KG 53. Between the period 7–8 July and 25–26 October 1944, I./KG 53 (and previously III./KG 3) launched 1,109 FZG 76s in 99 operations against London, Southampton, Gloucester and Paris.

By the beginning of November, KG 53 was regularly sending out formations in *Gruppe*-sized strength and, on occasion, two *Gruppen*. Despite the frequently appalling weather and the prospect of nightfighters, raids continued with determined regularity. Indeed, on 4–5 November, the *Geschwader* sent out a total of 50 sorties (some crews flew more than one mission per night), with London as the target (the aiming point was described as the 'north river [Thames] area between Charing Cross and West India Docks)'. To put the effect of the operation into context, of the 23 V1s plotted by the British defence network between 1909–1958 hrs, only three made landfall. One was destroyed by guns and another six fell into the sea off the Suffolk coast as a result of anti-aircraft fire. The remaining 13 simply disappeared. During the night, flying bombs came down scattered across Essex, Suffolk and Norfolk. None reached London. The most impact the Heinkel force inflicted was to damage two cottages in the village of Levington, between Ipswich and Felixstowe, where one person was killed and six injured. But this was for the loss of no fewer than five crews and several experienced officers.

Perhaps thankfully for the crews, during December, KG 53's scale of effort fell due to a shortage of fuel. On one occasion III./KG 53 was forced to cancel an operation because of a lack of fuel, with stocks only being sufficient for the next operation, and no stocks being held on the operating airfields. I./KG 53, in the meantime, had moved to Leck to initiate a major pulling back of the *Geschwader* to north German fields in the Schleswig area.

There was to be one last throw of the dice, however. Since August, the prime target for III./KG 3 and KG 53 had been London, with raids being mounted several times per week, despite the weather. However, with approval granted from no less a figure than Adolf Hitler, the *Luftwaffenführungsstab* decided to seek out another target in the north of England and selected the industrial city of Manchester. In the early hours of Christmas Eve, 50 He 111s launched V1s against the city. Thirty-one missiles crossed the coast between Skegness and Bridlington, coming down across Lancashire, Cheshire, Yorkshire, Derbyshire, Durham, Lincolnshire, Northamptonshire, Nottinghamshire and Shropshire, but only one fell within Manchester city limits. Eleven crashed 25 km from the city and six at a distance of 16 km. The remainder crashed even further away. Damage in Oldham was extensive, with 30 houses left requiring demolition and more seriously damaged. As rescuers went out into the town they found scenes of carnage. One woman was found dead in her bed, having been killed by a falling roof joist that had hit her head. Two baby boys were killed and four guests attending a wedding party were also victims. In all, 37 people were killed and 67 seriously injured in the Manchester attack.

Elsewhere, the effect of the V1s was negligible, even somewhat comical. One British Civil Defence worker watched as a lone flying bomb made its way inland up the Humber estuary until it started spiralling with its pulse jet still running, apparently suffering from some technical defect, and crashed into water a few moments later without exploding. South of the Humber, in Lincolnshire, another blew up in a stubble field, causing some superficial damage to the windows of a nearby farmhouse.

Following this raid, KG 53 operations were suspended for one week, but although the Luftwaffe persevered with the offensive until February, it was really just the last rounds. On the night of 3–4 January, the *Geschwader* is believed to have sent out a maximum effort, deploying 44 He 111s from all its *Gruppen* to strike at London. The Heinkels flew out in three waves, crossing the Dutch coast at Den Helder at around 2,000 m before heading over the North Sea at 200 m. It was a good night for the British anti-aircraft defences, with 12 V1s being shot down off the coast of Suffolk. Fourteen made it inland to fall widely across Essex and Suffolk. Just one bomb made it as far as London, coming down at Lewisham, but it caused no injuries. Four Heinkels failed to return, while a further three crashed on their return to base.

Between 7 July and 10 November 1944, a total of 1,310 V1s are believed to have been air-launched from He 111s, of which 1,176 were fired at London, 90 against Southampton, 21 against Gloucester and 23 against Paris. In addition, Manchester became a main target for one night as related above, seeing a further 30 bombs. One source lists the total number of flying bombs launched against Britain between September 1944 and January 1945 as being 1,030.

FLIEGENDER PANZERSCHRECK 88 mm ANTI-TANK ROCKET

The German view of the rocket as an anti-tank weapon had been a largely negative one throughout World War II. Despite its poor performance, the Russians used the 82 mm RS-82 rocket throughout the conflict, and the RAF's Typhoon squadrons employed rockets to some effect against armour, soft-skinned columns and trains in northwest Europe from the summer of 1944. The Luftwaffe, however, preferred to use heavy-calibre cannon and bombs in ground-attack operations, and had achieved considerable success with the former.

This attitude shifted to some extent in the spring of 1944 when, as an 'emergency measure', trials commenced at the bomb- and explosive-testing *Erprobungsstelle* at Udetfeld, near Gleiwitz in Upper Silesia, under the direction of an anti-tank pilot Major Herbert Eggers. Formerly of 13.(*Panzerjägerstaffel*)/JG 51, Eggers had been assigned to E.Kdo 26, an anti-tank weapons testing unit that had been established in January 1944 from 11.(Pz.)/SG 9 (see also Chapter One). Eggers and his team in E.Kdo 26 adapted the Army's 88 mm *Panzerschreck* ('Tank Fright' or 'Tank Fear') anti-tank weapon for airborne launching with electric discharge from Fw 190F-8 and F-9 ground-attack aircraft. The *Panzerschreck*, known officially as the *Raketenpanzerbüchse* 54, was a reusable, lightweight rocket-launcher intended for use by infantry and modelled on the American bazooka, examples of which had been captured in Tunisia in 1943.

The head diameter of the *Panzerschreck* was 8.8 cm, and it had a 4 cm diameter rocket motor and a total combined weight of 7.1 kg. The hollow-charge warhead, known as the *Puppehenkopf* (Doll's Head), was contained in a sheet metal casing. The burning time of the rocket was 0.04 seconds, and they were fired from three large, open-channel-section, sheet steel guide rails 1.5 m in length designed by Egger's team and known as AG-Ps (*Abchussgerät für Panzerschreck*). The rockets were contained within the AG-Ps and grouped in clusters of three. The projectiles were loaded from the front end of the rail and held in launching position by a spring catch. The cluster of rails was then suspended from an ETC 50 or ETC 71 bomb rack. In July and August 1944 Eggers used Fw 190F-8 Wk-Nr 580383 CM+WL, which was fitted with the latter rack, in order to test the rocket.

The penetration of the weapon was estimated at 140–160 mm at 60 degrees. Tests at Tarnewitz gave one hit in six on a 2 x 2 m target at 150 m.

In the early field trials at Udetfeld, 12 *Panzerschreck* rockets were fitted to an Fw 190 and fired at targets from the close range of 45–180 m. The low velocity of 120–135 m per second meant that it was necessary to approach a target at very close range and at a decreased speed of 490 km/h, which greatly endangered both aircraft and pilot, and for this reason only a small number of rockets were ordered initially. It was found that a direct hit by one rocket was sufficient to set a tank alight, but to achieve this required three aircraft with full loads.

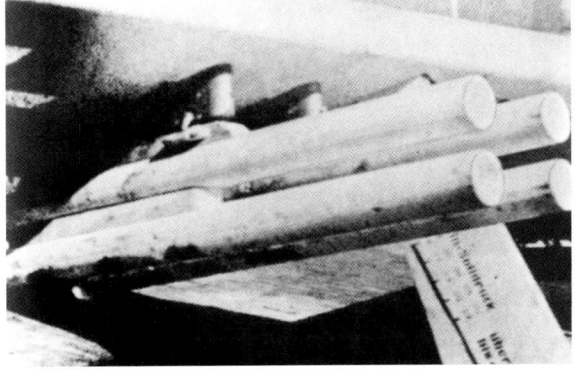

A mock-up of four wooden 'barrels' fitted beneath the wing of an Fw 190F-8 and used in initial assessment trials for the 88 mm *Fliegender Panzerschreck* anti-tank rocket. Low velocity resulted in tactical inefficiency and danger to both aircraft and pilot. (EN Archive)

It is believed that the first unit to use the *Panzerschreck* was 1./SG 10, which began to take delivery of the weapon in the autumn of 1944. During November this *Staffel* reported 23 Fw 190F-8s on strength and was based at Magyarmecske, 23 km southwest of Pécs in Hungary. Another early recipient of the rocket was 5./SG 77 under Oberleutnant Stephan Schmitt. The *Staffel* trained up at Udetfeld and then moved to Sarospatak, also in Hungary, in early October. In one of its first operations with the *Panzerschreck*, the unit attacked Soviet tanks close to the Hungarian–Rumanian border, but Schmitt's Fw 190 was hit by enemy anti-aircraft fire. The *Staffelkapitän* was killed, and he was subsequently awarded the Knight's Cross posthumously on 29 October. Other units understood to have taken delivery of the rocket in late 1944 included III./SG 3 at Frauenberg in Latvia, II./SG 2 and 8./SG 1, with further elements of SG 10 following in 1945.

The underwhelming performance of the early *Panzerschreck* prompted a second, revised version that used the ammunition of the original rocket as stocks were readily available. In this version, known simply as the *Panzerschreck II*, the 275 mm-long hollow-charge warhead was fitted with a flange to which a 350 mm cartridge tube with seven propellant powder charge rods was screwed. The 157 mm-long tail assembly carried the burner and four small, folding tail fins. The fins were each offset by two degrees, thus providing the spin necessary for stability.

In tests, the warhead was found to penetrate up to 160 mm of tank armour. In operational units *Panzerschreck II* rockets, which were known officially as PD 8.8 cm *Pz. Büchsenrohr*, were suspended on individual, underwing rails 328 mm apart in rows of six or eight rockets, but another method was known to be used that saw two sets of two rockets mounted above each other under a wing suspended from an ETC 71 rack.

The advantage of the *Panzerschreck* lay in the fact that it was no longer necessary to attack a tank from the rear. The development and use of such an improvised weapon gave impetus to the development of the more sophisticated *Panzerblitz* rocket.

Fliegender Panzerschreck	
Calibre	100 mm
Length	800 mm
Tail assembly span	230 mm
Weight with warhead	8 kg
Max speed	135 m per second
Range	135 m
Panzerschreck II	
Calibre	90 mm
Length	995 mm
Span	188 mm
Weight with warhead	7.2 kg
Maximum speed	240 m/sec (another source quotes 374 m/sec)
Range	200 m

PANZERBLITZ I, II AND III ANTI-TANK ROCKETS

In December 1944 a new rocket, developed following disappointment with the *Panzerschreck* as an anti-tank weapon, reached the *Schlachtgeschwader*. The first incarnation of a series of three variants developed by DWM, the *Panzerblitz* ('Tank Lightning') *I* carried a hollow-charge warhead identical to the *Panzerschreck* but fitted with a ballistic cap. The warhead carried a relatively small explosive content of around 590 g. A 323 mm-long cartridge tube, containing six propellant charge profile rods, was screwed into the 230 mm-long warhead. The 208 mm-long body attached to it had four slightly offset fins. The 127 mm-long burner had a nozzle opening of 19.8 mm, which widened up to 87 mm before reducing to 40 mm in diameter at the nozzle end.

The rocket weighed 7 kg. It was suspended from two round lugs that were arranged in one plane behind one another. Stabilisation of the projectile was attained by a stationary tail unit, with the four tail fins made of sheet metal, and this was considered to be very satisfactory. The rocket's trajectory was long and flat and its dispersion pattern was good. The all-burnt velocity was 320–340 m per second, with a burning time of between 0.4 and 0.8 seconds. This enabled an Fw 190 laden with 12 rockets to fire at a range of 200–300 m.

A one-time discharge device known as the *Einzelschussgerät Panzerblitz* (EG-Pb) was made by Curt Heber of Osterode, the firm also constructing launch racks for the R4M air-to-air rocket (see Chapter Three). The EG-Pb comprised a guide rail into which the lugs of the projectile slid from the front end when loading. The rocket was pushed back until the rear-sliding lug knocked against a stop. Simultaneously, a spring-loaded pawl was released that engaged the rear-sliding lug and prevented the projectile from slipping out accidentally. During launch, however, the thrust of the impulse had to overcome the resistance of the spring controlling the pawl. At the front and back ends of the rail, attachment points were fitted for suspension beneath the carrier aircraft. At the rear end of the rail there was also a terminal contact block connecting the ignition wires of the projectile. Following test-firing it was determined that the length of the rail needed to be no more than 700 mm.

The EG-Pb could be connected rigidly to an ETC 50 underwing rack using an intermediate carrier and offered little air resistance. The fairing of the intermediate carrier was sufficiently robust to resist the blast effect of the rocket on firing. In case of an emergency, the pilot could jettison the *Panzerblitz*, the EG-Pb and the intermediate carrier from the ETC 50 by operating the standard bomb release gear.

In trials, as many single rails as was practically and aerodynamically possible could be placed side-by-side beneath an aircraft's wing. Initially, projectiles were discharged as single rounds, but during the course of testing it proved effective to join several rails into a rail system. From this, the *Abschussschienegerät*-Pb (AG-Pb) was devised, which consisted of a system of six or eight rails held together by two cross-bars and fixed underneath the aircraft by four screws. The gaps between the rails was about 150 mm and depended on the type of tail unit used.

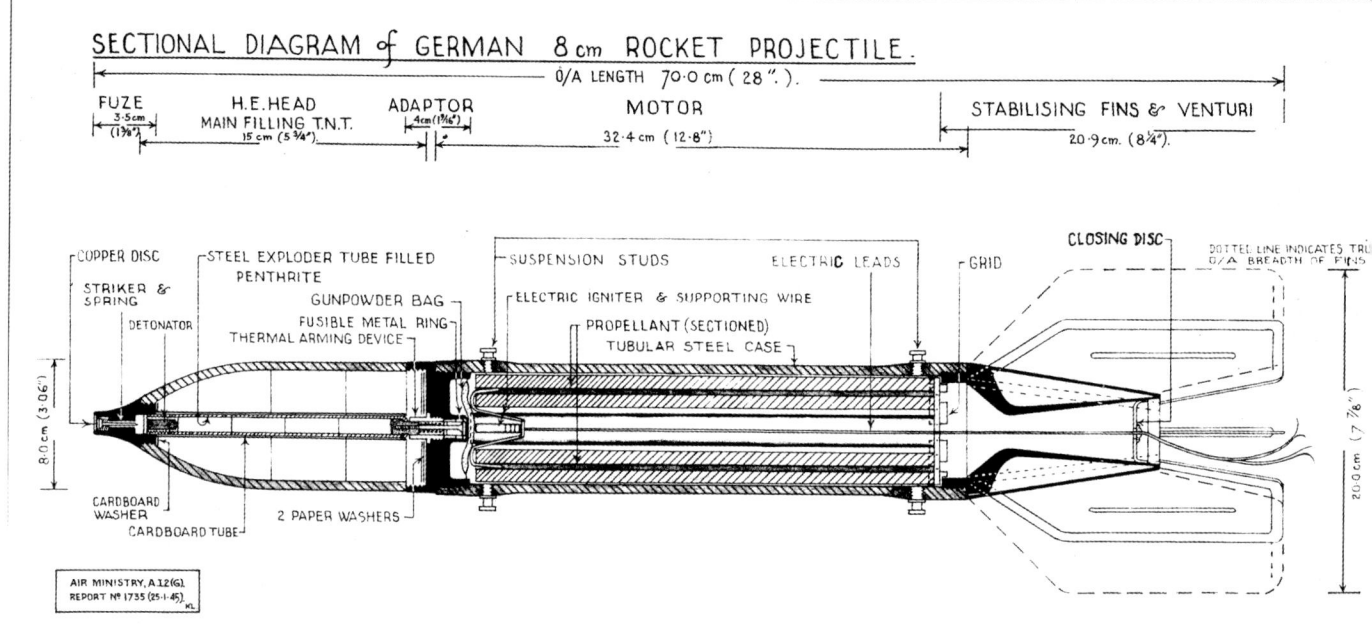

SECTIONAL DIAGRAM of GERMAN 8 cm ROCKET PROJECTILE.

O/A LENGTH 70·0 cm (28″.).

FUZE	H.E. HEAD	ADAPTOR	MOTOR	STABILISING FINS & VENTURI
3·5cm (1⅜″)	MAIN FILLING T.N.T. 15 cm (5¾″)	4cm (1⅝″)	32·4 cm (12·8″)	20·9 cm. (8¼″).

AIR MINISTRY, A.12(G)
REPORT Nº 1735 (25·1·45).
KL

British Air Ministry Technical Intelligence drawing from January 1945 of the 8 cm *Panzerblitz* rocket. (Forsyth)

In initial flights six *Panzerblitz I* were carried beneath each wing of an Fw 190, and they could be fired in salvos of three, six or 12. The loss in speed with racks fitted was around 15 km/h and when loaded with missiles, around 30 km/h. One aircraft so fitted was Fw 190F-8 Wk-Nr 733705 TX+PQ, which was assigned the prototype number V73 for tests with the *Panzerblitz*. Between 29 September and 6 October 1944, this aircraft was flown by Focke-Wulf test pilots Bernhard Märschel and Friedrich Schnier.

In order to determine the lead angle, the pilot of an Fw 190 would have to reduce his speed to around 490 km/h shortly before firing, if necessary by lowering his aircraft's undercarriage, but compared to the *Panzerschreck*, the *Panzerblitz* could be launched at twice the range – about 200 m. The projectiles could not be discharged all at once as lying close together, exhaust could cause interference. They were, instead, discharged in single rounds by means of a retarding relay or in a series with at least a 70 m interval between the individual projectiles by means of an automatic firing device. A projectile reached its maximum speed after only 0.8 seconds. In tests conducted at Tarnewitz, a strike rate of one hit in six was attained on a 10 x 10 m target at 250 m. The impact fuse was activated after about 50 m into the rocket's flightpath by the melting of a soft metal ring by heat from the propellant charge.

By February 1945 a total of 115 Fw 190F-8/Pb 1 aircraft were available, with 43,580 missiles having been manufactured in Czechoslovakia under the supervision of the SS. A monthly production rate of 16,000 missiles was targeted.

Generalmajor Hubertus Hitschhold, the *General der Schlachtflieger* from January 1944 to the end of the war, favoured the *Panzerblitz I* over the *Panzerschreck*, and recalled to British interrogators in 1945:

Fw 190F-8 wing underside fitted with 8 cm *Panzerblitz I* rockets on EG-Pb rail launch system.

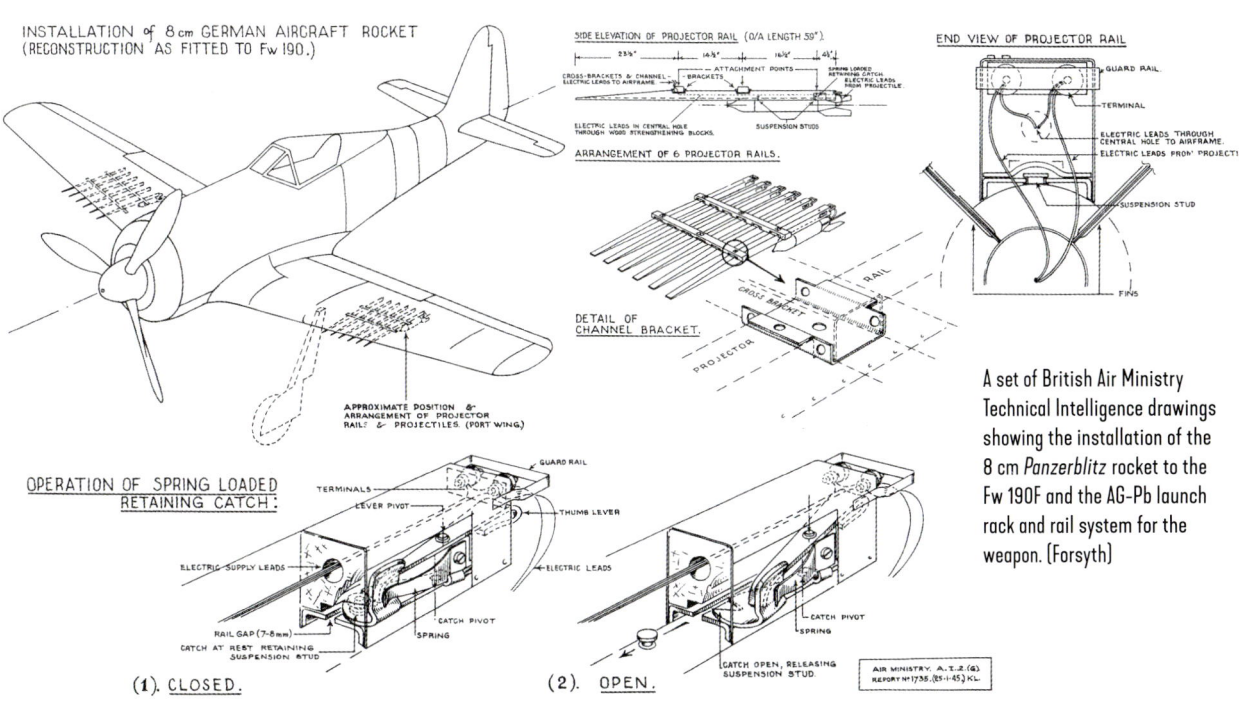

INSTALLATION of 8 cm GERMAN AIRCRAFT ROCKET (RECONSTRUCTION AS FITTED TO Fw 190.)

SIDE ELEVATION OF PROJECTOR RAIL (O/A LENGTH 59").

END VIEW OF PROJECTOR RAIL

ARRANGEMENT OF 6 PROJECTOR RAILS.

DETAIL OF CHANNEL BRACKET.

APPROXIMATE POSITION & ARRANGEMENT OF PROJECTOR RAILS & PROJECTILES. (PORT WING.)

OPERATION OF SPRING LOADED RETAINING CATCH:

(1). CLOSED.

(2). OPEN.

A set of British Air Ministry Technical Intelligence drawings showing the installation of the 8 cm *Panzerblitz* rocket to the Fw 190F and the AG-Pb launch rack and rail system for the weapon. (Forsyth)

It was possible to fire the rockets either in salvos of three, six or all 12 at once. The attacks were delivered at very close range, from 50 to 200 m, and depending on the defence, my pilots chose whether to fire all their rounds in one run or not. If the defence was weak, they made several attacks, and they always preferred to attack with several aircraft. It required three aircraft, firing their full loads, to ensure the knocking out of one tank – so one hit in 36. They had little training, however, but I think they improved things to something approaching one hit in 24. We were working on equipping the aircraft with more rockets so that each aircraft could reckon on one kill per sortie.

I never experienced any danger from my own rockets unless a tank blew up from a direct hit. The pilots were at least trained to turn away immediately after firing. Usually, every direct hit resulted in setting a tank on fire either immediately or very shortly afterwards, depending on where the hit had occurred. The crew of the tank was generally killed immediately, and that was attributed to either blast or splinters within the tank.

We did not try making steep dive attacks because of the short range which was required by the high dispersion of the rocket, and in a steep dive the tendency was to fire at too long a range, with a resulting inability to gain a hit. We attacked up and down wind as far as possible in a 20-degree dive. If tanks were parked up, we generally elected to use bombs.

We used the *Revi* sight with a 'ladder' graticule, and used our guns on the run-up because the Russian infantry usually rode on the outside of their tanks and put up sporadic small arms resistance. We also achieved good results with rockets against locomotives and soft vehicles, but I preferred using cannon against the latter.

A six-channel AG-Pb launch rack for *Panzerblitz I* rockets has been fitted to this Fw 190F of an unidentified unit. The rack was referred to by some Luftwaffe personnel as the *Gartenzaun* (Garden Fence). (EN Archive)

The first unit to use the *Panzerblitz I* operationally was III./SG 4 on the Western Front when it attacked enemy motor columns in the Strasbourg–Hagenau area on 7 December 1944. The *Gruppe* also took part in the Luftwaffe's New Year's Day attack on Allied airfields in northwest Europe. During the *Gruppe's* operations over Belgium that day, 9. *Staffel* pilot Feldwebel Rudolf Fye was shot down by USAAF P-47s and his Fw 190F-8 crashed near to the road that ran between Asch and Mechelen. It was equipped with *Panzerblitz* missiles, at least five live examples of which were discovered at the crash site, thus presenting Allied air and technical intelligence with a significant find.

In the period 21 January–20 February 1945, the Fw 190F-8s and F-9s of III./SG 4 accounted for the destruction of 23 Allied tanks and inflicted serious damage to 11 more, as well as causing the destruction of two armoured vehicles. This tally was achieved over 115 sorties flown in 16 operational missions in which the *Panzerblitz* was deployed.

Tactical doctrine eventually settled on the fitment of eight rockets per wing, which were fired in salvos of four or in pairs.

The Luftwaffe had wanted to equip three specially trained anti-tank *Staffeln* with rocket-firing Fw 190s in each *Schlachtgeschwader*, but by mid-April 1945 only 3 and 6./SG 1, 9./SG 2, 6./SG 3, 7., 8. and 9./SG 4, 1., 3. and 13./SG 9, 9./SG 77 and 13./SG 151 were equipped with the *Panzerblitz I*.

Panzerblitz I	
Length	705 mm
Maximum diameter	93 mm
Calibre	78 mm
Tail span	200 mm
Tail fin area	430 cm^2
Weight	6.54 kg
Maximum thrust	440 kg
Maximum speed	374 m/sec
Range	200–300 m
Rocket burn time	0.45 seconds

However, at this time the Soviet KV 1 tank had 90 mm armour and the American Sherman, 76 mm. Consequently, it was the low armour penetration depth of 90 mm achievable by the small hollow-charge that was behind the development of the *Panzerblitz II*. Influenced by the design and successful performance of the R4M air-to-air rocket, DWM's *Panzerblitz II*, officially designated as the R4 HL/8.8, mated a hollow-charge warhead to an R4M rocket body. The rocket weighed 5.5 kg and used the same propellant as the R4M, giving it a velocity of approximately 370 m per second.

For aerodynamic reasons the larger, 130 mm diameter warhead, which was capable of penetrating 180 mm of armour, was provided with an additional casing of metal sheet. Just behind the warhead, a clip of spring-steel plate ending in a guide key was mounted on the rocket motor which guided the projectile along the AG-Pb-type discharge rail as used for the *Panzerschreck*. To guide the folding tail unit of the R4M rocket motor, a strip of wood was fitted beneath the rear part of the rail.

Very few examples of the *Panzerblitz II* were produced, but some were tested on an Fw 190F-9. Longer term, it was intended to fit the missiles onto the planned Henschel Hs 132 jet aircraft. In February 1943, in response to a specification for a high-speed anti-shipping aircraft intended for deployment against the anticipated Allied invasion of Europe, the Henschel Flugzeugwerke at Berlin-Schönefeld offered a design centred around a small, mid-wing monoplane with twin end-plate fins and rudders, and a single turbojet mounted above the fuselage. The fuselage was to be of metal construction, while the wings were to be made of wood, with a plywood skin. The most radical feature was the cockpit, in which the pilot was to lie prone behind a glazed nose so as to be able to withstand forces of up to 10 g. This way, the Hs 132 would attack ships in a shallow dive, reaching speeds

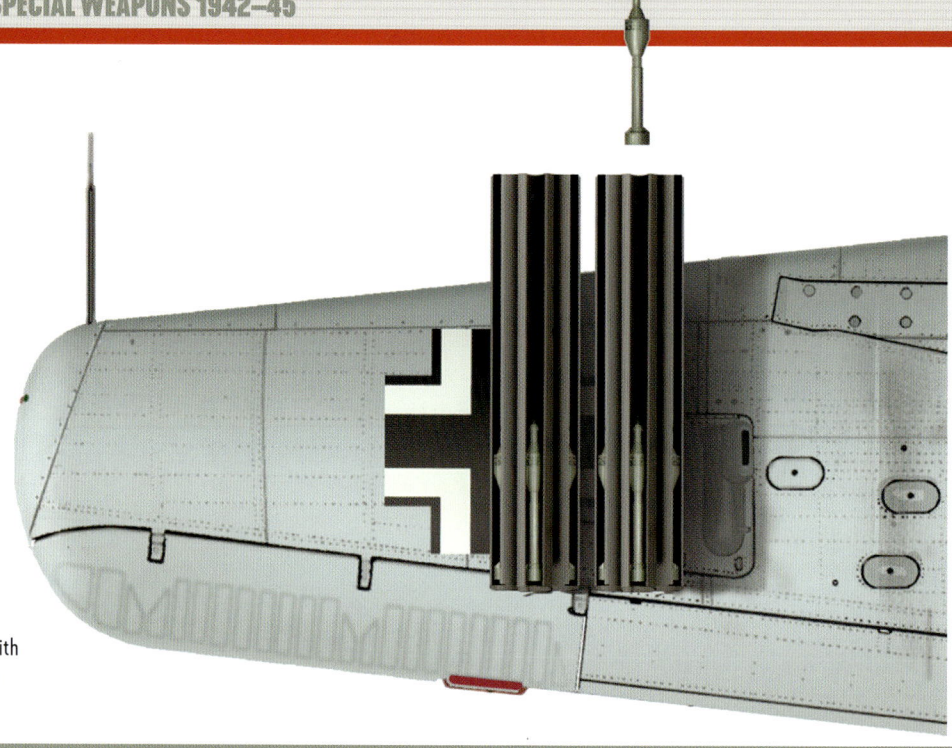

Fw 190F-8 wing underside fitted with 13 cm *Panzerblitz II* rockets on the AG-Pb rail launch system.

of up to 910 km/h, before dropping its 500 or 1,000 kg bombs at the target with the aid of a primitive computer, or firing a battery of rockets.

Three sub-types were planned, powered by either BMW 003 E-2, Junkers Jumo 004 B or Heinkel-Hirth HeS 011 engines, with an armament of two 20 mm MG 151/20 cannon – with an option for two MK 103s, if ordnance was reduced. The finished design bore a striking resemblance to the aircraft that would eventually emerge as the He 162, but Henschel only progressed as far as developing a wind tunnel model and producing the fuselage and wings of the first prototype.

Also considered suitable as a carrier aircraft for the *Panzerblitz II* was the proposed Junkers EF 126 *Walli* experimental, Argus pulse jet-powered interceptor project. But, like the Hs 132, the design never got further than the drawing board before the end of the war.

Panzerblitz II	
Length	815 mm
Calibre	130 mm
Weight	5.1 kg
Warhead weight	2.1 kg
Maximum speed	370 m/sec

The *Panzerblitz III* was also inspired by the R4M and had a highly sensitive 5.5 cm hollow-charge warhead based on the 75 mm HL.Gr. 43, which was shortened and converted to the new calibre. It contained an explosive charge of 0.290 kg. The result was an 85 mm-high hollow case, to which a 21 mm-long steel cylinder was welded. The new warhead, however, increased the length of the missile to 995 mm. The AZR2 detonator protruded 32 mm from the warhead. Although this reduced penetration of armoured steel to 120 mm at 60 degrees, it meant it was capable of being effective against the Soviet IS-2 Stalin tank with its 120 mm armour. Velocity was increased to 480–500 m per second, although this could be increased to 570 m per second with improved fuel.

Warheads for the *Panzerblitz III* were manufactured, but the defeat of Nazi Germany in May 1945 meant that the weapon never saw service.

Bü 181 *PANZERJAGDSTAFFELN*

In the closing weeks of the war a substantial number of Bücker Bü 181 Bestmann basic trainers were modified to carry *Panzerfaust* infantry anti-tank missiles with the aim of carrying out last-ditch missions to stem the advance of Allied armour into Germany.

It has remained unclear as to who originated the idea of fitting *Panzerfaust* to Bü 181s. However, it is believed that Oberst Adolf Dickfeld, the *General für Nachwuchs Luftwaffe* (General Commanding Luftwaffe Personnel under Training) and *Reichsinspekteur der Flieger-Hitler Jugend* (Reich Inspector of Hitler Youth Fliers), was responsible for organising the training of pilots to fly the type in the anti-tank role and for the formation of the so-called *Panzerjagdstaffeln* ('Tank Hunting Squadrons') that operated them during the latter half of April 1945.

The first trials were carried out at the NSFK *Reichssegelflugschule* (National Socialist Flying Corps State Glider School) at Trebbin, south of Berlin, probably around mid-March 1945. Shortly after the trials, an appeal was put out for pilots to fly light aircraft in the ground-attack role, with a stipulation that all volunteers were to have considerable experience in both low-altitude flying and aerobatics! This resulted in a nucleus of volunteer pilots at Trebbin who were predominantly instructors or members of the aircrew pool, the *Frontfliegersammelgruppe Quedlinburg*.

Training consisted of familiarisation flights firing the projectiles, concluding with three or four live-firing runs. For the final phase of training, the Bü 181s were fitted with four *Panzerfaust* 100m missiles, one mounted above and below each wing at about mid-span, positioned in a five-degree, nose-up attitude on makeshift wing racks.

As an infantry weapon, the *Panzerfaust* was a very simple design, weighing 5–10 kg and formed of a tube of low-grade steel, approximately a metre long and 4–6 cm in diameter. Attached to the upper-side of the tube were a rudimentary rear sight and trigger. There was no front sight, with the edge of the warhead being used instead. Inside the tube was a small charge of black powder for propellant. Fitted to the front of the tube by its wooden tail stem and metal fins was an oversized hollow-charge warhead, 15 cm in diameter and weighing 3 kg. It contained around 800 g of explosive.

Bücker Bü 181C-2 Wk-Nr 502167 of the 3. *Panzerjagdstaffel* based at Kaufbeuren in April 1945. The aircraft was photographed at Zurich-Dubendorf, having had its *Panzerfaust* wing racks removed shortly after being flown to Switzerland on 18 April 1945 by its defecting crew of Unteroffiziere Hans Ficker and Werner Diermayer. Note the firing sights fitted to the top of the cowling. (EN Archive)

The *Panzerfaust* 100m was the final version produced in quantity from November 1944 onwards. It had a nominal maximum range of 100 m. Some 190 g of propellant launched the warhead at 60 m per second from a 6 cm diameter tube. This version weighed 6.8 kg. The missiles were armed prior to flight and were aimed by means of two crude sights mounted on the Bü 181's upper engine cowling, one offset to port and the other aligned with the centre of the pilot's windscreen.

Many Bückers adapted to carry *Panzerfaust* had 50 cm x 50 cm metal plates fitted to both sides of the fuselage directly aft of the cockpit as a measure of resistance against blast and fire when the missiles were launched.

On 2 April 1945, orders were issued for the establishment of a number of *Panzer-jagdstaffeln*. By the end of the first week of April, sufficient personnel had been trained to form three operational units, the 1., 2., and 3. *Panzerjagdstaffeln*. Additionally, a 1. *Tiefangriffsgruppe Bücker 181* under the command of Hauptmann Hubert Jennes formed up in March 1945 in the Wittenberge area, south of Berlin, but whether this was an alternative name for 1. *Panzerjagdstaffel* remains unclear. Whatever the case, it is believed this unit, which served in the north of Germany, was formed with a strength of ten to 12 Bü 181s and operated under attachment to the reconnaissance unit *Nahaufklärungsgruppe* 8.

The 3. *Panzerjagdstaffel* was formed under the command of Oberleutnant Karl-Heinz Dragenscheck at Kaufbeuren im Allgäu, the home of *Flugzeugführerschule* (FFS) A/B 23, sometime around 7 April 1945. The airspace over Kaufbeuren airfield, situated about 30 km southwest of Landsberg, was still relatively free of American fighters and, to the benefit of the training programme, it possessed a large number of Bü 181 trainers. Training started in earnest on 14 April, with the cadre personnel, who had largely been recruited from the instructor staffs of various flying schools including FFS A/B 23, flying several ten-minute familiarisation flights with new volunteer pilots.

On 11 April, 8. *Panzerjagdstaffel* is known to have been based at Magdeburg-Süd with 12 aircraft and 22 pilots on strength, while the commander of 6. *Panzerjagdstaffel*, also at Kaufbeuren, reported that the strength of his *Staffel* was 'complete'. 3. *Panzerjagdstaffel* reported that the unit's strength was 'almost complete' and that its aircraft were being re-equipped. The *Kommandeur für Schlachtfliegerarbeitsstab* (Commander for Ground-Attack Planning Staff South) duly placed an urgent request at this time for 'recoilless anti-tank grenades' so that the training of the anti-tank pilots could commence.

On 18 April, 10. and 11. *Panzerjagdstaffel* arrived at Straubing, the former with 30 aircraft, while the latter was ordered to disband and surrender its aircraft and personnel to 10. *Staffel*. The latter then moved immediately to Münster. The same day, 1. *Tiefangriffsgruppe Bücker 181* transferred to Finow to commence operations, flying its first mission on 19 April in the Werneuchen/Müncheberg/Wriezen area. One crew did not return.

Despite the *Panzerjagdstaffeln*'s apparent departure from Straubing, one eyewitness recalled 'four or five Bü 181s daily' making low-level practise flights around the airfield, before the field was hastily destroyed in the face of advancing American forces on 25 April.

The tactics required for delivering the *Panzerfaust* missiles called for extremely low flying both to and from the target at maximum speed. Having selected his target, the pilot would, at a distance of approximately 500 m, pull up his aircraft to a height of 20–30 m and go into a shallow dive. Then, at a distance of 150–200 m from the target, all the projectiles would be fired at once by means of a cable situated between the pilot's and co-pilot's seats. At a speed of around 205 km/h, at a maximum distance of 200 m from the target, the pilot would then have to take abrupt evasive action to avoid flying into the explosions caused by the missiles. This was achieved by standing the nimble Bücker on its wingtip and performing the tightest of turns. Such a manoeuvre enabled a quick escape over the hedgerows before an unsuspecting enemy knew what had hit him.

8. *Panzerjagdstaffel* flew its first mission on either 11 or 12 April, during which one aircraft was hit by ground fire and was forced to land. It blew up shortly thereafter from the detonation of its own missiles. Franz Florian Winter had been an instructor with JG 101 at Stolp-Reiz and II./JG 108 at Wiener-Neustadt before joining 8. *Panzerjagdstaffel* on 9 April. He recalls that unlike most other units equipped with the Bü 181, his unit's Bückers carried only one *Panzerfaust* beneath each wing, but none on the upper side of the wing. Winter recorded:

Together with an Unteroffizier, I received my first operational order on 13 April 1945. We flew to the Halberstadt area, not at dawn or dusk, but rather at around 1400 hours. We flew just over the grass and over the American advanced tank units, since there was no longer any frontline. Over Klein Oschersleben we were jumped on by enemy fighters and there was an 'unequal' air battle. We were shot down. My comrade crashed on a meadow west of Klein Oschersleben. My machine smashed into a small coppice. I was trapped anxiously for 30 minutes because I thought that at any moment the aircraft would go up in flames. I was hauled from the machine by two Polish women.

ABOVE LEFT A Bü 181 fitted with two wing-mounted *Panzerfaust* 100m anti-tank missiles. The *Panzerfaust* were positioned in a five-degree nose-up attitude and launched from makeshift racks fitted at a central point on the wing. It is believed that some 150–170 Bü 181 trainers were converted to such a configuration. (EN Archive)

ABOVE The *Panzerfaust* missiles were aimed by means of two crude sights mounted on the fuselage of a Bü 181's engine cowling, one offset to port and the other aligned with the centre of the pilot's windscreen. (EN Archive)

Bü 181C-2 Wk-Nr 502167 of 3. *Panzerjagdstaffel*, Kaufbeuren, Germany, April 1945.

Panzerjagdstaffel, comprising 12 *Panzerfaust*-equipped Bü 181s, finally deployed from Kaufbeuren to Ringingen on 19 April 1945. It flew one of its first missions against American vehicle convoys in the Tübingen area that evening. Although no tanks were destroyed, the six Bückers that participated in the attack were able to account for a small number of trucks. No losses were suffered by the Germans, but at least two of their aircraft received light damage from the surprised crew of an American anti-aircraft gun.

To the north, on 20 April, Hauptmann Jennes and Oberfähnrich Peter Rambausek of 1. *Tiefangriffsgruppe Bücker 181* were reportedly hit by infantry fire during an early morning mission and were forced to land in an open field. They were picked up by German troops and were able to return to their base at Finow, where Rambausek and another Fahnenjunker-Unteroffizier promptly took command of another Bü 181.

This Bü 181 of 1. *Tiefangriffsgruppe Bücker 181* was photographed in March or April 1945. The aircraft has *Panzerfaust* missiles mounted on racks fitted to the upper- and undersurfaces of each wing. The aircraft is adorned with the unit's emblem depicting a crudely painted Mickey Mouse figure holding a *Panzerfaust*. The pilot seen here is Oberfeldwebel Paetzsch. (Chapman)

So as to maintain the element of surprise, and to afford the attacking aircraft the best measure of protection from Allied fighters, the Bü 181s mainly struck at dawn and dusk. During the day they remained hidden safely in trees at the edges of their landing grounds. By changing their locations virtually every day, the aircraft of 3. *Panzerjagdstaffel* remained undetected until the end of the war and were never once caught out in the open. This meant that their crews frequently spent cold and uncomfortable nights either in, or under, their aircraft since the unit's supporting supply vehicles often failed to find 3. *Panzerjagdstaffel* before it left for its next landing site.

One of 3. *Staffel's* last missions was undertaken during the early hours of 24 April from a small field near Immenhofen, close to Kaufbeuren. Some eight Bü 181s took part in the operation, the aim of which was perhaps indicative of a nation in defeat. The target on this particular day was abandoned Wehrmacht vehicles and Luftwaffe aircraft which may have held documents or other items of importance. In what was to be the last mission of his operational career, Oberfeldwebel Buchsteiner, for example, apparently attacked a stranded He 111 that had been abandoned by its crew in a meadow near Memmingen airfield.

However, as late as 29 April, the Bückers were known to be operating – at dusk that day, a column of 'soft-skinned' vehicles from the US Army's 10th Armored Division was advancing in the vicinity of the airfield at Schöngau. According to the Divisional history:

> Three light German planes took off from the field and flew over the column, dropping *Panzerfaust*s as bombs. One hit dangerously near 'Red' Hankins' vehicle and blew off his right rear tire. Fortunately, none of the occupants were injured when they were thrown from the jeep.

Panzerfaust-equipped Bü 181s received a special word of gratitude from XIII. SS-*Armeekorps* during fierce fighting near Nördlingen, Bavaria, in late April. SS commanders reported Luftwaffe 'light aircraft provisionally fitted with *Panzerfaust* under their wings' that came to their aid by carrying out daily, low-level attacks at both dusk and dawn. These attacks, which were aimed at American tanks and convoys, were described as being 'tireless and courageous'.

In total it is believed that some 150–170 Bü 181 trainers were converted to carry the *Panzerfaust*.

1. *Tiefangriffsgruppe Bücker 181* was disbanded at Eggebeck/Tarp with five or six aircraft on strength. On 4 May 1945, 3. *Panzerjagdstaffel* was disbanded at Reit-im-Winkel close to the Austrian border. On 7 May, in what was probably the last official mention of the *Panzerfaust* units, the Luftwaffe Quartermaster General reported that a *Behelfs-Panzer-Schlachtkommando* (Auxiliary Anti-Tank Detachment) was based at Gasteig, near St. Johann, with four Bü 181s on strength. 3. *Panzerjagdstaffel* surrendered to the US Army in Reit-im-Winkel on 9 May. The unit's surviving aircraft remained on the local glider field for some time after the war and provided a ready-made playground for local children until the aircraft were eventually carted away to the scrap yard.

CHAPTER FIVE

ANTI-SHIPPING WEAPONS

SD/PC 1400 X REMOTELY CONTROLLED GUIDED BOMB (*FRITZ X*)

Shortly after 1400 hrs on 9 September 1943, a formation of nine Do 217K-2s of III./KG 100 led by the *Gruppenkommandeur*, Major Bernhard Jope, took off from their base at Istres and headed south in cloudless skies over the Mediterranean at 7,000 m. Their mission was to attack capital ships of the Regia Marina which the Germans feared would be used by the Allies following Italy's surrender the day before. The Italian fleet, comprising the battleships *Roma*, *Italia* and *Vittorio Veneto*, along with six cruisers and ten destroyers, all under the command

A pristine *Fritz X* glide bomb resting on its timber delivery cradle. The bomb's cruciform tail unit, with struts, spoilers and control surfaces, is clearly visible. The interior of the bomb encased by the tail unit housed the venturi tubes and roll control gyroscope. The connection points for the flare unit can also be seen. (EN Archive)

of Admiral Carlo Bergamini, had set sail from La Spezia and Genoa for Allied ports in North Africa the previous day.

Each Dornier carried one of the Luftwaffe's new 1,570 kg PC 1400 X guided bombs, known as the *Fritz X*. Conditions for the attack were ideal as the Italian ships reached La Maddalena and made passage through the Strait of Bonifacio, between Corsica and Sardinia, towards the island of Asinara. At 1530 hrs an observer spotted a formation of aircraft approaching from the northwest. As the Dorniers made their approach, the German airmen readied themselves. 'I will never forget the imposing sight of the Italian war fleet as it loomed up beneath us,' Jope later remembered.

All bombs were released west of the Strait of Bonifacio. Those from the first wave fell short of their targets, having been released at an angle of 60 degrees rather than at 80 degrees. One from the second wave, however, released by Unteroffizier Klapproth aboard Jope's own aircraft, found the 46,215-ton *Roma* – Italy's most modern battleship, which had been in service for less than a year. The PC 1400 went through its quarterdeck near the engine room and two boiler rooms and exploded outside, just beyond the hull. The badly damaged ship dropped out of formation.

Five minutes later, another bomb, released from the Do 217 piloted by Oberfeldwebel Kurt Steinborn, smashed into *Roma*'s foredeck plating from a higher angle, causing another explosion, this time in the forward magazine near the B turret where the heavy calibre shells for the battleship's 381 mm main guns were stored. The 1,500-ton turret blew away high into the air. Steinborn recalled:

> My observer, Unteroffizier Degan, released the *Fritz X* guided bomb with a steady hand and switched on the automatic camera. As often practiced on the bombardier training course, Unteroffizier Degan now directed the bomb using the phosphorous flare. It was a direct hit amidships! I still remember that 42 seconds elapsed from the time the bomb was released until it hit. While the PC 1400 X was aiming for its target in free fall, it was tracked by my observer Degan with binoculars and readjusted using the radio set. One could clearly see that the *Dödel* detonated in the funnel amidships.

Indeed, with flames reaching 900 m into the air, the vessel eventually turned over, broke in two and sank. Of the 2,000 crew aboard, 1,552 perished in the fire or drowned, including Bergamini and his staff. Another bomb seriously damaged *Roma*'s sister ship, *Italia*, when it passed through the hull and hit the water close by. More than 1,000 tons of water poured into *Italia*'s badly ruptured hull, but the battleship managed to reach Malta under its own steam. Further successful attacks would be carried out against British and American warships over the coming days. The Luftwaffe had introduced a fearsome new dimension to its anti-shipping operations.

Built by Rheinmetall-Borsig at Berlin-Marienfelde, the PC 1400 X was the creation of Dr. Max Kramer of the DVL at Berlin-Adlershof. An insightful engineer with a degree in electronic engineering from the Technisches Universität in Munich, Kramer specialised in the modelling of laminar-flow dynamics. He had tried to respond to a call by Luftwaffe aircrew for a solution to a problem experienced when bombing ships. They had found that in early operations mounted by the *Legion Condor* against shipping during the Spanish Civil War, there was a time lag of 20–30 seconds for a bomb to reach a ship from altitude. This was enough time for a vessel to take evasive manoeuvres.

In 1938 Kramer set out to solve such a bombing problem by creating a *controlled* bomb. In one trial carried out at Brackwede in Westphalia in cooperation with Rheinstahl AG, he tested the use of radio-controlled spoilers fitted onto the cruciform tapered tail

surfaces of a 250 kg SC 250 bomb. Other tests were undertaken at the *E-Stelle* Rechlin on behalf of the RLM. The tests proved encouraging, and in 1940 the RLM gave the go-ahead for further trials applying a control procedure on the 1,400 kg SD 1400 armour-piercing bomb. The bomb, fitted with a special cruciform tail, was designated as the SD 1400 X, the 'X' denoting the tail design.

A Fritz X glide bomb suspended from beneath an He 111H, ready for air-launch tests by the DVL. The bomb is off-set to starboard and fitted to specially faired clasps. (EN Archive)

Testing was supervised by Generalingenieur Ernst Marquardt of the Ministry's *Abt.* E7 and one of his specialists, Dr. Theodore Benecke. The first airborne release took place on 21 June 1940 at the Luftwaffe's missile testing ground at Peenemünde-West on the Baltic coast when a *Fritz X* was dropped from 4,000 m. On 23 November, another bomb was dropped from 7,000 m and both events were deemed satisfactory. But the local weather conditions often precluded visual tracking in the air – a cloudless sky was needed for the operator in the carrier aircraft to be able to follow the bomb to ensure a direct hit on a target. To this was added the complexity of testing a myriad of sensitive individual components, and by the end of December 1941 only nine remotely controlled drops had been made from He 111s. That month the test programme relocated to the Regia Aeronautica's bombing range at Siponto, near Manfredonia in Foggia, where better weather was assured.

Meanwhile, at the end of 1941, the Luftwaffe had established *Lehr- und Erprobungskommando* (E.Kdo) 21 at Schwäbisch Hall as a dedicated test command for the *Fritz X.* By the late summer of 1942, the unit was under the command of Major Ernst Hetzel, former *Gruppenkommandeur* of III./KG 100 which was also based at Schwäbisch Hall between April and July 1943 equipped with new Do 217s.

From Siponto, at what became known as the *Erprobungsstelle Süd'* between March and May 1942 under the supervision of the E7 *Gruppenleiter*, Dr. Hans Bender, 20 remotely-controlled drops were performed from 6,000–8,000 m and guided to target

An SD 1400 X *Fritz X* remotely controlled guided bomb is launched against Allied merchant shipping by a Do 217K-2 of III./KG 100 over the Mediterranean in the autumn of 1943.

crosses. In a test-drop carried out on 8 August of that year, the *Fritz X* succeeded in penetrating a 120 mm-thick armoured bulkhead.

Tests did not always prove successful; the high velocity in free fall of such a large bomb meant that it became difficult to control. It was not until the availability of a high-speed wind tunnel at the DVL in early 1942 that the creation of an improved spoiler system could be undertaken. Prior to this, if the bomb was released from a height of less than 4,000 m, there was insufficient time for a bomb-aimer to make corrections to the trajectory.

In all, of 100 bombs to be dropped in trials, 49 hit their target or struck close enough to cause serious damage.

In essence, the SD 1400 X remained faithful to the armour-piercing bomb on which it was based, the front part of the weapon retaining its shape and size, but with the addition of four short wings. The electronic guidance system containing the radio receiver, power source and reference gyroscope, was housed in the rear section of the bomb, which was fitted to the central warhead section with a ring of studs and nuts and made of cast magnesium alloy. Control spoilers and operating solenoids as well as an all-metal dive brake assembly were added to the four tail fins and intended to limit terminal velocity to 280 m per second. One section of the dive brake was electrically insulated to enable it to be used as a radio aerial.

In its operational form, the specifications of the remotely controlled, wire-guided PC 1400 *Fritz X* were as follows:

Length	3,262 mm
Span	1,350 mm
Diameter	562 mm
Total weight	1,570 kg
Warhead weight	1,150 kg
Tail unit weight	120 kg
Explosive weight	300 kg
Fuse	AZ 38B (impact 0.045 secs; first delay 0.1 secs; second delay 0.2 secs)
Velocity	After 4,000 m drop = 250 m/sec.
	After 8,000 m drop = 290 m/sec.
Self-destruct fuse	AZ 80
Bombing altitude	4,000–8,000 m
Maximum penetration at 6,000 m – 130 mm armour plate	

The means of controlling and guiding the *Fritz X* lay in the FuG 203/320 *Kehl/Strassburg* radio control system. The *Kehl* FuG 203 transmitter for fitment into the carrier aircraft had been developed by Telefunken and named after a district in Strassburg, the Franco-German city on the Rhine. The transmitter operated on one of 18 frequencies between 48.2 and 49.9 MHz, separated by a 100 kHz transmitter which transmitted orders to the FuG 230 radio receiver as frequency modulations on a radio frequency carrier, with the bomb requiring a set of four orders: right, left, up and down. Before take-off, bombs would be pre-set to one of the frequencies, with a corresponding setting made in the carrier aircraft.

The system did suffer from problems associated with moisture and condensation, especially during longer range, high-altitude missions, but these were countered by incorporating a special heating system in the carrier aircraft that directed hot engine exhaust over the weapon controls. But this installation meant that the aircraft had to be specially fitted with both *Kehl/Strassburg* and the heating system, thus limiting their number.

It was planned to deploy the *Fritz X* from the greatly anticipated but troublesome He 177 bomber, but delays in the production and introduction of that aircraft saw the specially developed Do 217K-2 sub-variant used as an interim measure. One *Fritz X* was suspended beneath the wing, between the fuselage and the starboard engine nacelle. In order to climb above 6,000 m whilst carrying the heavy bomb, the Do 217K-2 had extended wings, along with a fully glazed and rounded nose section from which the pilot and bomb-aimer had better visibility for releasing, guiding and tracking the *Fritz X*.

From the end of 1942, E.Kdo 21 supplied trained crews, armourers and technicians to III./KG 100 at Schwäbisch Hall, but delays in production of the planned first batch of 1,000 *Fritz X*s at Rheinmetall-Borsig in Berlin-Marienfelde and in the conversion of Do 217s meant that operations did not commence until July 1943, when III./KG 100 moved from Schwäbisch Hall to Istres.

A *Fritz X* glide bomb fitted beneath the fuselage of an He 177 probably of KG 100. The ground-clearance for a bomb loaded in such a way was minimal and potentially dangerous. The bombs were usually carried on ETC 2000/ XII D 1 wing racks. (EN Archive)

A standard *Lotfe* 7d bombsight was used to aim the bomb, which was released from between 5,200 and 7,000 m at an approach speed of between 250 and 270 km/h. The bomb-aimer used a *Knüppel* (joystick) to control and adjust the steering. A straight and level bombing run of two-and-a-half minutes was required prior to bomb release. Immediately after release, the aircraft was throttled back, pulled up sharply for eight to ten seconds, and then levelled off in straight and level flight. This manoeuvre allowed the bomb sufficient time to get ahead of the aircraft so that the observer could aim it for its entire flight.

The bomb could not be controlled during the first 15 seconds after release; on the 16th second, the bomb-aimer took control. No evasive action could be taken until after the bomb had been seen to explode. The *Fritz X* carried a flare in its tail section that enabled the observer to follow its fall visually, while a supplementary electric light performed the same function for nocturnal drops. It was estimated that the *Fritz X* could be guided to within a margin of error of only 50 m from an altitude of 7,000 m. The bomb took 42 seconds to reach the ground from 7,000 m and 38 seconds from 6,000 m. The lowest height for satisfactory release was 4,000 m.

Operations commenced with the PC 1400 X on 21 July when III./KG 100 sent three Do 217K-2s loaded with the bombs to attack ships in Augusta harbour, Sicily, but this and following missions against other targets in Sicilian waters are believed to have been without success.

A few days after the successful attack on *Roma*, KG 100 struck again when its II. and III. *Gruppen* under Hauptmann Molinus and Major Jope, respectively, carried out a number of attacks against the Allied landings at Salerno. In the pre-dawn darkness of 11 September, one *Fritz X* almost hit the US Navy light cruiser USS *Philadelphia* (CL-41). The vessel did not sustain damage, although several members of its crew were wounded when the bomb exploded just 15 m away. Ten minutes later, another bomb struck the forward gun turret of *Philadelphia*'s sister ship USS *Savannah* (CL-42). The ensuing explosion in the lower handling room ripped a hole open in the vessel's deck, killing the crew of the turret and a damage-control party. Another large hole opened up in the ship's bottom and forced a seam in its side. For a period the warship was without power and settled by the bow until its forecastle was almost awash. However, the light cruiser survived and reached Malta. A total of 197 crew had been killed in the attack.

Two days later, in the early afternoon of 13 September, a *Fritz X* dropped by a lone Do 217 from III./KG 100 hit the Royal Navy light cruiser HMS *Uganda*. The bomb coursed through seven steel decks before exploding in the water beneath the vessel's keel. The hull was left with a large hole and 1,400 tons of sea water flooded in. Sixteen sailors were killed in the attack, but like *Savannah*, *Uganda* survived.

The Luftwaffe continued operations with the *Fritz X*, albeit sporadically after September 1943. Indeed, the weapon would not be used again until a failed mission against Plymouth on 30 April 1944, and after that in only in a handful of attempted attacks against the Allied landing fleet off Normandy and in the Bay of Biscay in July and August 1944. On most occasions the Do 217s were effectively scattered by Allied air defences. Ultimately, only one ship, *Roma*, is known to have been sunk by the *Fritz X*, with six others believed to have been damaged.

A monthly production rate of 300 *Fritz X*s was planned, rising to 750, but this was never achieved once manufacture began in April 1943, and yet the bomb remained in production until December 1944. It is believed somewhere between 1,500 and 2,000 were built.

HENSCHEL Hs 293 REMOTE-CONTROLLED GLIDE BOMB

The man chiefly responsible for the design and development of the Hs 293 was Professor Herbert Wagner, an Austrian and a career aeronautical scientist who was born in Graz in 1900. After service in the Imperial Germany Navy during World War I, Wagner commenced studies at the Technisches Hochschule in Berlin. In 1924, after graduating, he joined the Rohrbach Flugzeugbau, and three years later began an academic career at the University of Danzig as a Professor of Aeronautics. In 1930, while a professor at the Technical University of Berlin, he developed an early axial turboprop engine.

In 1934, Wagner joined the Henschel Flugzeugwerke as a manager, but moved to Junkers the following year, where he worked on the development of the EF 61 high-altitude bomber for the RLM as well as early Junkers jet engine projects at the company's Magdeburg plant. He later became a board director, but left Junkers in May 1939 to return to Henschel at Berlin-Schönefeld, where he became involved primarily in rocket development projects, but also, under encouragement from the DVL, in the development of missile guidance systems. This was concurrent with a desire within the RLM to explore guided missiles for deployment against shipping.

The Hs 293 remote-controlled glide bomb, complete with Walter rocket engine. (Goss)

Do 217E-5 Wk-Nr 5552 6N+HP fitted with an Hs 293 remote-controlled glide-bomb and flown by Leutnant Horst Preißer of 6./KG 100 from Istres, France, in September 1943.

ABOVE An Hs 293 is launched over the Baltic during trials from He 111 carrier aircraft PH+EI attached to the *E-Stelle* Peenemünde's Test Section E4, which was responsible for the development and testing of remote-controlled guidance systems, radio communication and the use of television as a means of weapon control. PH+EI was fitted with extremely advanced equipment intended for television-guided launch trials with the Hs 293D. (EN Archive)

RIGHT An operator aboard an He 111H-12 keeps his eyes focused on an He 293 after launch and uses the command joystick to visually control the bomb in its gliding descent. (EN Archive)

Wagner put forward a proposal for a steerable, air-launched, glide bomb which would make use of the FuG 203b/230 *Kehl/Strassburg* radio control system as used on the *Fritz X*. The RLM accepted Wagner's proposal and designated it the Hs 293. His first design, the Hs 293 V1, was only a drawing-board product, essentially a glider version of the standard SC 500 bomb. The V2 version was actually produced in the winter of 1940, leading to an ensuing production run of around 100 unpowered, pure glide-based missiles. These were used to conduct test-drops from an He 111 in the late summer of 1940, but there were regular guidance failures caused by malfunctioning valves and electronics unable to cope with vibration and temperature fluctuations experienced between ground-level and the intended operational altitude.

It was also realised that for a glide-weapon to be effective without power, it would need to be launched from a sufficiently high altitude to allow time for a bomb-aimer/observer aboard the launch aircraft to gain control. Nevertheless, on 18 December 1940, Professor Wagner managed to steer an unpowered glide bomb over a distance of 3.5 km across the Peene estuary to make a direct hit on its target – a haystack.

The third, limited, pre-series production of 100 missiles saw the fitment of a Helmuth Walter HWK-109-507 rocket motor to provide power boost at launch. The first air drop, carried out on 16 December 1940, failed because the left and right directives had been erroneously

This is understood to be Hs 293A Wk-Nr 241339 rested atop one of its packing crates. (EN Archive)

interchanged. Subsequent tests conducted from Peenemünde-West against the remains of a beached merchant ship proved generally successful, although on one occasion the remote control system failed and a missile crashed between the platforms of the works railway station at the neighbouring army test centre of Peenemünde-East, fortunately without any fatalities. Most electrical problems were solved by replacing the switching valves in the radio control receiver with Siemens magnetic relays.

In November 1941, it was decided to establish a dedicated Luftwaffe Hs 293 test unit, to be known as *Versuschsstaffel* 293 at Gärz under the command of Hauptmann Franz Hollweck, with which to conduct detailed operational assessment and training with four He 111Hs. Hollweck was an experienced naval aviator, having served with 10.(*See*)/LG 2, KG 30 and 1./Kü.Fl.Gr. 606. By March 1942, the unit had been redesignated E.Kdo 15, and it had worked closely with Wagner and the technicians at Peenemünde-West.

The *Kommando*'s aircraft carried out training flights so that crews could master approach and jettison procedures, as well as take-offs and landings with asymmetric configurations using an Hs 293 suspended on only one side of the aircraft. Flight control was accomplished by ailerons and elevators, with the bomb-aimer rolling the missile clockwise or anti-clockwise until the target was on the missile's vertical (yaw) axis before applying a pure pitch order to climb or dive so as to align its flight axis with the target. Bomb-aimers, however, did complain that they experienced difficulty in following bombs during flight. This issue was dealt with effectively by fitting flares in the tail of the bomb which would not be visible to ground or surface anti-aircraft gunners.

According to Diplom-Ingenieur Max Mayer, who worked as a test pilot and specialist in unmanned missiles at Peenemünde and who later managed *Abt.* E2, which was concerned with missiles and rocket-propelled aircraft:

> With well-trained controllers, impact accuracy, as with the SD 1400 X, lay within a 5 x 5-m box. With this missile it was possible,

One of the early prototype Hs 293s, possibly from the V2 series, beneath an He 111 carrier aircraft. At this stage the missile was unpowered and lacked the familiar warhead shape with its trimming weight. By July 1941, tests were being carried out with the improved Hs 293 V3 series, but this also lacked its rocket engine and its minimum release altitude was 1,000 m. (EN Archive)

HENSCHEL Hs 293
REMOTE-CONTROLLED GLIDE BOMB

The Hs 293 was one of the most sophisticated and complex weapons in the Luftwaffe's arsenal.

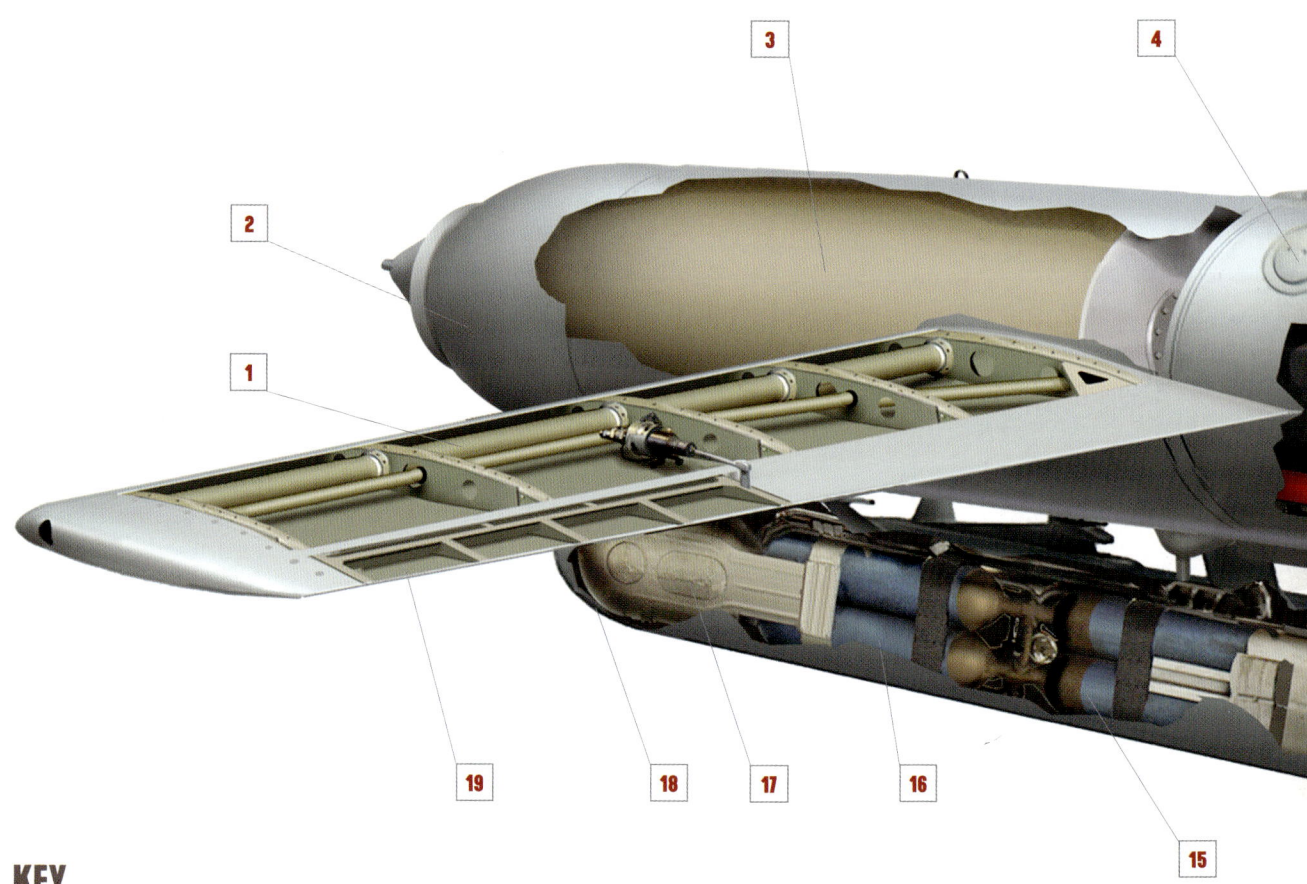

KEY

1. Main wing spar
2. Warhead casing
3. Trialen 105 500 kg high-explosive warhead
4. Hot-air intake (from parent aircraft)
5. Guidance filter unit
6. DC generator
7. STARU E230 Straßburg radio command receiver
8. Main antenna support
9. Antenna support
10. Elevator
11. Tail tracer flare assembly
12. Guidance tuning window
13. Radio destruction device
14. Rocket combustion chamber
15. Air pressure tank
16. Z-Stoff tank
17. T-Stoff tank
18. Walter HWK-109-507B rocket motor
19. Aileron

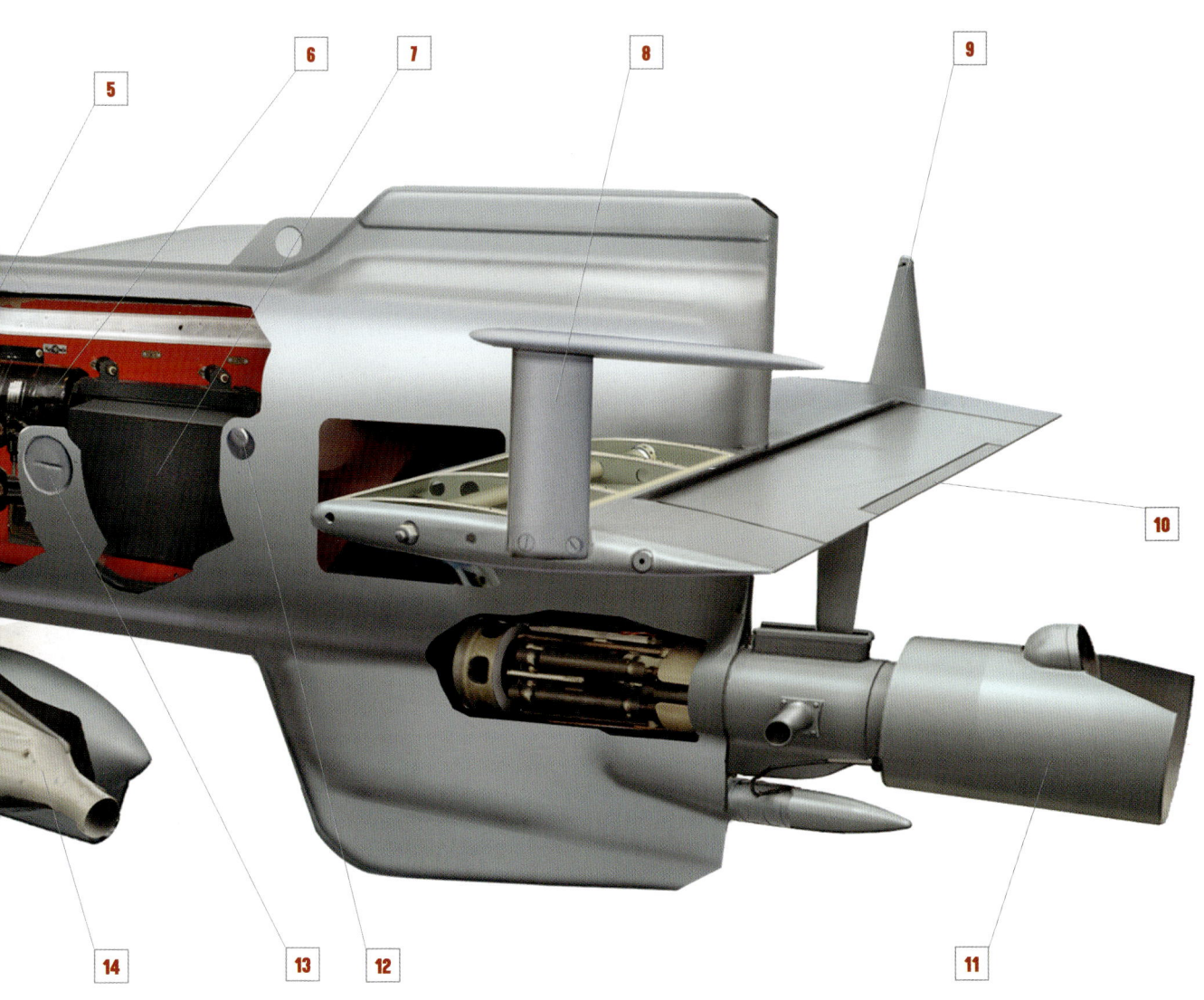

depending on visibility and the tactical opportunities offered in the target area, to operate it in a very wide range: from a minimum altitude of around 400 m and from 3.5–5 km distance, right up to over 10,000 m altitude and up to 18 km distance from the target.

Eventually, in November 1942, E.Kdo 15 received its first Do 217s, the aircraft with which it was intended to go operational using the Hs 293. Presumably with training deemed to have reached a satisfactory level, in April 1943 the *Kommando* was disbanded and its crews absorbed by II./KG 100, with the promoted Major Hollweck succeeding Major Fritz Auffhammer as *Kommandeur*.

The Henschel-designed weapon was formed from a principally aluminium, stressed skin, spot-welded body. It carried a 500 kg Trialen 105 warhead in the forward part of a small, cylindrical fuselage, itself forming part of a small 'monoplane' with rectangular wings and tailplane. The aft part of the missile, which was made from light monocoque alloy, contained the control gyroscope, radio receiver, batteries and battery-driven motor generator. The missile weighed 800 kg in total and was fitted with ailerons for longitudinal control and lateral steering, as well as an elevator operated by a motor. The rocket motor was suspended from beneath the fuselage by a three-legged support, with the thrust being transmitted to the missile body by a spring-loaded catch fitted on the rear support leg. The thrust axis of the motor was inclined at 30 degrees to the horizontal and the support was adjustable so that thrust could pass through the bomb's centre of gravity.

After launch, the small Walter rocket motor accelerated the speed of the Hs 293 to around 600 km/h. The motor functioned through the decomposition of highly concentrated hydrogen peroxide into a high-pressure steam to produce an initial thrust of 600 kg, decreasing to 400 kg, the catalyst being an aqueous solution of calcium permanganate known as *Z-Stoff*. The motor had an empty weight of 517 kg and carried 68 kg of propellants; it held 7.1 litres of hydrogen peroxide and 2.36 litres of calcium permanganate. Cutting out after about 12 seconds, the bomb then coasted towards its target in a shallow dive guided by the bomb-aimer in the parent aircraft. The bomb-aimer moved a joystick to merge the flare marking the path of the bomb over the target image. On paper, the Hs 293 had a range of about 11 km.

For night operations, a tracer flare or a searchlight fitted with a coloured filter housed in the rear fuselage would aid direction and control.

The Hs 293 was intended to be launched from the He 111H, Do 217E-5 and K-2, the He 177, the Fw 200 and the Ju 290A-7 on the basis that such aircraft were fitted with *Kehl/ Strassburg*. The first aircraft to deploy the weapon in action were the Do 217s of Major Hollweck's II./KG 100, which usually carried out attacks at night from 1,000 m at 320 km/h. Before take-off, the bombs would be pre-set to one of the frequencies of the *Strassburg* receiver, with a corresponding setting activated on the *Kehl* transmitter in the launch aircraft.

Attacks with the Hs 293 were conducted in both daylight and at night under conditions of good visibility and a near full moon. 'Official' tactics adopted by units armed with Hs 293s saw missions undertaken by a formation of 12 to 15 aircraft, in turn broken down

ABOVE The unusual sight of an Fw 200 laden with Hs 293 missiles. From 1940 the Condor was viewed as the Luftwaffe's main long-range anti-shipping aircraft but, ultimately, operations never took place using the type with the Henschel glide bomb. This aircraft, Fw 200C-5 Wk-Nr 0226 DP+ON, was used for carrying out trials with the Hs 293 at the *Erprobungsstelle* at Karlshagen in 1943. (EN Archive)

LEFT Do 217E-5 6N+NP of II./KG 100 is loaded with an Hs 293A below its starboard wing. Note the large, faired carrying rack. The successes achieved by the Hs 293 did not reflect the time spent on development, the cost involved and the deployment effort expended. (EN Archive)

into *Kette* elements of three bombers. The lead aircraft in each *Kette* would be equipped with FuG 200 *Hohentwiel* air-to-surface vessel search radar – a low-UHF band system which had a range of some 60–80 km when used to locate individual ships. It also had less impact on aircraft speed because of its small antennae array.

The lead aircraft would make the initial attack against a nominated target ship, while the other two aircraft positioned themselves to attack regardless of which way the ship was turned. If an attack was being made against a convoy, the leader of the whole formation assigned each *Kette* a specific ship to attack. The aircraft flew as low as possible to within striking distance of the objective. A climb was then made to a height of approximately 600 m and position was taken on the port quarter of the ship at a range of 10–12 km. In order to avoid anti-aircraft fire, the attacking aircraft headed for the port quarter of the ship and then, at a range of about six kilometres, made a 30-degree turn to port. As soon as the aircraft steadied down after the turn, the bomb was released.

The bomb-aimer waited until the Hs 293's Walter rocket motor started up and then took over radio control. He first brought the bomb up to his own eye level and then turned it to starboard toward the target. He would constantly endeavour to keep the bomb in a direct line of sight from his aircraft to the ship, holding it in a set glide throughout its flight. This was done because it was extremely difficult to judge at any moment how far the bomb was from its target. The only satisfactory means of controlling it so that it did not over-shoot or under-shoot, therefore, was for the observer to hold it directly between his eye and the target.

On some occasions a stopwatch was used to give a rough idea of how far the bomb had travelled, but this practice was not common. In order to assist the bomb-aimer and to give him as much time as possible, the pilot reduced the speed of the aircraft from 360 to 300 km/h as soon as the bomb had been released. The aircraft, therefore, flew on a straight course at a speed of 300 km/h while the bomb travelled in a curve towards the ship at an average speed of around 600 km/h. These tactics were adopted so that the bomb reached the targeted ship before the bomb-aimer had lost sight of it as a result of the launch aircraft passing too far beyond the vessel.

One undesirable feature of such tactics was that it was necessary for the launch aircraft to maintain a steady course during the flight of the bomb. Evasive action during this period was impossible, and thus the Do 217, and later the He 177, became vulnerable targets to both fighter attack and anti-aircraft fire.

The Hs 293's baptism of fire came on 25 August 1943 when 12 Do 217E-5s of II./KG 100, escorted by seven Ju 88C-6s of 15./KG 40, attacked Royal Navy ships off the Spanish coast. The sloops HMS *Egret* and HMS *Bideford* and the frigate HMS *Waveney* suffered varying levels of damage inflicted by missiles. *Bideford* had one crewman killed and 16 injured. Two days later, *Egret* received a direct hit from an Hs 293 launched by II./KG 100 30 miles west of Vigo and sank with the loss of 198 lives. The Canadian destroyer HMCS *Athabaskan* was heavily damaged in the same attack.

From then, until the end of November 1943, Hs 293s would account for 19 Allied naval and merchant vessels sunk or damaged, comprising sloops, cargo ships, a cutter, a destroyer, a cruiser, a landing ship, a hospital ship and a minesweeper.

On 21 November, the Luftwaffe decided to launch an attack against convoy SL.139/ MKS.30, inbound from Sierra Leone and Gibraltar, using the problematical He 177A-5s recently assigned to II./KG 40 and fitted with Hs 293s. This mission would be typical of many. A force of 25 He 177s, led by Major Rudolf Mons, the *Gruppenkommandeur*, took off from their base at Bordeaux-Mérignac in western France and flew in close formation at just ten metres above the sea, with escort provided by eight Ju 88C-6s of ZG 1 for a part of the way. The Heinkels arrived over the convoy, some 675 km northeast of Cabo Finisterre, about four-and-a-half hours later. The formation was already depleted by this point since two of the Heinkels had abandoned their mission due to problems with their *Kehl III* equipment.

By this stage the weather had deteriorated, with the cloud base down to 300–400 m. Nevertheless, commencing at 1700 hrs, the bombers attacked individually, with one part of the formation, led by Hauptmann Alfred Nuss of 6./KG 40, concentrating on the

4,045-ton *Marsa* and the 6,065-ton *Delius* that were straggling 3.5 miles behind the convoy. The Hs 293s were launched from altitudes of between 400–600 m. Four He 177s fired eight missiles at the former vessel, leaving it burning, although just one sailor from its 50-man crew was killed. *Marsa's* captain, Thomas H. Buckle, watched as three Heinkels came towards his ship, the first two each releasing an Hs 293 off the starboard beam, then a second bomb off the port quarter, while the third bomber approached from astern on the port side. When 4,000 yards abeam, it released two Hs 293s some 45 seconds apart, but both of these bombs appeared to fail and:

> . . . fell straight into the sea with their rocket motors smoking. We were attacking the aircraft with our 12-pounder, and as the bursts seemed to me to be close to the aircraft, this may have caused the bomb-aimer to have lost control to some extent.

As a fifth Heinkel came in:

> . . . I brought my stern around, but as it travelled round the stern I lost track of the bomb and consequently I was unable to take any further avoiding action, and it struck the water between the davits of the port lifeboat, exploding in the engine room near the main discharge.

Buckle also described how, when the Hs 293 impacted with the surface, it sent up a column of water some 10.5 m high:

Six He 177s of II./KG 40 gathered by the damaged hangars at Bordeaux-Mérignac in western France in late 1943 or early 1944. Five of the bombers are ready for a mission over the Atlantic, each armed with two Hs 293s. Success in any great measure with the Hs 293 evaded the crews of the troublesome He 177. (EN Archive)

The detonation did not appear to be particularly loud. The blast, however, was terrific. I was walking from the starboard to the port side of the bridge at the time, endeavouring to trace the bomb's path, when a gunner and the Second and Third Officers were blown in through the door for which I was heading. The ship was hit near the water line. I could not see the extent of the damage to the shell plating because it was underwater, but the deck was indented for the full length between the two boat davits, and the engine room flooded so rapidly that I think there must have been a hole in the ship's side.

Marsa was last seen the next day drifting.

Moving at just 7.5 knots, *Delius* received damage to its bridge. The vessel's Chief Officer, Gordon Marshall, described events as follows:

At 1615 hrs the bomb struck the ship. The starboard wing of the glider-bomb hit the foremast; this turned it inboard and downwards, then it hit No. 3 derrick and exploded on the port side of the hatch. The explosion was extremely violent, and debris was thrown to a tremendous height. I was knocked over by the blast and was dazed for a few minutes. The master on the bridge, the lookout on the monkey island and the assistant steward in the saloon were all killed instantly. The explosion shattered the bridge, chartroom, wireless room, etc., leaving them a complete shambles. Fortunately, the engine room was undamaged.

Delius was brought back under control and managed to rejoin the convoy, even delivering its cargo to its assigned destination.

Two frigates, HMS *Calder* and HMS *Drury*, were also attacked with Hs 293s, but they evaded the bombs by rapid manoeuvring, defensive fire and the letting off of flares. A single He 177 attempted to attack the destroyers HMS *Watchman* and HMS *Winchelsea*, but failed in its attempt.

Defence for the convoy also came from the Canadian anti-aircraft ship HMCS *Prince Robert*, which arrived at the height of the attacks. The vessel zig-zagged across the wake of the convoy, firing regularly at the German aircraft and adding to the barrage already in progress.

During their attack, which lasted for about 40 minutes, the He 177s, launched 40 bombs, 25 of which failed. II./KG 40 returned to Bordeaux, having lost three of its He 177s, with a fourth aircraft crashing on landing, suffering 45 per cent damage, and three others sustaining lighter damage from the convoy's anti-aircraft defence.

In total, 37 Allied vessels were damaged, sunk, scuttled or written off as a result of being hit by Hs 293s between August 1943 and August 1944, when maritime bomber operations began to recede as the German war effort flagged and units were pulled back into the Reich. A few luckless operations were conducted against bridges being used by Allied forces in France in August 1944 and, towards the end of the war, against bridges over the Oder. However, the Hs 293 proved inadequate as a weapon against strong land-based structures.

Primary and secondary sources vary considerably on the numbers of Hs 293s manufactured, some stating that as many as 5,923 examples of the standard Hs 293A-1 were built from November 1942 to the end of 1943, with a further 3,500 following in 1944, most constructed at Henschel's Niederschöneweide factory in southwest Berlin. It is also stated that by July 1944, some 11,400 Hs 293s of all types had been completed, with many later variants being A-0 or A-1 conversions. More conservative estimates put the figure at closer to 1,000.

There was also an extensive plan for further development. The Hs 293C was fitted with the cone-like nose of the Hs 294, with radio (C-2) and wire-controlled (C-3) sub-variants; the D was fitted with a camera in the nose and a tail antenna for improved guidance using television, control transmission and reception in tests by Fernseh GmbH. It was to be launched on an experimental basis from a Do 217 or an He 111, but in early development the television picture was found to be poor at distances greater than four kilometres from the target and the equipment was never completed. The Hs 293F was a tailless design and the G was intended for high-altitude releases.

The 2,176 kg Hs 294 'torpedo nose' depth-bomb version, dating from January 1943, was intended to be launched from either the He 177A-5 or Ju 290A-7 and to run underwater. It had a 1,250 kg warhead carrying 650 kg of explosive and two rocket boosters to carry a heavier overall weight, as well as options for a fuse to function after a pre-determined period of underwater travel registered by a log – there was a nose switch for impact fusing or a self-destroying fuse. A total of 300 units were planned for production between early 1943 and the autumn of 1944.

The Hs 295, started in April 1944, was an enlarged version with an armour-piercing warhead initially intended for radio-control but later for wire-control for launch from an He 177A-5.

The Hs 293 in a planned 'H' variant was also considered for deployment against enemy bomber formations in late 1943–early 1944, first as a radio-controlled weapon and then for control by wire, for launching from Do 217K-2/U-1 or Do 217M variants. The bomb was to be launched with a remote fuse. It also had *Neptun* R-1 tail warning equipment, but its Walter rocket motor with its hydrogen peroxide and calcium permanganate mixture was unsuitable at the low temperatures encountered in high-altitude combat operations. The Wilhelm Schmidding firm of Bodenbach had worked on a competitor motor, the 109-513, for the Hs 293 which used methanol and compressed gaseous oxygen. This motor was seen, potentially, as a viable alternative, but the growing strength of Allied fighter escort for bomber formations meant that any employment of Hs 293s in such a role was unrealistic.

Nevertheless, in July 1943, personnel from E.Kdo 25 and JG 11 visited a detachment of KG 100 at Garz on the Baltic coast, where crews were being trained on the weapon, in order to explore the possibility of deploying the Hs 293 against bombers using either Fw 190s or Bf 109G-5s as guidance aircraft. A captured Luftwaffe signals mechanic interrogated by the Allies in September 1944 recalled how he had heard a fellow technician relating how a specialist *Sonderkette* (three-aircraft flight) was using the

Hs 293 VARIANTS

1. The Hs 293A-1 was the most widely produced version of the series.

2. The Hs 293C was fitted with the 'torpedo' nose of the larger Hs 294 series. The Hs 293C-3 was the wire-guided version, evident from the wire bobbins fitted at the wingtips.

3. The Hs 293D was fitted with a camera in the nose for terminal guidance and a Yagi antenna in the tail to improve data transmission and reception.

4. The Hs 294 was the 'torpedo' version of the series, with a special conical warhead designed to travel underwater to the target and strike the ship under the waterline. Due to its increased weight, it had two rocket boosters.

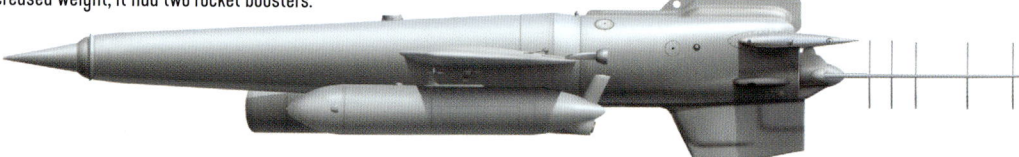

5. The Hs 295 was an enlarged version of the family, with a heavy, armour-piercing warhead.

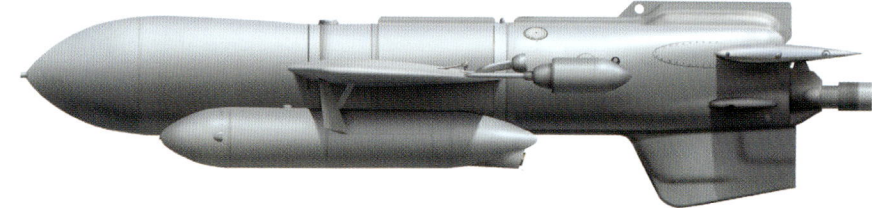

Hs 293H for anti-bomber purposes. The former commander of E.Kdo 25, Horst Geyer, explained this to the author:

> The idea was to fit this 'flying bomb' onto a Ju 88 and fly at a considerable distance behind a bomber formation in the same way as the weapon was used to attack ships, thus avoiding the bombers' defensive fire. The carrying aircraft would have flown at a distance of about three kilometres behind the formation and the fighter, flying closer, would then have steered the bomb directly towards its target. One day, an Hs 293 was delivered to our airfield and we fitted it onto one of our Me 410s. An unfortunate incident occurred whereby the aircraft was parked in a hangar undergoing routine maintenance. The Hs 293 had been positioned in front of the aircraft and left unattended momentarily while pre-flight checks were conducted. An engineer had been checking the ignition, but had applied the wrong procedure and set off the fuse. The missile ignited, blew up the hangar, killed five ground personnel and destroyed the Me 410. No trials were performed and this idea was eventually shelved.

Aside from trials with the Me 410, the *Erprobungskommando* is believed to have tested the Hs 293 operationally against enemy bombers on at least one occasion when, on 29 November 1943, two Do 217s of the aforementioned *Sonderkette* attempted to deploy them during a raid on Bremen. After this attempt, the *Kommando* reported:

> Especially important is the fact that further testing of the Hs 293 with only three aircraft (Do 217), of which 50 per cent on average are serviceable, promises little success. It is therefore requested that the reinforcement of the *Sonderkette H* be arranged with A.O. Bombers.

Henschel Hs 293A-1	
Length	3,818 mm
Span	3,100 mm
Total weight	975 kg
Warhead weight	500 kg
Explosive weight	295 kg
Propulsion unit weight	134 kg
Propulsion unit thrust	6,500 kg per second
Fuse	E1/AZ 38B
Starting velocity	Airspeed of carrier aircraft
Maximum side acceleration	3 g

Velocity at impact	150–240 m per second, depending on drop altitude and range
Release altitude	500–7,000 m
Target range at release point	4.4 km to 14 km

AERIAL TORPEDOES

Almost from the outbreak of war, considerable expectation surrounded the planned deployment of the aerial torpedo. The bomber *Staffel* 7./KG 26 had an He 111H-4 fitted with torpedo equipment as early as the beginning of 1940, but the weapon's introduction to service had been hampered because of indecision as to whether the Kriegsmarine or the Luftwaffe should be responsible for general torpedo development. Notwithstanding that, Major Martin Harlinghausen, a *Legion Condor* veteran and an aerial torpedo specialist who served as Chief of Staff to X. *Fliegerkorps* in the Mediterranean from December 1940, had been a driving force in promoting and developing the He 111 bomber as a potential torpedo-carrier. Hitler's instruction that torpedo trials with the Ju 88 should be suspended (presumably because emphasis was placed on that aircraft's role as a *Schnellbomber*) worked in the He 111's favour in this regard.

In the spring of 1941 the *Torpedo Schul- und Erprobungsstelle* was established at Grossenbrode on the Baltic coast. Then, in December 1940, a cadre of operationally experienced officers from the minelaying unit KGr. 126, including Oberleutnant Josef Saumweber, a specialist from the *Fliegerwaffenschule* (*See*) (Maritime Air Weapons School) at Lobbe, was assigned to II./KG 26. After operations over England in the summer of 1940, 1./KGr. 126 is understood to have conducted its first experimental torpedo flights in November. The KGr. 126 cadre would form the core torpedo element within II./KG 26 in the Luftwaffe's campaign in the Mediterranean, and would be supported by a team of torpedo mechanics and a specialist workshop at Comiso with an initial batch of just six torpedoes.

Throughout the war, the Luftwaffe would deploy two standard types of air-launched torpedoes. The LT (*Luft Torpedo*) F5b was powered by compressed air. Weighing 750 kg, it had a charge of 250 kg, was five metres in length and required a minimum depth of water of 16–20 m. Its range was 3,000 m and speed was 33 knots. The torpedo's gyro would start to run as it was launched from the carrier aircraft, and the engine was started by the pressure from the water pushing back a flap as it went below the surface.

An LT F5b plunges into the water to run close to the surface moments after being released from He 111H-5 Wk-Nr 3891 BK+CD at the Luftwaffe torpedo training school at Grossenbrode. (EN Archive)

A 'train' of LT F5b torpedoes is towed past He 111H-6 1H+FN of 5./KG 26 in either Greece or Italy in 1942. Note the cockpit glazing of the Heinkel has been covered with a tarpaulin as protection from the heat of the sun. (EN Archive)

The second type was the Italian-manufactured F5W 'Whitehead' torpedo, which was also powered by compressed air, weighed 900 kg and had a charge of 195 kg. Slightly longer than the F5b, it had a speed of 36 knots. The F5W was generally considered to have been the better of the two weapons, mainly because of its sensitivity. The F5b's main weakness was that when hitting a target at more acute angles, detonation was not always guaranteed, whereas the F5W, while not without faults, was generally satisfactory, although crews were encouraged to watch the track of a torpedo to make sure it was running properly. Unlike the F5b, however, the F5W had no stabilising fins and a much more complex air rudder system.

One Luftwaffe torpedo-bomber pilot was of the opinion that one of the principal difficulties associated with the F5b lay in the anti-dive and rolling rudder. If the torpedo entered the water at a slant, due to the aircraft not having been on an even keel at the moment of release, the rudder did not always compensate sufficiently, with the result that the torpedo did not run true.

It is believed that the first aerial torpedo operation in the Mediterranean was mounted on 4 February 1941 when three He 111s of II./KG 26, along with a formation of 16 Ju 88s from LG 1, attacked a convoy as it departed Benghazi for Tobruk. These early missions were mounted as armed reconnaissance, but the aircraft used them as potential opportunities to trial torpedoes. The 4 February mission was not a success as no ships were believed hit, and an He 111H-5 of 4./KG 26 was lost after it was forced to make an emergency landing and then burst into flames.

There was a further blow a few days later, on the 19th, when the He 111H-5 flown by Oberfeldwebel Joachim Conrad was shot down near Benghazi, probably by ship-mounted anti-aircraft fire. On board the aircraft was the torpedo specialist Oberleutnant Saumweber, and it is believed that the Heinkel had been attempting to make a torpedo attack on a convoy leaving the port, which was unsuccessful.

On 16 March, two torpedo-carrying He 111s flown by Hauptmann Kowaleski and Leutnant Karl-Heinz Bock took off from Comiso to attack an enemy group west of Crete. Reaching their target area shortly before 1855 hrs, they spotted enemy ships in the twilight against a darkening horizon about 15 minutes later and adopted a parallel course, five kilometres from the targets. The Heinkels then turned in to attack, and at a height of between 25–30 m, closing from 3,000 to 2,000 m range, the first aircraft launched its torpedoes at a battleship. The ensuing explosion was '100 m high'. The second Heinkel launched and the crew observed a black 'detonation cloud'. Under heavy defensive anti-aircraft fire, the two He 111s turned for home.

The German crews believed that they had hit two ships – the battlecruiser HMS *Repulse* and the battleship HMS *Nelson* – and in doing so heralded the real baptism of fire of the aerial torpedo as a Luftwaffe anti-shipping weapon. In fact they had not, and this mistake subsequently proved very costly to the Italian fleet, which engaged in naval operations under the false assumption that KG 26 had rid the Mediterranean of a significant British naval threat.

At this point, operations ceased temporarily because of a shortage of torpedoes and warheads, but by 22 March stocks had been replenished and an He 111 narrowly missed hitting a Brazilian merchant vessel some 130 km northwest of Alexandria. The bomber then attacked the ship with its machine guns. Two days later, four He 111s were active against convoys and claimed two steamers hit by torpedoes.

Operations by torpedo-carrying He 111s in the North African and Mediterranean areas during mid- to late 1942 were continued by II./KG 100 based at Athens and Kalamaki, in Greece, augmented by elements of I. and II./KG 26 passing in and out of the theatre. Also moved to Grosseto, in Tuscany, at the end of 1941 was I. *Kampfschulgeschwader* (KSG) 2, which had been established in October of that year at Grossenbrode from the *Stab* of the *Torpedoschule und Erprobungsstelle der Luftwaffe* under Oberstleutnant Karl Stockmann, who had previously been *Kommandeur* of Kü.Fl.Gr. 406, together with his head of training, Hauptmann Werner Klümper.

An armourer appears to be inspecting his handiwork – two LT F5b torpedoes – hung beneath this He 111H-6 of I./KG 26 in preparation for a mission over the Mediterranean. The torpedoes are fitted to faired PVC 1006B torpedo-carriers. Note also that the bombsight housing beneath the nose of the Heinkel has been covered and the nose machine gun has yet to be mounted. The He 111 carries the *Vestigium Leonis* ('the trace of the lion') emblem of KG 26. (EN Archive)

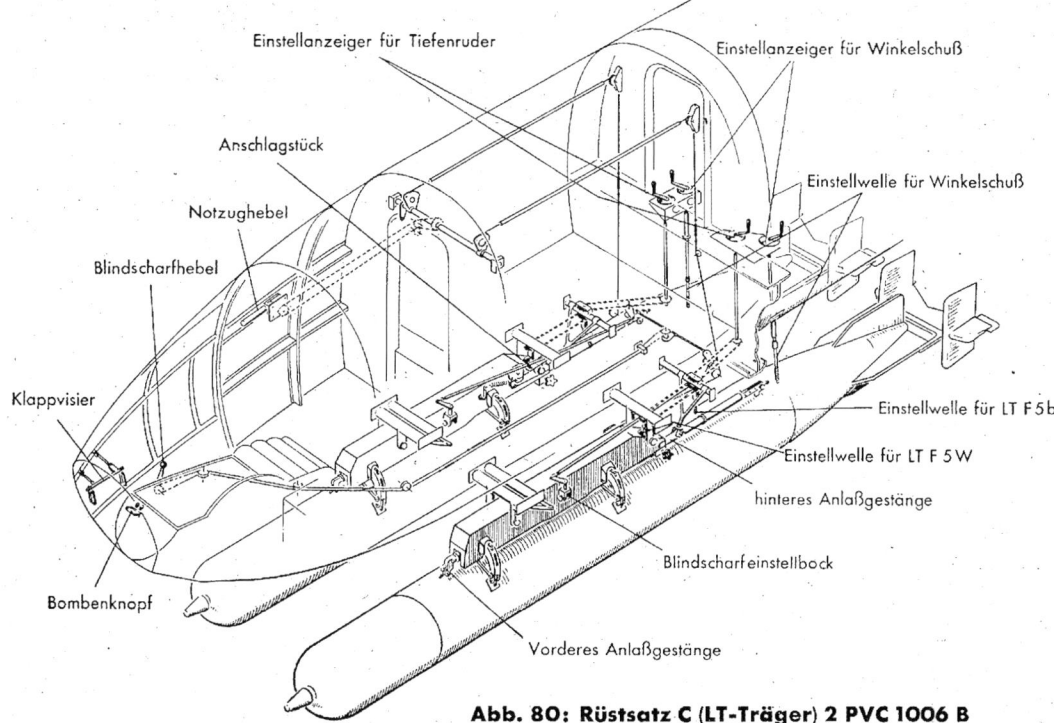

Einstellanzeiger für Tiefenruder

Einstellanzeiger für Winkelschuß

Anschlagstück

Einstellwelle für Winkelschuß

Notzughebel

Blindscharfhebel

Einstellwelle für LT F 5 b

Klappvisier

Einstellwelle für LT F 5 W

hinteres Anlaßgestänge

Blindscharfeinstellbock

Bombenknopf

Vorderes Anlaßgestänge

Abb. 80: Rüstsatz C (LT-Träger) 2 PVC 1006 B

The control, winching and mounting system for a pair of LT F5b or F5W torpedoes on PVC 1006B racks as contained within the official He 111 H-6 handbook. (EN Archive)

Grossenbrode's position on the Baltic Coast meant that the winter weather, which frequently brought snow and ice, hampered Klümper's efforts, and so a decision was made to move the school to Grosseto, shortly after which KSG 2 followed, the unit having several He 111s on its strength. Oberst Harlinghausen, who was by this stage the Luftwaffe's *Lufttorpedo Inspizient*, had recognised the advances in Italian torpedo design and wanted to combine and centralise German and Italian efforts in that area. The *Geschwader's* I. *Gruppe* was formed of two *Staffeln* drawn from the *Torpedoschule's* 1. *Lehrgang* (Intake). By the time it moved to Italy, I./KSG 2 had already carried out a few operations against enemy convoys heading to Murmansk. A II. *Gruppe* was established in June 1942, and all of these units would move from Grossenbrode to Grosseto.

During the initial period at Grosseto, crews flew a course of practice flights, without torpedoes, learning how to approach targets and to interpret and work on reconnaissance situation reports. Then flights were conducted in full daylight using practice torpedoes, each crew making around five such flights usually at an 'attack' altitude of 40 m. Then followed similar training at dawn and dusk when daylight was limited, representing what would be the optimum tactical attack conditions, and finally a series of night flights. Crews would then progress to perfecting the art of *Zangenangriffe* ('pincer attacks'), which were intended to be carried out against larger merchant vessels and warships in *Staffel* strength, where torpedo-bombers would 'swarm' around a target vessel, preventing any effective evasive manoeuvring. The training course would conclude with instruction in large formation flying and attack methods.

Abb. 1: Lösen des Torpedos

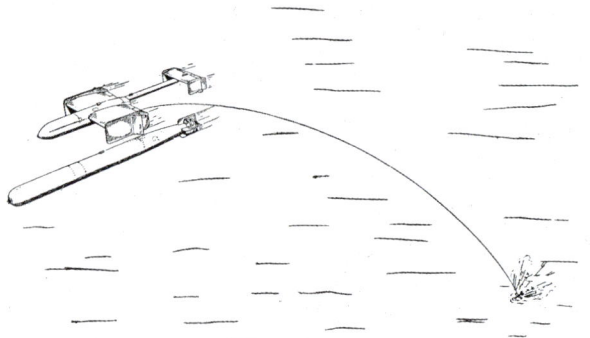

TOP This He 111B has been loaded with a Blohm und Voss L10 ahead of a test flight. With the torpedo having been fitted to the aircraft, the hydraulic trolley will shortly be towed away. The early variant of Heinkel suggests that the aircraft had been specially assigned for test flights. (EN Archive)

ABOVE RIGHT An LT F5b training torpedo with a dummy warhead fitted to a Blohm und Voss L10 glider carrier that hangs suspended just off the ground from the hook of a hydraulic hoist. Visible under the starboard wing of the L10 is the small torpedo release drogue. (EN Archive)

ABOVE Handbook illustration issued by the RLM's Chef der TLR, Fl.-E9 from October 1944 showing the out-spooling of the small surface activation drogue and the ensuing moment of release of an LT F5b torpedo. (EN Archive)

It seems that other crews and units did not receive such structured training. Indeed, at least one crew from I./KG 26 reported that no training in formation flying or attack was given at all, with the type of attack being decided by the *Gruppenkommandeur* at 'the last moment' in accordance with the sailing formation adopted by a convoy. German crews also monitored the work and experiences of their Italian counterparts.

By July 1942, KSG 2 had some 15 He 111H-6s on strength, plus two He 111H-5s, as well as eight Ju 88s and a Do 217. The unit also had a stock of 50 LT F5b and 14 F5W torpedoes.

German torpedo design embraced inventive concepts. These included the Blohm und Voss L10 *Friedensengel* (Angel of Peace) glider torpedo-carrier of late 1944. This took the form of a small faired and winged glider body that was clamped to the top of a standard LT F5b. This enabled release of the torpedo in a long glide at 310 km/h from a higher (2,500 m, compared to the usual 200 m) and greater distance from a target (8,500 m, compared to 3,000 m), effectively giving it greater 'range' and allowing the carrier aircraft to break away at a safer distance. As the L10 was launched, a small drogue would be reeled out from its body using a ten-metre cable, and

Ju 88A-17 'White' or 'Yellow B' of I./KG 77 loaded with two LT F5b torpedoes in what is probably a posed photograph in southern France in the summer of 1944. Note the 'meander' over-water camouflage pattern which extends over the entire fuselage and engine nacelles. (Goss)

when this touched the water the torpedo would be released. The production run of the L10 is not clear. A more sophisticated version, the L11 *Schneewittchen* (Snow White) was developed, but little is known about any deployment.

The Ju 88 was initially outfitted to carry torpedoes in numbers during the first half of 1942. Development had commenced the previous year with the V46, and led to the Ju 88A-4 LT (*Luft Torpedo*) sub-variant. This differed from a standard A-4 principally in the replacement of the two external underwing ETC bomb racks inboard of the engines with PVC torpedo racks and a dedicated torpedo sight in the cockpit, from which it was possible to input the torpedo run control settings. The necessary control mechanism was housed in a fairing that extended from the nose of the aircraft back towards the wings and the racks. The latter, which were larger and deeper than the ETC racks, were able to carry one LT F5b torpedo each.

Initial operations with the Ju 88A-4 LT were considered sufficiently successful to justify a small run of production models, specifically the Ju 88A-17, in which the most significant feature was the removal of the ventral gondola. The torpedo load was the same as the A-4 LT.

The first Ju 88 unit to receive training and equipment for torpedo operations was III./KG 26, originally based at Banak, in Norway, under Hauptmann Ernst Thomsen, where the unit had converted to the Junkers from the He 111 in June 1942. Having headed south to Grosseto shortly thereafter to receive torpedo training, the *Gruppe* went into

Ju 88A-4 BF+YT, based at Gotenhafen-Hexengrund, is fitted with a Blohm und Voss L10 *Friedensengel* glider torpedo-carrier, to which is suspended an LT F5b dummy-head training torpedo, denoted by the red and white warhead. (EN Archive)

Possibly a Do 217K-2 of KG 100 carries a rare configuration of Blohm und Voss L10 torpedoes beneath each wing, probably for an operational test. (EN Archive)

action from the Italian airfield following the Allied landings in North Africa in November 1942 – it attacked enemy shipping off Algiers on the very first day of the campaign.

The Ju 88 would prove to be a formidable torpedo-bomber, giving rise to several notable *Experten*. One such individual was Hauptmann Günther Trost, *Staffelkapitän* of the training unit 12./KG 26 in late 1943 who was awarded the Knight's Cross on 12 November that year for his accomplishments as a *Torpedoflieger*. Tröst had been with III./KG 26 since June 1942, flying missions against the Murmansk convoys from Banak. Relocating to Grosseto in early November, he and his *Gruppe* flew many torpedo missions against Allied convoys off North Africa. Such was Tröst's capabilities that he was assigned for a period to the staff of the *Bevollmächtigten für die Lufttorpedowaffe* (Plenipotentiary for the Aerial Torpedo Arm). He ultimately carried out 42 combat missions against enemy shipping in the North Sea and Atlantic, and 47 against convoys in the Mediterranean.

Following the Allied landings at Anzio and Nettuno in late January 1944, III./KG 26 was moved from Montpelier to Piacenza to be closer to launch attacks on the enemy beachhead, but the supply of torpedoes fell away quickly and so the Ju 88s resorted to flying high-altitude missions at dusk dropping fragmentation bombs. Losses were high.

Through the spring of 1944, the Ju 88 torpedo-carriers of KGs 26 and 77 continued to be a thorn in the side of Allied shipping in both the Mediterranean and in the Arctic, albeit with diminishing success because of the continuing increase in enemy tonnage, warships and aircraft. Typical of such activity was the mission flown by around 20 Ju 88A-17s of I. and III./KG 77 in the early hours of 1 April against the American convoy UGS-36 en route from Hampton Roads to Port Said. The bombers struck the convoy west of Algiers. Torpedoes narrowly missed the *Marion McKinley Bovard*, passing 75–100 yards from its side, but the 7,180-GRT Liberty ship *Jared Ingersoll*, carrying stores and petrol, was hit and badly damaged. Fire broke out on board the vessel and it was abandoned by the crew, who were rescued by the destroyer escort USS *Mills* (DE-383). These were costly operations, for three Ju 88s were lost in the action as a result of defensive fire, including one shot down by *Jared Ingersoll*.

On the night of 11–12 April, another formation of around 20 Ju 88s from I. and III./KG 77 attacked again. This time they targeted the 102 merchantmen, and their 19 escorts, of convoy UGS-37 bound for Port Said. The Junkers reached the convoy 56 km east of Algiers just before midnight and had to make their attack through dense anti-aircraft fire. One torpedo narrowly crossed astern of the *Horace H. Lurton* and another hit the Edsall-Class destroyer escort USS *Holder* (DE-401) amidships on the port side, causing two large explosions. Despite fire breaking out on board and serious flooding, the crew remained at their posts and continued to fight off the Ju 88s. *Holder* was eventually towed to Oran for repairs. No fewer than seven Ju 88A-17s had been lost in the attack.

With the coming of the Allied landings in Normandy on 6 June 1944, III./KG 26 and I. and III./KG 77 quickly left their southern French bases for the invasion front. They were immediately engaged on the night of the 6th in attacking the landing fleet off the Cherbourg Peninsula. For the next three months, these *Gruppen* mounted determined, if sporadic, operations against enemy warships and supply vessels, but with little success. On the night of 13–14 June, two torpedoes launched by Ju 88s of KG 77 hit the B-Class destroyer HMS *Boadicea* as it escorted a westbound convoy off Portland. One of the torpedoes detonated the ship's magazine and it sank rapidly, with only 12 survivors from a complement of 182.

On the 18th, the Luftwaffe sent up no fewer than 69 aircraft carrying torpedoes and mines to strike at enemy shipping. The 1,760-GRT Canadian-built British supply ship *Albert C. Field*, carrying 2,500 tons of munitions and 1,300 sacks of post bound for Normandy, was struck by a torpedo launched by a Ju 88. According to the vessel's First Officer:

The third prototype Blohm und Voss L 950 torpedo glide-carrier falls away from He 111 DF+OV at 1,000 m during launch tests over the Baltic from Gotenhafun-Hexengrund on 24 April 1942. The rocket motor trigger arm is visible hanging from the lower fuselage of the L 950. This carrier was seen as a progressive development of the L 10 that enabled launching from greater height and longer range, but only seven examples were made before the project was abandoned. (EN Archive)

> At 2340 hrs on 18 June 1944, when in position off St. Catherine's Point steering a course south at six knots, we were struck by one torpedo from an aircraft. The weather was fine and clear, visibility good. No one saw the track of the torpedo, which struck our starboard side, amidships, on a bulkhead separating our two holds. The explosion was dull, like a very heavy depth charge, no water was thrown up and no flash was seen.

Albert C. Field sank within three minutes approximately ten miles southeast of Anvil Point, Dorset.

On 8 August, Hauptmann Siegfried Betke of 9./KG 26 was awarded the Knight's Cross and became a 'torpedo ace'. His record for anti-shipping missions stretched back to 1942 and operations against British shipping off the southern coast of Ireland and the west of England with 9./KG 77. He later flew against Malta and Allied convoys in the Mediterranean, as well as conducting bombing missions over North Africa. After a period of torpedo training on the Baltic coast in late 1943, he returned to the Mediterranean with 9./KG 26,

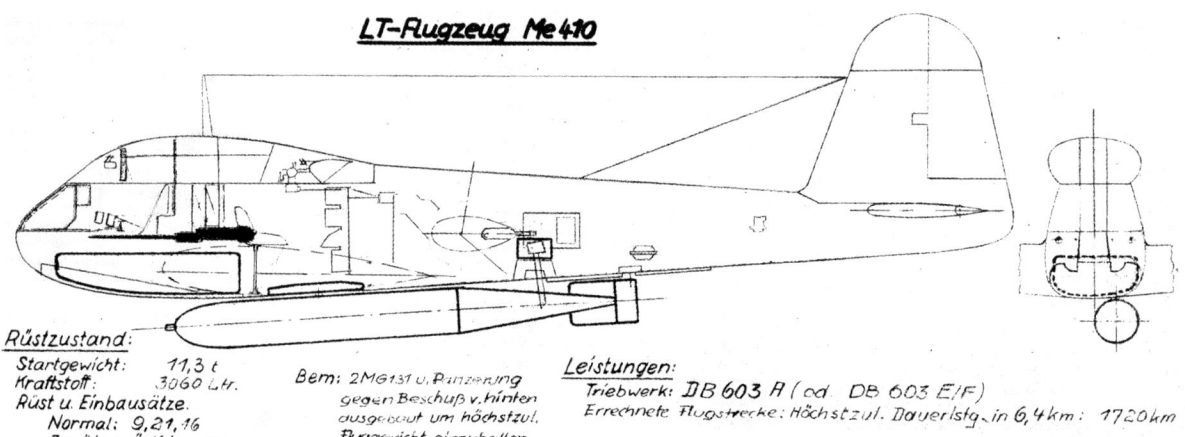

LT-Flugzeug Me 410

Rüstzustand:
Startgewicht: 11,3 t
Kraftstoff: 3060 Ltr.
Rüst u. Einbausätze.
  Normal: 9,21,16
  Zusätz. möglich: 14

Bem: 2MG131 u. Panzerung
gegen Beschuß v. hinten
ausgebaut um höchstzul.
Fluggewicht einzuhalten.

Leistungen:
Triebwerk: DB 603 A (od. DB 603 E/F)
Errechnete Flugstrecke: Höchstzul. Dauerstg. in 6,4 km: 1720 km

TOP Heinkel He 177 V31 Wk-Nr 550202 TM+IF releases a Blohm und Voss L10 torpedo over the Baltic off Gotenhafen in early 1944. (EN Archive)

ABOVE This Messerschmitt drawing shows the planned adaptation of the Me 410 as a torpedo-carrier, fitted with a standard LT F5b beneath the fuselage and offset to port, plus defensive cannon and machine gun armament. Range was projected at 1,720 km. (EN Archive)

RIGHT A Blohm und Voss L10 torpedo carrier is raised from a hydraulic trolley to the underside of an Me 410 for tests. (EN Archive)

flying the Ju 88A-17 against convoys off North Africa, before being relocated to France for attacks against the Operation *Overlord* landing fleet. Betke became a recipient of the Knight's Cross for his success and leadership of torpedo operations. By war's end he had flown 190 combat missions and sunk six merchantmen, with damage inflicted on many others.

In the south of France on 15 August, after Allied forces had landed along the coast between Cannes and Hyères in Operation *Dragoon*, individual Ju 88s from III./KG 26 attempted to attack the landing fleet with torpedoes, but failed.

All the torpedo *Gruppen* flew on until the end of the war, operating mainly against convoys in the Far North, but worsening weather, poor serviceability, a lack of torpedoes, torpedo failures and heavy losses conspired to erode their effectiveness significantly.

BOMBEN TORPEDO

In the spring of 1943, Fliegerstabsingenieur Dr. Theodor Benecke, head of the *Lufttorpedo* development section in the RLM's *Technisches Amt*, in collaboration with the *Forschungsanstalt Graf Zeppelin* at Stuttgart-Ruit, attempted to overcome the complexity and expense in the manufacture of the torpedo as an aerial weapon. Moreover, Benecke wanted to produce a series of weapons which would have the effect of a torpedo, but which could be released from a fighter-bomber such as the Fw 190. Hence the concept of the *Bomben Torpedo* (BT – Torpedo Bomb) was born, in which, by the omission of a torpedo motor and control gear, simplicity could be optimised while cost and production time were minimised. As a joint US Army and US Navy post-war review of German ordnance summarised:

> If this simplified weapon were launched so that the greater portion of the distance to the target was covered through the air, as with an ordinary bomb, the initial speed of launch would be retained over nearly all the range. The projectile would enter the water just short of the target and carry on in the direction of its flight in air by reason of its momentum in the same way as does a torpedo. To prevent it from going too deep before detonation, a relatively flat angle of entry into the water is necessary. Such a weapon was developed in Germany in the closing months of the war. It combines the characteristics of the bomb to travel a long distance in a short time interval with the characteristics of a torpedo in that underwater travel eliminates range errors.

The BT series was developed in 200 kg, 400 kg, 700 kg and 1,400 kg versions, all of the same shape and steel construction. The bombs were formed of three sections: the two-section warhead and the tail section. The forward section of the warhead was in the shape of a truncated cone and the rear section was cylindrical, this part also containing the magnetic fuse which was such that it reacted to the variable magnetic field of a ship, but this aspect never proved satisfactory by war's end.

Three large fins were mounted at 120 degrees apart around the weapon's tail, which ensured good stability while it was in the air. When the bomb struck the water the tail section was jettisoned.

It was essential that the bomb did not ricochet off the water, even in flat angles of entry, and that it should continue its path without any deviation. However, in the case of ogival

GENERAL ARRANGEMENT DRAWING OF *BOMBEN TORPEDO* (BT) 400

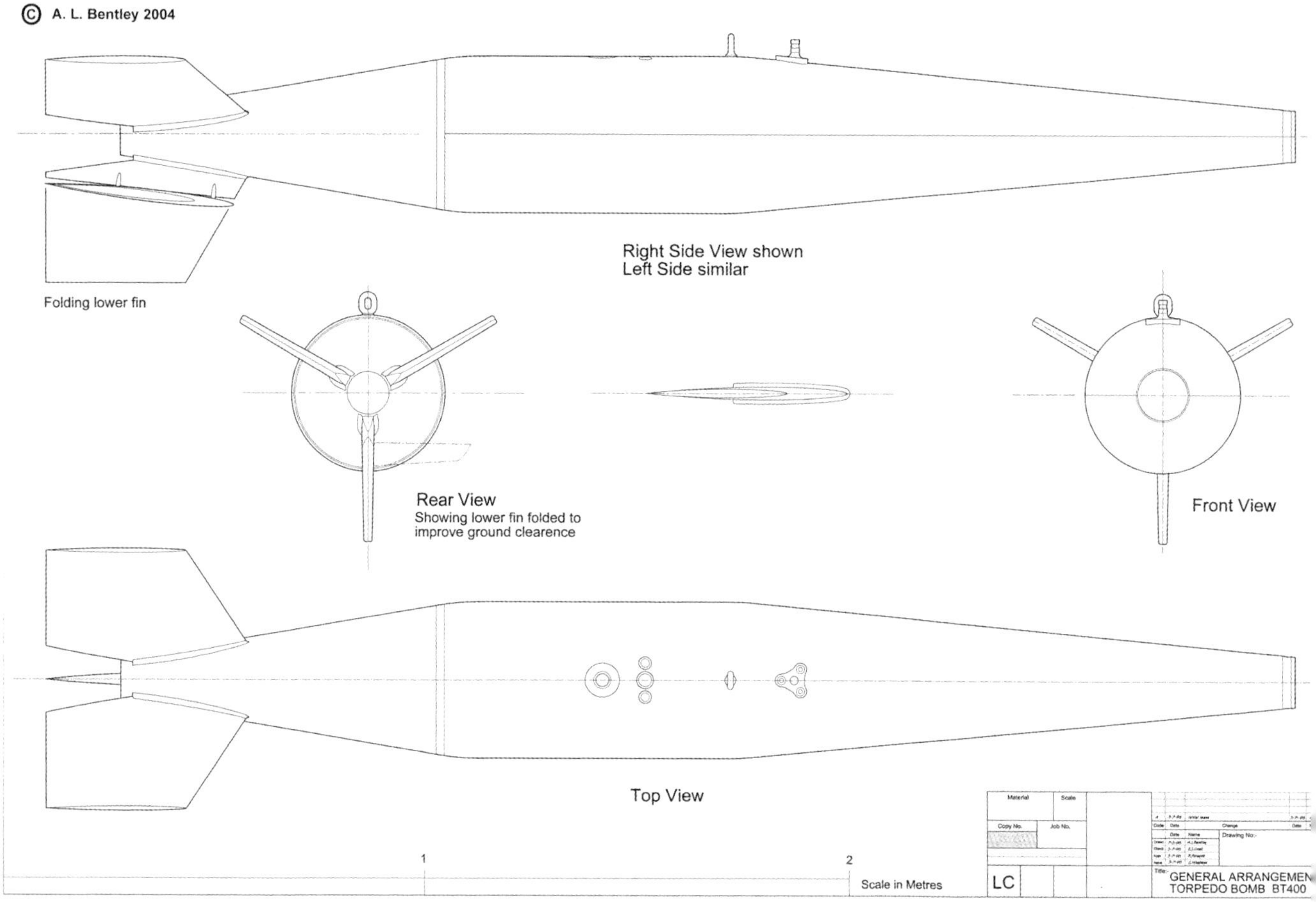

© A. L. Bentley 2004

Folding lower fin

Right Side View shown
Left Side similar

Rear View
Showing lower fin folded to
improve ground clearance

Front View

Top View

Material   Scale

Copy No.   Job No.

LC

Title: GENERAL ARRANGEMEN
TORPEDO BOMB BT400

1   2

Scale in Metres

BELOW This Messerschmitt drawing shows the planned adaptation of the Me 410 as a *Bomben Torpedo*-carrier. The ordnance shown here appears to be two *Bomben Torpedo* series suspended from the fuselage. Range was projected at 1,320 km. (EN Archive)

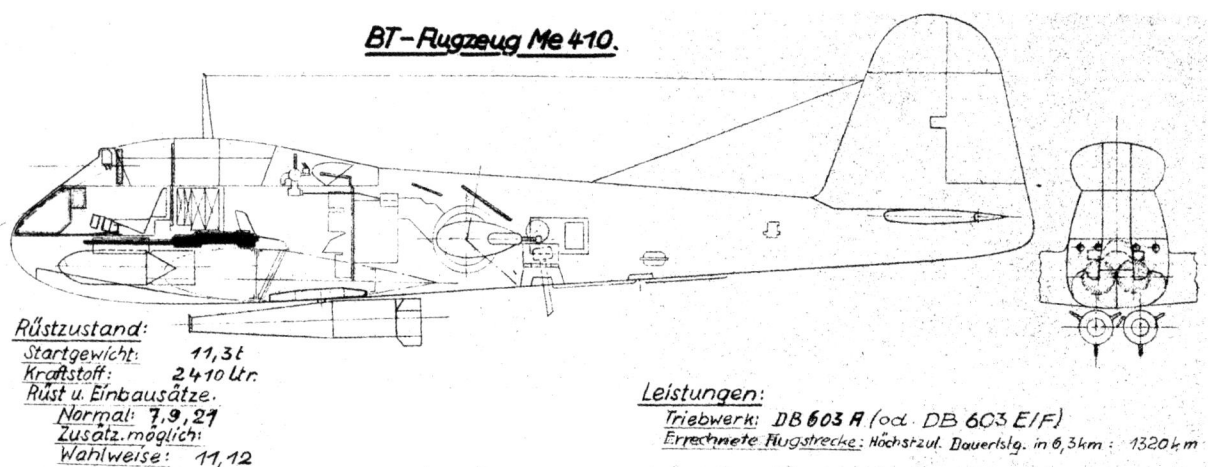

BT-Flugzeug Me 410.

Rüstzustand:
Startgewicht: 11,3 t
Kraftstoff: 2410 ltr.
Rüst u. Einbausätze:
   Normal: 7,9,21
   Zusatz. möglich:
   Wahlweise: 11,12

Leistungen:
Triebwerk: DB 603 A (od. DB 603 E/F)
Errechnete Flugstrecke: Höchstzul. Dauerlstg. in 6,3 km : 1320 km

noses, such as that proposed for the BT 1400, a ricochet would occur if striking the water at a flat angle. In the case of the BT 700, which used a flat nose, ricochet could be avoided.

The BT series was unpowered and bombs were to be carried on an ETC 502 fuselage rack. Work did commence on a BT 1000 which had a rocket motor installed in its tail section, but this plan was dropped as it proved impractical to operate.

BT bombs were built only in small numbers due to problems with developing a suitable method of aiming and release. It is possible some examples were delivered to III./KG 200, but as far as is known, they were never used.

	BT 200	BT 400	BT 700A	BT 700B	BT 1000	BT 1400	BT 1850
Length	2,395 mm	2,946 mm	3,500 mm	3,358 mm	4,240 mm	4,560 mm	4,690 mm
Diameter	300 mm	378 mm	426 mm	456 mm	480 mm	620 mm	620 mm
Tail span	560 mm	710 mm	1,100 mm	850 mm	1,305 mm	1,160 mm	1,160 mm
Weight	220 kg	435 kg	780 kg	735 kg	1,180 kg	1,510 kg	1,923 kg
Warhead	100 kg	200 kg	330 kg	320 kg	710 kg	920 kg	1,050 kg

BLOHM UND VOSS BV 143 AND BV 246 EXPERIMENTAL GLIDE BOMBS

The BV 143 was a winged torpedo, with a flying weight of around 1,000 kg, intended for release from an aircraft at around ten kilometres from its target – a distance impossible for conventional aerial torpedoes. Developed in 1939 by the team of Dr. Richard Vogt, Dr. Zeyns and Flugbauführer Jürgen Cropp at Blohm und Voss in Hamburg-Finkenwerder, it comprised a cigar-shaped body with unswept wings and a cruciform tail unit. Launch trials commenced at Peenemünde in September 1940, but there followed a series of crashes and failures.

After release, the weapon's HWK 109-501/2 rocket motor started up, accelerating the bomb to approaching 450 km/h, and it maintained a fixed course by means of a gyroscopic device that adjusted the control surfaces. After release from the aircraft, it approached the target in a flat glide, and when two metres from the surface of the water, a trigger arm hanging from the lower fuselage was jolted by the sea to activate a rocket motor. This was intended to bring the missile into level flight, where it would speed into the target ship above the waterline.

Limitations were evident in the weapon's ability to change height in level flight due to the length of the trigger arm. Design of an electric altimeter device that may have solved the problem was shelved. The BV 143 never saw operational use.

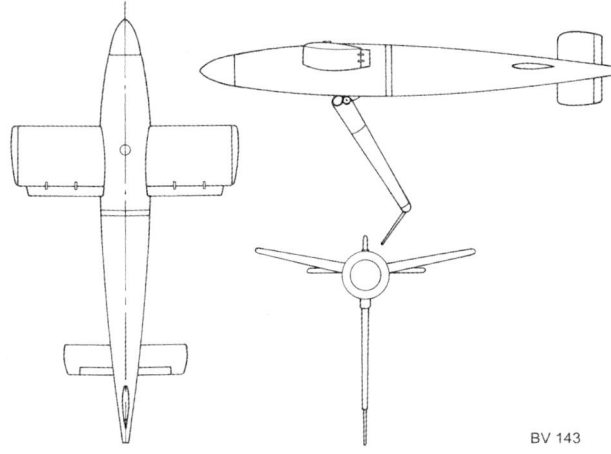

Plan view of a BV 143 air-launched glide bomb. (EN Archive)

BV 143

BV 143 rocket-assisted winged torpedo loaded to an He 111.

An He 111 loaded with what appears to be an early variant of a BV 143. (EN Archive)

Peenemünde's test pilot and unmanned missiles specialist Diplom-Ingenieur Max Mayer recounted:

Even optical or electrical fine-precision altimeters, regarded as promising and even used in trials, did not bring the desired results. I myself conducted several flight-tests with these devices, flying in an He 111 and a Do 217 over the Peene estuary right down to treetop height – in other words, to within about five metres proximity to the sea surface. These trials, however, still provided no conclusive evidence regarding the indispensable reliability of the BV 143's automatic 'pull-out' mechanism.

Similar to the BV 143, in that it was designed to approach the target in a glide from some distance, the BV 246 *Hagelkorn* (Hailstone) was an aerodynamically clean, unpowered glide bomb dating from 1943. Like the BV 143, the fuselage of the BV 246 was cigar-shaped, tapering finely towards the rear where there was

The long, cigar-shaped BV 143 air-launched glide torpedo on a purpose-built trailer. Visible in the photographs is the folded trigger 'arm' that would drop down and upon touching the surface activate a rocket motor. (EN Archive)

a cruciform tail unit, although this was changed on later developments to a twin fin arrangement. The warhead was filled with Amatol-39 high-explosive – a mixture of 50 per cent TNT, 40 per cent Ammonium Nitrate and ten per cent Hexogen.

A distinct feature was the high-mounted, narrow chord, high aspect ratio wing, with a span of 6.4 m, made from pre-stressed lightweight concrete that enabled the *Hagelkorn* to attack its target from a distance of 210 km at a speed of around 450 km/h if released from an altitude of 10,500 m, or a distance of 200 km if the bomb was released at an altitude of 7,500 m.

A gyroscope relayed control signals to the rudder, and these signals were, in turn, modified by a FuG 103 direction-finding device tuned to a radio beam from the carrier aircraft, planned as an He 111 or a Ju 88. From the early summer of 1944 at least four Fw 190s were delivered to the *E-Stelle* Karlshagen (formerly Peenemünde-West) also for test-launching the Bv 246.

The project generated little interest in the RLM and the *Erprobungsstellen* due to known Allied countermeasures directed at German radio navigation aids. However, between 3 July 1943 and 5 July 1944, 119 bombs were

Plan view of BV 246 *Hagelkorn* air-launched glide bomb. (EN Archive)

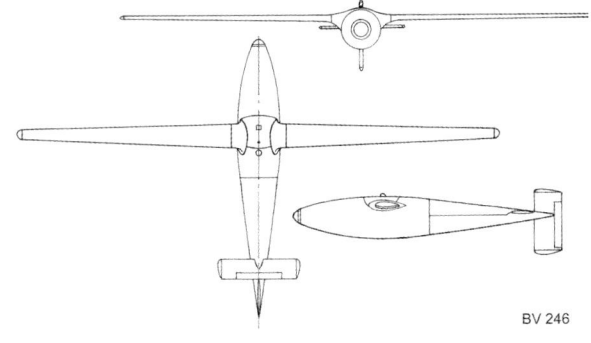

BV 246

He 177A-3 fitted with a BV 246 *Hagelkorn* guided glide bomb.

The remains of a later variant of the BV 246 *Hagelkorn* air-launched glide bomb found abandoned by British troops on an airfield in northern Germany. This version has a double-finned tail assembly. (EN Archive)

tested and the weapon was stockpiled in some quantity. A number of versions were produced, the most important of which was the BV 246B long-range glider bomb, but other option variants included functioning as a container for additional fuel supplies, testing UHF radio equipment, television guidance, modified radio equipment using a changed modulated frequency and a fully radio-controlled version.

Tests continued to be undertaken against a range of targets, including exploring potential for land-based operations. The Bv 246 saw very limited deployment in such a capacity with KG 101, which had 60 bombs and achieved mixed results, but any likelihood

of wider use was stymied by protracted development, lack of resources, bureaucracy and, eventually, the worsening war situation. Some of the final work included tests for an ultra-short wave passive homing device for use against ground transmitters that was installed in the missile's modified nose and acted on the gyroscopic control equipment for the rudder and elevator.

BV 246B	
Length	3,525 mm
Wingspan	6,408 mm
Wing area	1.47 m²
Tail span	1,200 mm
Tail fin area	0.29 m²
Tailplane area	0.32 m²
Fuselage diameter	542 mm
Overall height	880 mm
Weight	730 kg
Warhead	435 kg
Speed at target	450 km/h
Range	210 km

SB 800 RS *KURT WASSEROLLBOMBE*

Developed in late 1943 by Rheinmetall-Borsig in cooperation with Dr. Herbert Wagner of Henschel, having been inspired by the RAF raid on the Möhne and Eder dams with a 'bouncing bomb' on the night of 16–17 May 1943, the *Kurt Wasserollbombe* ('Rolling Water Bomb') was a spherical, hydrostatically operated, aircraft-dropped weapon designed to operate like a skipping stone over a smooth water surface and intended for deployment against ships, power plants, lock gates and harbours.

Working simultaneously for both Rheinmetall-Borsig and Wagner, scientists at the DVL examined the feasibility of ricocheting discs in a vertical plane across water by firing projectiles from an air gun across a 100 m-long tank that was also capable of creating specially disturbed surface conditions. Spins of up to 300 rpm were imposed before launching in an attempt to improve stability.

The somewhat ambitious concept behind *Kurt* was first trialled at the *E-Stelle* Travemünde in late 1943. The initial design took the form of a spherical steel ball of two halves containing 300 kg of explosive. There were two hydrostatic Krupp fuses. Behind this was a cylinder 1,150 mm in length and 420 mm in diameter containing a rocket engine that delivered a burst of power lasting 2.8 seconds. Just forward of the rocket exhaust was a box tail built of 6 mm thick plywood, measuring 275 x 900 x 1,400 mm to provide stability. Tests were not encouraging: initial releasing at 700 km/h in a seven-degree dive from only 20 m proved to be too fast when skimming across the surface. Subsequent tests were carried out using an Fw 190, a Ju 88 and an Me 410 against targets on land, but the Trialen 105 explosive was found to be too sensitive. It was replaced by a compound of 50 per cent TNT, 30 per cent ammonium nitrate and 20 per cent aluminium dust.

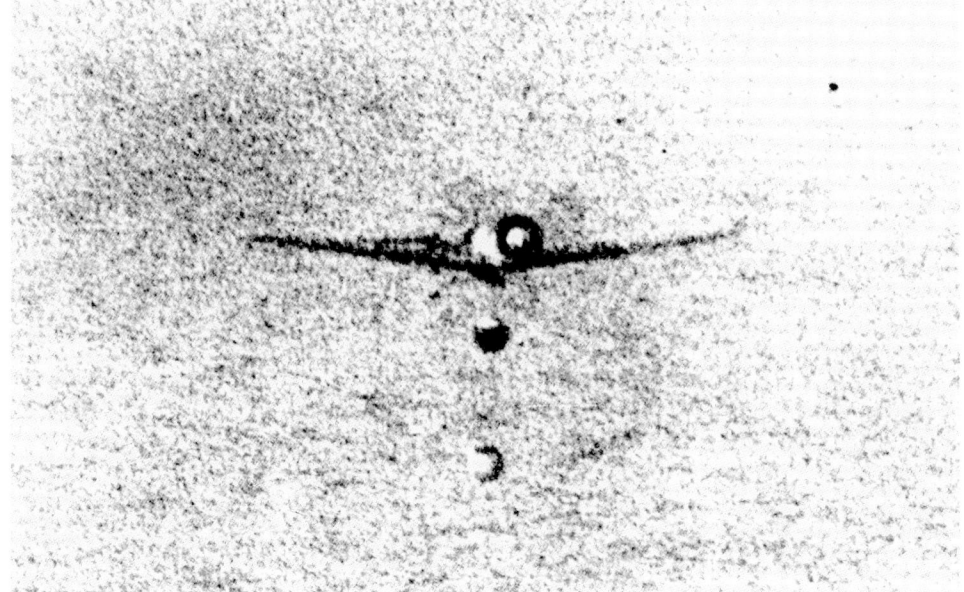

An Fw 190 flies low over the water to release an SB 800 RS *Kurt Wasserollbombe* during tests. The spherical steel bomb is already separating. (EN Archive)

The SB 800 RS *Kurt* impacts with the water, the two halves of the bomb breaking apart as they 'skip' across the surface. (EN Archive)

Experiments took place simultaneously with a new *Kurt II* in the wind tunnel at the *Aerodynamische Versuchsanstalt* (Experimental Aerodynamics Institute) at Göttingen. This version was built in four sections carrying, from the front, a warhead housed in a long cone with a maximum diameter of 750 mm, then a 100 mm-wide band carrying two suspension arms, then a section containing the jettisoning mechanism for the rocket engine and, finally, a small section for the rear suspension arm with the 608 mm-long rocket engine at the tail.

Work stopped in August 1944 when efforts were directed to more advanced projects.

CHAPTER SIX

RADICAL MEASURES

As the Allied strategic bombing offensive against the Third Reich intensified in early 1943, so its effects were increasingly felt by all who lived and worked in the Homeland. The impact of the bombing upon the military, industry, utility services, transport, education and homes, combined with the infamous exhortations of the Minister of Propaganda, Joseph Goebbels, in February 1943 for Germans to engage in *Totaler Krieg* (Total War), produced an understandable response.

In an interview with the author, Horst Geyer, the former commander of E.Kdo 25, recounted:

> E.Kdo 25's main brief was to develop and test new and effective weapons with which to bring down heavy bombers.

Stabsingenieur Öhlkers demonstrates a selection of large containers, shells and warheads to the visiting senior officers of Generalleutnant Galland's staff during their visit to E.Kdo 25 and *Sturmstaffel* 1 at Achmer in November 1943. (Forsyth)

ABOVE Generalleutnant Galland walks past a line-up of Me 410s belonging to E.Kdo 25 at Achmer in November 1943. The two aircraft nearest the camera are fitted with 3.7 cm *Flak* 18 cannon, which the unit was evaluating as a potential weapon against bombers, while the furthest machine carries underwing W.Gr. 21 mortars. Accompanying Galland from the right are Hauptmann Geyer, Hauptmann von Kornatzki, Oberleutnant Eugen Reebman (E.Kdo. 25), Oberst Hannes Trautloft, Oberst Günther Lützow (commander of 1. *Jagddivision*), Oberstleutnant Edu Neumann (*Stab.* G.d.J.), Stabsingenieur Öhlkers (E.Kdo. 25) and Oberleutnant Franz Frodl (E.Kdo. 25). The *Kommando* received proposals from the military, the Air Ministry and civilians in various capacities and occupations on how to combat the enemy daylight bomber threat. (Forsyth/Geyer)

We tried many things, but the ideas did not always originate from within. We received many letters and proposals from civilians, from companies and manufacturers, from other branches of the armed services and also from the Luftwaffe testing centre at Rechlin; 'Why don't you try this, or that?' – and so on. All suggestions were investigated, and if something looked hopeful, then we proceeded with trials. We were basically free to do what we liked, buy what we liked, design what we liked and test what we liked. But it fell to me to report everything to Oberst Galland, the *General der Jagdflieger*, and the *Erprobungsstelle* at Rechlin.

Nothing was deemed to be too imaginative, bizarre or beyond consideration.

CABLE BOMBS

One weapon to be trialled in the Luftwaffe's daylight campaign against USAAF bomber formations was the towed bomb. The concept had been developed at the *E-Stelle* Rechlin, with experiments commencing in August 1943 using an He 111 to 'tow' both box- and ring-ended SC10 10 kg bombs through the air. Trials were conducted over open moorland using steel cable of 2.5 mm diameter, and 60 m in length trailing down to 21.5 m, the Heinkel flying at a speed of 350 km/h. A second test was made with high-tensile strength, carbon steel piano wire of 1 mm diameter, 100 m in length, trailing down to between 25.5–32.5 m.

In order to prevent the trailing bomb from striking the fuselage or tail of the carrier aircraft during the first test flight on 24 August, the Heinkel flew at 220 km/h and the practice bomb was fitted to a steel cable and at first fed out from the fuselage manually to a length of two metres. This 'proceeded flawlessly', and after some further observation, the cable was unspooled to a length of 60 m. Again, there were no problems, and the cable was reeled back into the aircraft by hand.

Subsequent experiments were conducted using piano wire. It was found that when the aircraft changed course, the wire at first swung about erratically, but after five to ten seconds of continued flight, it straightened again. Recovery and reeling back to the aircraft went without problem at speeds of between 220–250 km/h, except on one occasion when the bomb was caught in the slipstream of the left-hand engine, flailing in the air some 20 m from it and only narrowly avoiding striking the aircraft. In other tests with piano wire,

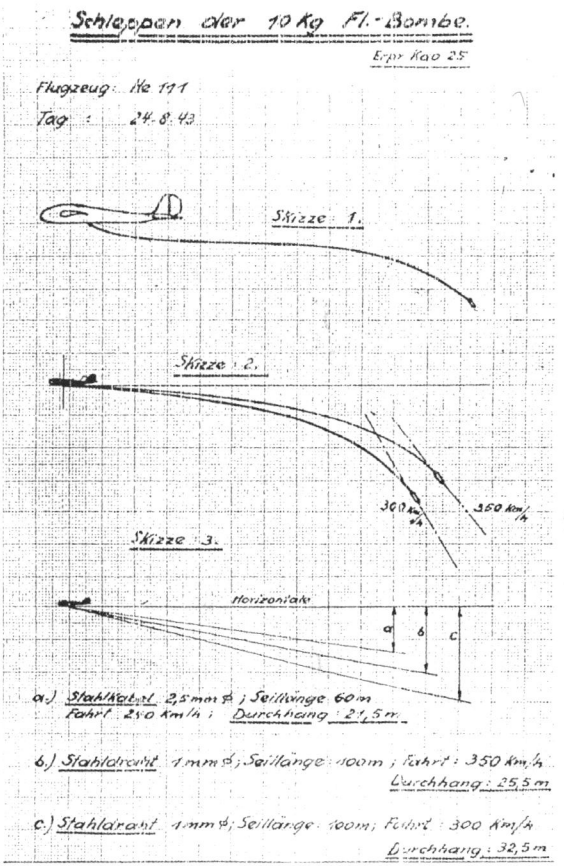

A diagram prepared by E.Kdo 25 dated 24 August 1943 showing comparisons in the use of a towed 10 kg bomb' carried on a 60 m-length of 25 mm steel cable flying from an He 111 at 240 km/h against lengths of 1 mm steel wire, 100 m in length at 300 and 350 km/h. The diagram also shows the varying amounts of reach with the three types set against the horizontal. (Forsyth)

the bomb was suspended to a length of 200 m before it was judged to be 'still'. But after just a few seconds the cable swung to one side, before swinging out into ever-increasing arcs until there was a continuous circling motion of between 50–70 m in diameter. At this, the wire was cut and the bomb fell onto the moorland. At the end of August 1943, despite the erratic and hazardous nature of these initial experiments, there was sufficient belief in the principle of towed bombs to develop an automatic reeling and cutting device based on a cable drum for installation within the carrier aircraft.

Curiously, it was representatives from the German police and postal ministry who suggested that steel cable offered an opportunity for aerial deployment. They foresaw such cables being 'dragged' into enemy bomber formations and 'dropped' onto bomber engines. Thus, in late September 1943, Rechlin delivered an experimental 10 kg 'sharpened' cable bomb to E.Kdo 25 at Wittmundhafen for fitting to an Fw 190 with the objective of deploying such a weapon against the bombers. Rechlin had originally tested the 2.5 mm twisted-steel cable bomb with a view to using it against high-tension electric power cables and telephone lines in enemy territory, but such plans were abandoned when the OKL voiced concern that the appearance of German aircraft in the sky trailing sharpened steel cables could incite the enemy to adopt similar measures over Germany.

To assist in E.Kdo 25's cable bomb trials, which commenced in early November 1943 under the name *Gerät Schlinge* ('sling' device), the *Erprobungsstelle* also furnished the *Kommando* with the salvaged wing section of a B-24 Liberator on which it could conduct ground tests.

The technical personnel of the unit devised a means in which cable of 100 to 400 m in length could be stowed into a specially adapted cylindrical metal container, with the 10 kg 'bomb', effectively used as just a weight and to provide momentum, left outside the casing. The whole apparatus was then attached to a fuselage-mounted ETC 50 bomb rack. The *Kommandoführer*, Horst Geyer, recalled:

> At first we carried out tests with a very small 'bomb' – about the size of a man's fist, with no charge or blast – attached to a length of 400 m, two- to three-millimetre twisted steel cable, which was extremely sharp. You could easily cut your hand on it.

It was planned to approach an enemy formation from the front and about 500 m above. The 'bomb' would be freed on impact with a bomber via means of a 'weak link' in the cable and the container jettisoned. The fighter would then exit flat over the bombers and subsequently be available to operate in a conventional role. One limitation was the fact that once the cable had been released, it could not be reeled in again – if a release was made in error, a danger existed if the cable had to be dropped straight down onto friendly territory.

Having moved to Achmer from Wittmundhafen in October, E.Kdo 25 conducted further tests, with Horst Geyer flying several trial flights against the wing of the Liberator in order to assess the damage the cable would inflict. These tests proved disappointing, and Geyer recalled:

> Some tests were made with a weight and others without, but approach and correction became very difficult. Rechlin then sent over the wing of an old B-24 Liberator, which was placed on a specially constructed wooden cradle. I flew several trials against this wing in an Fw 190 to assess the damage inflicted, but the cable just kept swinging about and didn't hit the target. The 400 m cable was carried in a cylindrical container beneath the Focke-Wulf's fuselage and was opened at a height of 500 m on the approach to the target. The bomb came free on impact with the target and the cable was released later whilst over open countryside. The device was made so that it could be fitted to virtually any aircraft.

Stabsingenieur Öhlkers of E.Kdo. 25 shows a selection of cable bombs to Oberst Hannes Trautloft at Achmer in November 1943. Standing next to Trautloft is Oberstleutnant Edu Neumann, who had led JG 27 in North Africa and the Mediterranean prior to being appointed to the staff of the *General der Jagdflieger*. Oberleutnante Franz Frodl and Eugen Reebmann of E.Kdo. 25 look on. (Forsyth)

Undaunted by the results of the trials with the *Schlinge*, E.Kdo 25 reported the weapon operationally ready in the first half of October. There, it seems, further work stopped until 11 December 1943 when the Eighth Air Force despatched 37 B-17s and 86 B-24s to bomb aircraft industry targets at Emden under strong fighter escort. As the B-24s of the 44th BG approached the target, an Fw 190 trailing a length of steel cable 'with a weighted object on the end' was seen to make a head-on approach towards the formation, followed by a shallow dive from slightly above. The German fighter was then seen to release the cable, which impacted with a B-24, entwining itself around the bomber's nose. The cable injured the bombardier and the navigator. Shortly after this attack, the bomber's right-hand side bomb-bay door inexplicably blew in and was torn away in the slipstream. USAAF technical personnel later assumed this was as a result of the cable weight smacking against the aircraft.

The steel wire tangled around the nose of a B-24 of the 44th BG was photographed upon the bomber's return to England following the raid to Emden on 11 December 1943. The wire was probably dropped from an aircraft belonging to E.Kdo. 25. The bombardier and navigator aboard the Liberator were wounded. (Forsyth)

The Liberator was able to return to base and the cable was removed and taken away for scientific analysis. This showed the cable to have been 0.15 inches in diameter and made up of five wires wound around a single core wire. The individual wires were square (0.047 inch each side) and the pitch on the outside wires approximately ¾ inch. Chemical analysis tests showed the wire to contain the following elements:

Carbon	0.55 per cent	Manganese	0.50 per cent
Silicon	0.20 per cent	Nickel	less than 0.01 per cent
Sulphur	0.034 per cent	Chromium	0.10 per cent
Phosphorus	0.031 per cent	Molybdenum	not found

Two days after the raid on Emden, German radio broadcasts were heard proclaiming that this new weapon had been used against the American formations 'with devastating effect'. A more grounded assessment was given by Geyer in his report on the *Schlinge* for the first half of December 1943:

> This device has been used twice so far on operations, firstly with two aircraft and on a second occasion with three aircraft. This showed that when attacking from the front, contact became difficult as a result of the enemy taking evasive manouvres. In one case the cable was apparently pulled over an engine, and the engine came to a stop shortly

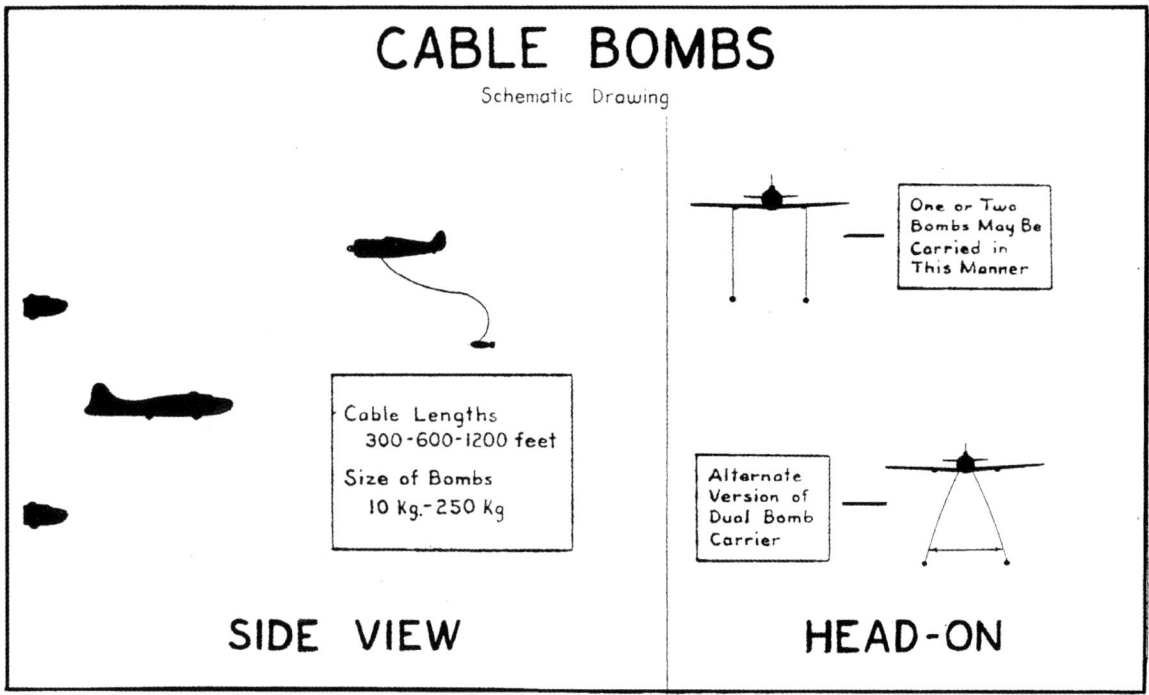

CABLE BOMBS
Schematic Drawing

One or Two Bombs May Be Carried in This Manner

Cable Lengths
300 - 600 - 1200 feet

Size of Bombs
10 Kg. - 250 Kg

Alternate Version of Dual Bomb Carrier

SIDE VIEW

HEAD-ON

A drawing produced by USAAF intelligence officers in early 1944 based on reports from aircrews who had experienced or witnessed Luftwaffe cable bomb attacks. (Forsyth)

thereafter. In a second instance a severe cut was seen across the centre of the wing of a Boeing. Further attacks and observations could not be made because of the commencement of combat with the accompanying enemy fighters. More accurate assessment can only be expected after further operations which must be carried out at least in *Schwarm* strength.

The Eighth Air Force reported further sightings of Fw 190s and Ju 88s trailing cables through bomber formations on at least three occasions in December 1943 and January 1944 during raids to Bremen and Oschersleben. However, USAAF Intelligence was not perturbed, and reported:

The conclusion to be reached after a study of reports is that although the attacks with cable bombs are becoming more frequent, they are not particularly dangerous. Even though large bombs may be carried, the question of aiming them restricts their effectiveness – plus the fact that aircraft trailing these cables must come into range of the bombers' guns, thereby making themselves very vulnerable targets. Even when the bombs reach their target, their effectiveness, so far as is known from the single attack in which a plane was hit, is relatively light. The restriction of aiming, vulnerability of carrier aircraft, and limiting of its manoeuvrability seem to indicate that at present, the bomb-on-cable tactic will not be a successful countermeasure against Allied bomber formations.

At Achmer, Geyer reported:

> Contrary to the hoped-for expectations, as has been shown with the third and fourth operations against the enemy, it is extremely difficult to pull the rope over the enemy machine. Particularly when making minor corrections in approach, the rope describes such large swings and deviations that only in rare cases can it be pulled over the target. During the last missions, after frontal approaches closing to up to ramming range, cables were cut on three occasions. In all three cases when the cables broke, the enemy aircraft suffered no particular effect. Fixed weapons cannot be used during a cable attack as the processes for cable and fixed weapons are different.
>
> Due to the extraordinary difficulties recognised so far in getting the cable to the target at all, and the very questionable effect on the target, it is requested that field tests be terminated following negative results. The existing aircraft can then be converted immediately to testing the 'Liesel' device.

Indeed, after mid-January 1944, further experiments were stopped. However, theoretical proposals for cable bombs were still being drawn up by the DVL at Adlershof in March of that year, when Dr. Ingenieur Ulrich Schmieschek proposed using Bf 109 fighters to which would be fitted steel cable of 1,000 m in length carrying 50 kg or even 200 kg bombs for use against enemy formations. The bombs were to be ignited electrically by the pilot at the optimum moment of approach towards a *Pulk*. Six months later, Dr. Ingenieur Walter Wundes of the Gothaer Waggonfabrik of Gotha put forward a proposal for a method to destroy an enemy bomber by means of:

> . . . a bomb towed on the end of a wire cable towards the path of an oncoming enemy aircraft. Upon the impact of the enemy bomber with the cable, the explosive charge would be forced by its momentum towards the bomber and explode. The wire cable would be released from the tow aircraft at the moment the enemy aircraft makes contact with the cable.

During a post-war interrogation, the former *General der Jagdflieger*, Generalleutnant Adolf Galland, stated that two unconfirmed victories had been claimed using cable bombs. However, experiments had been stopped because:

> . . . the bombs tended to trail behind the Fw 190 rather than hang down, because the bomb swung about too much and because the fighter aircraft had to come very close to the bombers to achieve victories.

The dropping of steel nets on bombers was also considered, but never adopted. Geyer believed that the weight of such nets would have slowed the carrying aircraft down considerably, thus making it an easy target for enemy escort fighters.

CHEMICAL WEAPONS

In February 1944 following a proposal from Oberst Edgar Petersen, the *Kommandeur der Erprobungsstellen*, and another from a member of the *Kommando* itself, E.Kdo 25 explored the feasibility of the spraying of fouling chemicals onto the engines and windscreens of enemy bombers. The proposals became the subject of much debate and investigation, and resulted in a salvaged engine from a shot-down USAAF bomber being sent to the chemical firm of I.G. Farben, which was instructed to conduct experiments with prospective chemicals. Horst Geyer recalled:

> One member of the unit had contacts with the I.G. Farben company and he worked with them on trials designed to foul and clog up an aircraft engine using certain chemicals, but they found that the quantity of chemical needed even to 'kill' one engine was too great, so to have brought down a four-engined bomber would have been impossible.

Indeed, by June 1944, Geyer noted that all efforts in this direction had thus far proved without success. The *Kommando* also consulted Dr. von Harz of the research laboratories at the chemical and weapons company Dynamit Nobel A.G. in Troisdorf. He advised the officers of E.Kdo 25 that the agents available did not possess sufficient energy to destroy an engine. It had been suggested that the use of ozone could damage the engines of enemy aircraft, but Geyer conceded that procuring ozone was extremely difficult, and that in any event, there were no devices available able to carry and spray the required quantities of the substance, other than at the DVL at Berlin-Adlershof, where von Harz recommended any further testing should be carried out.

Geyer further recalled that:

> Tests were also carried out at Rechlin with chemicals designed to spray over cockpit and gun turret Plexiglas – to adhere to it and to mask it, but not, necessarily, to destroy it. Civilian laboratory researchers analysed fragments of windshields from shot-down B-17s and B-24s in an effort to determine the manufactured composition of the Plexiglas. They subsequently developed certain types of chemicals in liquid and powdered form which could be dispersed over the glass. Rechlin then asked us to conduct trials using an Fw 190, and we found, that depending on which kind of chemicals were being used, it was not necessary to use large quantities.

Geyer remembered a group of chemists visiting the *Kommando*'s airfield at Parchim one day in the summer of 1944 to deliver a sample of one such 'white liquid' that was duly sprayed over a large piece of Plexiglas:

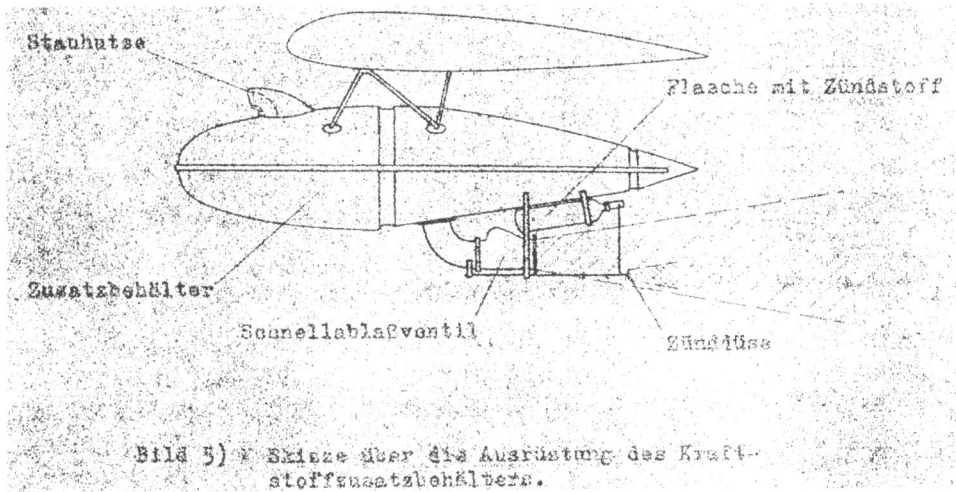

Stauhutze

Flasche mit Zündstoff

Zusatzbehälter

Schnellablaßventil

Zündüse

Bild 3) * Skizze über die Ausrüstung des Kraft-
stoffzusatzbehälters.

A profile view of the underwing tank which it was proposed to fit to either an Fw 190 or a Ju 88, and from which would be blasted volatile explosive fuels directly into the atmosphere and targeted at enemy bombers to rupture their airframes. This proposal was based on a standard Luftwaffe 300-litre drop tank, to which a smaller tank containing explosive mixture was fitted beneath and at the rear end of the larger tank. The contents of the main tank passed through a quick-drain valve and an ignition nozzle. (Forsyth)

As soon as the liquid hit the glass, you went blind – you couldn't see anything in front of you. But I wasn't sure about it and I sent a report to Galland, warning him that, if necessary, American bomber crews would attempt to break their windshields if they were sprayed, thus nullifying the effect. Galland understood what I was saying. But Göring also grew worried about the idea and instructed Galland not to pursue it, since he was concerned that the enemy would employ the same methods against us.

Geyer did remember one Fw 190 being fitted with underwing tanks with valves designed to eject a chemical spray. The valves were of an open-or-shut, one-use only, jettisonable type and were intended to be used following a head-on attack against an enemy bomber formation. The Fw 190 would make a standard approach, using cannon, pass over the formation and then the pilot was to open the valves to the tanks to spray the American aircraft. No such operations were ever carried out, however.

NITROPENTASCHNUR

During the summer of 1944 E.Kdo 25 assessed the possibility of dropping *Nitropentaschnur* (15 m-long strips of detonating cord) on enemy formations in clusters. *Nitropentaschnur*, more widely known as PETN cord, was made from Pentaerythritol tetranitrate. This form of explosive compound, which is similar to nitroglycerin, was first manufactured in 1894 by the Rheinisch-Westfälische Sprengstoff A.G. of Köln, with production commencing in 1912. It was used by German forces during World War I, and in World War II it formed a component in ammunition used in the Luftwaffe's MG FF/M series of cannon as well as in the high-explosive M-shell.

The rear fuselages of a small number of E.Kdo 25's Fw 190s were fitted with specially designed containers manufactured by the Max Baermann engineering firm at Köln-Dollbrück. These containers could carry 20 lengths of 15 m-long cord. The plan was

At the end of July 1944, *E.Kdo 25* was redesignated JGr 10 under the command of Major Georg Christl, a Knight's Cross-holder, seen to the extreme left in a field cap, who had previously commanded III./ZG 26. *Jagdgruppe* 10 was redesignated again as 'JG 10' and continued in its weapons development role until December 1944, by which time, in reality, it was little more than an *Einsatzschwarm* (Operational Flight). JG 10 moved to Erfurt-Bindersleben in March 1945 and was disbanded soon thereafter. Also in this photograph are Adolf Galland and Horst Geyer in conversation at centre. (Forsyth/Geyer)

that a frontal approach would be made against a bomber *Pulk*, and as the fighter passed through the enemy formation, the container would be released and the cords dropped down onto the bombers, the flight of each piece of cord being controlled by two small parachutes. They would either become wrapped around propellers or strike the metal of the enemy machines to hook and embed themselves into their panels with the aid of small, sharp-clawed 'anchors'. The fuses were set to detonate at seven seconds from impact. Such contact with an enemy aircraft would result in detonation.

Initial test flights against static targets on the ground demonstrated flawless opening of the containers and discharge of the *Nitropentaschnur* cord, but it was also found that the fuse delay was too short and it was recommended that the period be extended to 20 seconds. Furthermore, it was believed that a load of 20 cords would be insufficient and would have little effect, simply scattering harmlessly through an enemy formation. In cases where the anchors managed to embed themselves in the metal of the static test targets, it was observed that while the outer metal skin was punctured, the cord had failed to entwine itself around any parts, and thus would fail to detonate.

In March 1944, E.Kdo 25 transferred to Parchim, and during the ensuing summer it underwent changes in structure. In June the *Kampfstaffel* moved to Tarnewitz briefly, before heading to Finow. At the end of July, Horst Geyer had been reassigned to take command of E.Kdo 262, the Me 262 jet fighter test and evaluation unit at Lechfeld, while E.Kdo 25 was redesignated JGr 10 and placed under the command of Major Georg Christl, a Knight's Cross-holder who had previously commanded III./ZG 26.

In September 1944, Christl noted that:

. . . further experiments with *Nitropentaschnur* has revealed new factors. Dropping tests with 60 m-long detonating cords are inconclusive due to technical deficiencies with the parachute system. When the length of the cords are extended from 15 m to 60 m, the number of cords carried in the container has to be reduced from 20 to seven, thus meaning that the additional cords would have to be transferred to suspended external holders in a further development. Since the fundamental problem of looping the cord around aircraft components has still not been resolved, it does not seem promising to continue with further tests based on the [negligible] measure of success so far achieved.

'MIST OF FIRE' – INCENDIARY CLOUDS AND THE AIR AS A WEAPON

Among the more inventive proposals to combat enemy bombers was one from Dr. Wendland of the design department at the Focke-Wulf aircraft company at Bad Eilsen dated 8 March 1944 in which he proposed to destroy enemy bomber formations by artificially generated gusts, or squalls, of air. The gusts were to be created by the 'combustion of fuels in the atmosphere'.

Wendland envisaged replicating the levels of natural energy which were contained in the 'storm clouds of nature' – around 6.5 kilocalories to a kilo of air – which created updrafts with speeds of more than 25 m per second. If, however, combustion could occur without excess air, enormous levels of energy – up to 680 kilocalories to a kilo of air – could be produced. Wendland proposed producing such energy by blasting volatile explosive fuels directly into the atmosphere from specially designed external tanks, one of which he proposed fitting to a Ju 88, or smaller versions under each wing of an Fw 190. According to Wendland:

The ignition of such fuels in the atmosphere would produce updrafts of tremendous strength. Aircraft which have less resistance to gust, for example bombers, would suffer extreme flows of wind on their wings of sufficient strength to cause rupture of the airframe.

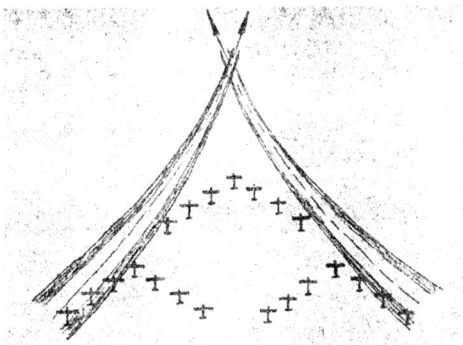 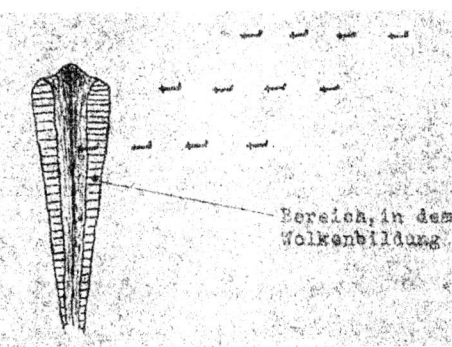

By the spring of 1944, no proposal to combat the ever-increasing formations of Allied heavy bombers from pounding the cities and industrial centres of the Third Reich was dismissed out-of-hand. This diagram, produced by a design department employee at Focke-Wulf in March of that year, illustrates a bizarre plan to bring down, or at least destabilise, enemy bomber formations by artificially generated gusts, or squalls, of air. (Forsyth)

However imaginative Wendland's proposal may have been, it did not progress beyond report stage.

Nine months later, on 24 December 1944, an equally imaginative proposal landed on the desk of the aircraft designer Professor Willy Messerschmitt from the *Autobedarf Lagerlechfeld* – the cover name given to Lechfeld airfield, where Me 262 test-flights were being conducted. The unidentified writer, clearly of some technical and scientific mind, had been inspired by fanciful rumours of a new secret weapon apparently known colloquially as 'exploding mist'.

His idea was for one or more Me 262 jet fighters to make a frontal approach on an enemy bomber formation and about 500 m above it, towing winged *Anhänger* (trailers). One such trailer would house a pressure-sealed container carrying 700 kg of a light gas fuel. Using a pyrotechnic propellant, the fuel could be atomised in five seconds through nozzles in the wings of the trailer. According to the writer, with a ratio of 1 kg of fuel to 15 kg of air, an area of explosive air/fuel mixture spanning a cross-section of 20 m² and a length of 1 km could be created in the air at a height of 8,000 m. Once the outflow from the trailer ceased after five seconds, the trailer ignited automatically. If a suitable degree of atomisation and fuel (such as dissolved acetylene) could be achieved after three seconds, then 'a powerful pressure wave would emit a front of flame'. The writer further stated that, 'This should destroy a large number of the bombers, or the *Pulk* would have to break up, making it easier for individual fighters to approach.'

The report calculated that the explosive effect of this reaction would extend for about 200 m. The Me 262 would not be placed at undue risk because the releasing of the trailers would be well ahead of the target and, combined with the jet's superior speed, meant it would be able to break away and escape quickly. But questions remained as to exactly how much pressure could be expected in the 'pressure wave', what would be the most favourable fuel, and how much pressure was needed to destroy a bomber. The answers to these questions would come from consultation with thermodynamicists and combustion and ballistics specialists. The writer stated that:

> The starting point is the assumption, based on the observation of the effects of air mines, that 1,500 kg of explosives would damage a bomber at a distance of 100 m.

It is not known if the proposal was considered further by Messerschmitt.

FLAMMENWERFER (FLAMETHROWERS)

Flamethrowers were first employed by the Imperial German Army during World War I for use against fortified positions and bunkers. Following a proposal received from Leutnant Dr. Peter W. Stahl, the Technical Officer of I./KG 51, in February 1940, trials were conducted at Tarnewitz with a rearward-facing *Flammenwerfereinbau* (flamethrower installation) installed in an He 111 which would emit burning oil from specially fitted

pipes. The installation was intended as a defensive weapon against fighter attack, the thought being that pursuing enemy aircraft would fly into the large, dense, soot-black cloud of smoke produced by the flamethrower and become engulfed, blinding the pilot.

The trials were carried out with a Bf 109 flying in the role of the attacker, and although the tests were effective, resulting in the Messerschmitt's windcreen being caked in soot at 150 m, officially, the idea went no further.

Undeterred by the lack of interest in his idea, Stahl pressed ahead and installed a flamethrower in Ju 88A-1 Wk-Nr 6093 9K+EH of 1./KG 51 at Villaroche, in France, in the late spring of 1940. An oil tank was installed within the rear of the fuselage, close to the tail wheel, to which were attached two vent pipes, while a third pipe was fitted to an enlarged port wingtip. When applied, the results were visually spectacular, and on one occasion, on 19 August 1940, Stahl may have even contributed to the downing of a Spitfire over the south coast of England. The specially configured Ju 88 later took part in operations in the Soviet Union in 1941, assigned to *Stab* I./KG 51, and on at least one occasion defended itself successfully using the flamethrower during an attack by Russian fighters.

Ju 88A-4 Wk-Nr 1050 9K+FB was similarly fitted with a flamethrower device by the *Stab*

TOP A Ju 88, possibly A-1 Wk-Nr 6093 9K+EH of 1./KG 51 based at Villaroche in France in the spring of 1940, makes a dramatic low-level flypast as the crew activates the *Flammenwerfereinbau*. (EN Archive)

ABOVE Ju 88A-4 Wk-Nr 1050 9K+FB of *Stab* I./KG 51 was fitted with a *Flammenwerfereinbau*. It is seen here in January 1942 with a rudder adorned with ship kill markings and at the time of flamethrower trials at Wiener Neustadt when flown by Hauptmann Heinrich Hahn. (EN Archive)

Drawing of a *Gero II*
installation in an Fw 190F.
(EN Archive)

Flammenwerfer "Gero"
Schlachtflugzeug Fw 190 F
X./XI./

I./KG 51 in the autumn of 1941 and flown by the *Gruppenkommandeur*, Hauptmann Heinrich Hahn. Tests with the flamethrower were later conducted by Hahn at Wiener Neustadt, but this aircraft was destroyed in a crash on 3 February 1942 in which Hahn and a crewman were killed.

In 1942 work commenced officially with a *Flammenwerfereinbau* in a Do 217E, which comprised an adapted army flamethrower and, in a fashion similar to Stahl's installation, a purpose-built tank fitted in the fuselage from which fuel was pumped towards the tail, where it was passed out under pressure and ignited. It is possible that this was the first installation in the *Gero* series, becoming the *Gero I*. The fuel used for these experiments was 'Aeroflame', an extremely volatile compound of petroleum and benzine which had undergone trials at the Army's gas and warfare testing establishment at Raubkammer, northwest of Münster. Perhaps not unexpectedly, the Dornier involved in the ground tests caught fire and it is not known if any air tests took place.

It was not until August 1944 that another *Flammenwerfereinbau* was trialled in Hs 129 Wk-Nr 142001, intended as a ground-attack weapon. This development had its origins in discussions between Diplom-Ingenieur Friedrich Nicolaus, the chief designer at Henschel, and members of his team, and Hauptmann Eggers and Leutnant Scholz of *Pz.Jä.St./*JG 51 in August and September of the previous year. The plan was to install a tank, possibly with a 300-litre capacity, inside an Hs 129 in a similar way to the earlier installation in the Do 217, but with a four-metre-long pipe that could be raised and lowered hydraulically. The fuel compound was pressurised with nitrogen and ignited with a thermite cartridge. Unfortunately, no further information is known regarding experiments with the Hs 129.

Further ground-attack trials took place utilising a variation of the *Gero*. First developed in October 1944, the *Gero II* was produced in three variants – the *IIA*, *IIB* and *IIC* – for mounting beneath an Fw 190F-8. The device comprised a jettisonable tank suspended from a centreline ETC 504 rack. As with the earlier systems, a highly flammable petroleum-based fuel would be expelled under pressure through a nozzle angled downwards. While there were only minor differences between the *IIA* and *IIB*, the *IIC* was a shorter and lighter installation.

In February 1945, Focke-Wulf was asked to make ready an Fw 190F-8 with a specially protected rear fuselage and tail area for an anticipated *Flammenwerfereinbau* made by Arado and Norddeutsche Dornier, but such work is not believed to have taken place before war's end.

Gero IIA and IIB	
Overall length	3,120 mm
Diameter	470 mm
Fuel capacity	375 litres
Empty weight	230 kg
Loaded weight	540 kg
Gero IIC	
Overall length	1,920 mm
Diameter	450 mm
Fuel capacity	164 litres
Empty weight	173 kg
Loaded weight	300 kg

A 1,000 kg SC 1000 *Grossladungsbombe* loaded onto the centreline bomb rack of an Fw 190. This bomb was usually carried by an ETC 2000 or strengthened PVC-1006 or ETC 500 rack for use against non-armoured shipping and land targets. Note the fin lowest to the ground has been removed to enable clearance during movement by the aircraft on the ground. (EN Archive)

SONDERSTAFFEL EINHORN

Some of the activities of the *Sonderstaffel Einhorn* (Special Squadron Unicorn) merit inclusion in this book because arguably, and comparatively, they can be viewed as a radical measure.

Over the past 40 years Luftwaffe researchers Günther W. Gellermann, Geoffrey J. Thomas and Nick Beale have unearthed documentary evidence that the notion of deploying Fw 190 fighter-bombers laden with very heavy bombs of more than 1,000 kg against large,

'solid' targets such as bridges had its origins in the mind of former glider pilot Oberleutnant Karl-Heinz Lange of I. *Luftlandegeschwader* (Glider Assault Wing) 1 who had been wounded in Sicily. In late 1943, it was Lange's belief that a small unit of ideologically radicalised pilots could inflict great damage against an anticipated Allied invasion fleet by flying *Totaleinsatz* ('total commitment' operations) in manned glide bombs. Put another way, these would be *Selbstopfer* (self-sacrifice) missions. Lange put his idea forward to the OKL as a serious proposal.

It was, obviously, an extreme and contentious suggestion that ran against the fundamental creed of the German officer corps. But these were extreme times, and there was an adequate number of young pilots within the ranks of the Luftwaffe who were sufficiently imbued with National Socialist fervour to execute Lange's radical measure as a way of defending both their principles and their homeland.

But there was one problem: in early 1944, no such manned glide bomb existed to carry out the task. Notwithstanding this, a *Staffel* was established on 3 February 1944. At a conference just over two months later, on 17 April, the Technical Officer of KG 200, Fliegerstabsingenieur Hans-Christian Tilenius, suggested that Fw 190s could crash 2,500 kg bombs into enemy ships, although it was recognised that to sink a ship more effectively or conclusively required holing it below the waterline. Tilenius' suggestion, however, went some way to compensating Lange.

In the meantime, Lange was able to assemble a small group of volunteers, many of whom had, like him, experience on gliders. The group was assigned to II./KG 200 at Dedelstorf and later embarked upon a crude form of training with JG 103 at Stolp-Reitz, although by this point it had been agreed that, realistically, the heaviest ordnance that could be carried by an Fw 190 was a 1,100 kg bomb.

By late August 1944, the group had acquired the somewhat intriguing name of *Sonderstaffel Einhorn* and as new *Staffelkapitän*, former reconnaissance pilot Oberleutnant Robert Schuntermann. With the prospect of any operations against shipping targets by heavily laden Fw 190s forgotten as the Allied armies advanced through France covered by an 'umbrella' of fighters, the unit was to strike at ground targets with the heaviest ordnance in cooperation with III./KG 51. In October, the *Staffel* was at Achmer, where, equipped with Fw 190F-8s, it worked on its ordnance procedures involving, by this time, 1,750–1,880 kg SC 1800 *Minen* bombs filled with 1,000 kg of Trialen explosive. These weapons were intended for use against buildings with no-delay fuse settings or against large merchant vessels, where a high pressure wave destructive effect could be created.

Einhorn's Focke-Wulfs were fitted with ETC 502/*Bombenträger* 2000 centreline bomb racks and had special tyres capable of taking the abnormal weight. For reasons of safety to both air- and groundcrews, the heavy fighter-bombers would line-up for take off spaced 20 m apart between wingtips and with 40 m separating following aircraft. *Einhorn*'s Fw 190F-8s are believed to have been deployed for the first time on 28 September against the road and rail bridges at Nijmegen following the initial Allied crossing of the Waal on the 20th. The War Diary of *Luftflottenkommando* 3 noted:

Seven Fw 190s were deployed as heavy fighter-bombers to attack the bridges at Nijmegen. One aircraft is missing. The effects of the attack can only be assessed after evaluation of the air reconnaissance flown on 28 September is available.

Results showed that one direct hit was scored on the road bridge and another on the rail bridge. A second mission against the bridges by '*Sonderstaffel Einhorn* with crews of III./KG 51 for a *Zerstörangriff* at dusk' was slated for 30 September. The following month the *Staffel* was transferred south to Verona, in northern Italy, where it remained under Schuntermann's command until December. It is believed the *Staffel* carried out a few night sorties using small numbers of Fw 190s with 1,000 kg bombs against facilities of the Allied air forces, as well as roads and bridges.

The *Staffel* returned to Germany in December 1944, where it was attached to *Nachtschlachtgruppe* (Night Ground-Attack Group) 20 before being rolled into 11./KG 200 under Hauptmann Erwin-Pieter Diekwisch in February 1945. It flew missions against the Remagen bridge, dropping SC 500 and SC 1000 bombs, and then conducted attacks on Allied vehicles east of the Rhine using various types of 'AB' ordnance (*Abwurfbehälter* – bomb canisters packed with smaller anti-personnel bombs). Such missions continued as 11./KG 200 pulled back to northern Germany, from where it carried out its last missions against western Allied and Soviet troop and armour targets in the spring of 1945 until the German surrender.

CHAPTER SEVEN

THE AIRCRAFT AS A WEAPON

MISTEL COMPOSITE BOMBER

Alongside its arsenal of technologically advanced jet fighters and bombers, the Luftwaffe could call upon an altogether more unconventional yet equally ingenious weapon from late 1944. The *Mistel* composite bomber (also known by the Germans as *Vater und Sohn* ['Father and Son']) had been designed and developed in 1943 by engineers at Junkers as a weapon to be used against 'difficult', heavily armed, armoured or strengthened targets, such as aircraft carriers, battleships, large fortifications and bridges. The design involved the placing of either a Bf 109 or Fw 190 fighter on top of a Ju 88 airframe that had its cockpit section removed and replaced with a large, hollow-charge

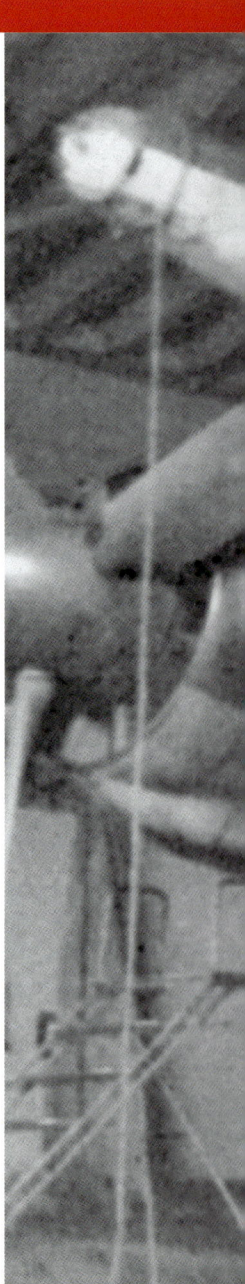

An enormous SHL 3500 hollow-charge warhead, known as the *Elefantenrüssel*, has been fitted to the bulkhead of a Ju 88 via purpose-built lugs after being hoisted into position by chains. The warhead was 1.8 m in length and contained 1,700 kg of explosive. (EN Archive)

Bf 109F-4 CI+MX, which was used in some of the early Bf 109/Ju 88 *Mistel* trials at the DFS Ainring in late 1943/ early 1944, lifts away from a manned Ju 88A-4. Separation was activated by explosive attachment bolts being released by the pilot of the Bf 109. There was a risk of structural damage to the fighter during separation and, therefore, it was recommended that the Messerschmitt should always be trimmed nose down. (EN Archive)

warhead carrying a 3.5 ton payload of high-explosive. The concept was for the pilot of the fighter upper component to fly and control both aircraft to the target, whereupon the lower component would be released, on course to, hopefully, achieve a direct hit.

The composite was constructed in such a way that the Bf 109/Fw 190 rested on two three-legged supports, fastened with spherically aligned bolts. The supports were attached to points on the front and rear spar joints, as well as to frame nine of the Ju 88's fuselage. The third attachment was made in front of the tailwheel well of the fighter via a collapsible strut, attached to frame 20 of the Ju 88's fuselage, which enabled the fighter's incidence to be increased during separation. The fighter was fastened to the supports by means of explosive bolts. On separation, only the bolt keeping the collapsible strut straight was detonated by the pilot. The subsequent increase in the angle of incidence of the fighter operated a switch to command the detonation of all three bolts holding the aircraft.

The Ju 88's engines were throttled mechanically from the fighter via a linkage system, which during the release sequence was also separated explosively. The throttle levers fitted to the fighter were much longer than those normally found in the Ju 88 in order to overcome the higher friction forces associated with the modified throttle system. Two dual-function instruments, indicating engine manifold pressure and engine speed (rpm), were fitted to the fighter to monitor the performance of the Ju 88 engines. The electrical connections between the two aircraft were made with two multi-pin shear connectors. Their halves were secured by locking wire to prevent inadvertent separation.

A directional control system enabled the *Mistel* combination to fly solely under control from the fighter. The system could be operated in two conditions: 'automatic' and 'cruise'. In cruising flight, control movements were measured by potentiometer and transmitted electrically to the servos coupled to the flying control surfaces of the Ju 88. In the 'automatic' flight condition the Ju 88 could be controlled by switches fitted to the fighter's control column and instrument panel. The switches were arranged to send directional and lateral control commands to the Ju 88 in the same way as if the pilot were physically controlling the aircraft.

The *Mistel* warhead, the *Schwere Hohlladung* (Hollow Charge) SHL 3500, was developed under great secrecy by the Lauchammer firm at Riesa some 55 km east of Leipzig. It was designed to pierce the armoured steel of a battleship or blow open a wall of reinforced concrete such as that found on heavy gun emplacements, command bunkers, factories and power stations. The 1,700 kg of explosive and detonator for the main charge was placed at the rear of the warhead, with a cone-shaped cavity to the front of it, 1.8 m in diameter. The cone was lined with a layer of soft metal, either aluminium or copper, with four electrical crush fuses positioned at the tip of a 2.75 m probe, protruding from the front of the warhead and known as the *Elefantenrüssel* ('Elephant's Trunk'). Soft metal was important, since a harder metal would prevent the hollow charge action from functioning properly. When this probe struck the target, the fuses would trigger the detonator behind the explosive charge. The warhead could not be armed until the carrier aircraft's landing flaps had been retracted. This safeguard prevented the warhead from exploding in the event that the upper aircraft was unintentionally separated during an aborted take-off.

After firing, the charge – a mixture of 70 per cent Hexogen high-explosive and 30 per cent Trinitrotoluol – would focus all its force on the soft metal liner which then became liquid and projected forward in a fine jet. Travelling at more than 20 times the speed of sound, the jet could drill a hole through eight metres of armoured steel or 20 m of reinforced concrete. Once through the outer layer of a target and subsequently confined within it, the jet of metal would vaporise anything in its path.

The length of the probe could be varied to trigger the charge at the optimum distance from the target. When used against armoured steel it could be as long as 2.75 m, but for less well protected targets, the length was considerably shortened and at least three variants of probes, of varying length, are known to have been developed. A stand-off probe was necessary in order to allow time for the soft metal liner to form itself into a thin jet before impact. Broadly speaking, the greater the distance between the charge and the target at detonation, the thinner and deeper the hole drilled; the closer to the target, the wider and shallower the penetration. The entire detonation procedure took place within one 10,000th of a second.

Towards the end of 1943, the *E-Stelle* Rechlin arranged for static tests of a hollow-charge, similar to the one intended for use on the *Mistel*, to be conducted against the 25,000-ton French battleship *L'Ocean*, anchored in the approaches to the naval port of Toulon, in southern France.

The four-ton charge was directed at the vessel's two main gun turrets. Additional 10 cm steel armour plating had been fitted to the 'target' to make it more representative of modern warships. When the charge was detonated it shot through the additional armour, through the 30 cm armour of the first gun turret, passing through the turret and out the opposite side (which was of similar thickness) and through the armour plating of the second turret. The result was an effective total penetration of 28 m into the ship.

Other static trials with the hollow-charge against structures made of reinforced concrete were undertaken in East Prussia, where the warhead blasted its way through some 18 m of concrete.

It was intended that such a warhead could be fitted to a purpose-converted Ju 88 with relative ease by trained specialist Luftwaffe armourers in the field. To facilitate this, a Ju 88 airframe would have its crew compartment removed at the aft bulkhead, this process

Smoke billows skywards following a trial detonation of an SHL 3500 hollow-charge warhead on the hulk of an old French navy warship, possibly the battleship *L'Ocean*, at Toulon in late 1943. (Forsyth)

being carried out at a Junkers *Mistel* conversion facility. Four spherically aligned quick-release bolts were fitted that allowed the crew compartment to be re-installed for training purposes or ferrying to an operational airfield.

Then, once delivered to an operational unit, and at the time a mission was ordered, the crew compartment would once again be removed and the warhead fitted, again using the quick-release bolts. This process – lasting approximately one day – required a team of six mechanics, two armourers and a crane capable of lifting four tons. After the warhead was attached, the composite had to be towed to the take-off position because the pilot of the fighter could not operate the brakes of the Ju 88.

The all-up weight of an operational *Mistel* combination fitted with a warhead was about 20,000 kg, some 7,000 kg heavier than a normally loaded Ju 88, and thus approaching the load limits. This placed heavy demands on the undercarriage which needed to be completely redesigned. In fact, no such modifications were carried out, and the undercarriage was extremely prone to collapse on take-off. A safe take-off could only be attempted from concrete runways that were considered to be in perfect working order. Even the smallest hole or imperfection on a runway's surface could have dire consequences for the machine and its pilot. Later in the war, operational *Mistel* were fitted with larger tyres that were rated as high as 23.4 tons for take-off. Once in the air, a combination fitted with a warhead was impossible to land, so the only choice left to the pilot in an emergency was to jettison the complete lower component.

Apart from a weak undercarriage, another problem was the time lag between the fighter's pilot operating a control and the auto-pilot relaying this to the bomber. The Ju 88 always had a tendency to swing on take-off, and this tendency was magnified by the delay in the time needed to correct. It was not unknown for the combination to swerve off the runway on take-off, and it was impossible to fly the *Mistel* in tight formation due to the same problem. Despite this, most *Mistel* pilots found the dual controls easy to handle and responsive – although it was still a daunting machine to have to come to terms with. Feldwebel Rudi Riedl was a flying instructor who was assigned to 6./KG 200, a *Mistel* unit, in late 1944. He recalled his first impressions:

When I saw my first *Mistel*, which was a training variant, I thought, 'How the Hell am I going to handle this monster?' I had to sit in the cockpit for a long time to get a feel for the machine – to understand how it worked and to come to terms with how high up I was. It's a very unusual feeling to be sitting six metres above the ground! Eventually, however, after having checked out the instrumentation and controls, I began to realise that perhaps it was not going to be so difficult after all. But when I later saw the operational variant with that warhead mounted, I was absolutely astounded. I thought, 'How am I going to handle this? There is no other crew except me!'

Riedl made a total of ten training flights, which was considered to be the standard number prior to embarking on operations. Six of these flights were aiming exercises in a training variant in which there was a pilot in the Ju 88, following which the complete composite returned to base. The remaining four flights involved separation exercises, during which there were occasional accidents. At the moment of separation, the natural inclination of the pilot in the upper component was to push the stick forward, causing, on occasion, the propeller blades of the fighter to strike the cockpit of the Ju 88, with fatal results. Riedl remembered:

> Starting and taxiing in the *Mistel* was hard. It was a real beast. Visibility was very restricted. Because of the height and angle at which the machine sat on the ground, you could only see the end of the runway when the tail came up and you were ready to lift off. Also, often when manoeuvring and turning into the take-off position, the tail wheel was known to come off, and that was when there were accidents; machines slewed off the runway. But once the machine was airborne, there was no real problem. A little sluggish perhaps, but that was all. It has to be remembered that we were all

These Luftwaffe armourers are probably receiving instruction on the engineering of the *Mistel* from the civilian technician seated in the centre of the group and looking over his shoulder at Junkers' Nordhausen assembly plant. Note the support frame fitted to the top of the Ju 88's fuselage for the upper component fighter. The removal of a Ju 88 cockpit and replacement with a warhead took a team of eight, with a crane, around a day. (EN Archive)

Mistel 1 Bf 109F PI+MI/Ju 88A-4 'White 2' of 2./KG 101,
Saint-Dizier, France, June 1944.

experienced pilots – former instructors – not novices. In flight, the *Mistel* handled comparatively well, very like any other twin-engined bomber. The Junkers technicians who worked on matching the controls of the upper and lower components did a fantastic job.

The *Mistel* was first used operationally to devastating, albeit sporadic, effect in June 1944 when aircraft of the first unit to be so-equipped, 2. *Staffel* of the bomber operational trials *Geschwader* KG 101, attacked shipping off the Normandy coast. One *Mistel* narrowly missed the Royal Navy frigate HMS *Nith*. A member of the crew remembered that the composite caused:

> . . . an enormous explosion alongside the ship. Our seaboat was turned outboard at its davits ready for an emergency and the wing of the plunging aircraft cut the boat in half, which gives you some idea of how close the Ju 88 was when it hit the water.

The *Mistel* blew in *Nith*'s starboard side amidships, causing it to heel over to port, and the entire length of the ship was raked by steel fragments. Steam pipes in the boiler room burst and the main generator was put out of action. For a time, the ship was without electricity. *Nith*'s second in command recalled grimly, 'I just remember the awful sight of maimed bodies, blood, and flesh.' Ten sailors were killed and 26 wounded.

A lack of parts and the time needed to train pilots in adverse war conditions prevented widespread deployment on the Western Front for several months, but on 22 March 1945 *Mistel* were called upon to destroy pontoon bridges over the Rhine near Oppenheim over which the US Third Army was crossing. Earlier attacks by fighter-bombers and Ju 87s had failed to make an impact, and so, as a last measure, the Luftwaffe turned to the few available Ju 88/Fw 190 composites of II./KG 200 based at Burg. During the late afternoon, five

Mistel were prepared, and the airfield cleared of personnel as a safety measure, but only four composites managed to get airborne, the fifth having broken down on the ground. They were accompanied by five Ju 88 and Ju 188 *Beleuchter* (illuminator) from 5./KG 200, which led the way to the river. At the controls of one of the *Mistel* was former blind-flying instructor Leutnant Alfred Lew of 6. *Staffel*:

Following a wide left turn, our formation headed towards the Rhine. Initially, we were at 1,500 m, but soon climbed to a cruising altitude of 2,000 m. The approach flight took two-and-a-half hours. As it began to turn dark, so I saw the River Rhine glittering below us in the moonlight. Meanwhile, our *Beleuchter* aircraft had dropped their signal flares, but still, I could not recognise the bridge. I flew a full circle to orientate myself, but as I did so, I ran into heavy American anti-aircraft fire. In order to locate the target, I descended to 1,500 m and then – Bang! – I was hit. My *Mistel* lurched to port and went into a spin on its back. As I was no longer able to control the machine, I decided to separate my Fw 190 while simultaneously diving away from the *Mistel*. At 300–400 m, I finally managed to regain control over my aircraft and got away from that terrible Flak, heading east towards Burg.

In the darkness and flying on instruments, my return flight became quite hairy and I flew off course. I found myself approaching the River Elbe. Since the Russians were quite close by at that stage, it was dangerous to cross the river. Fortunately, I reached the Elbe at Torgau. Now my experience as a flying instructor became very useful and I managed to find my way home. At 2200 hrs, I landed safely at Burg. One *Mistel* pilot, another former flying instructor, did not make it back from the mission. The success of the operation was virtually nil.

Feldwebel Willi Döhring of 6./KG 200 gesticulates to the photographer in a gesture intended to reflect the airman's superstition of taking a picture immediately before a flight. In the background, two of the unit's operational *Mistel 2* combinations line up for take-off from Burg, with the engines of the lead Fw 190 and Ju 88 already running up. The pilots wear life vests for an over-water flight. (EN Archive)

As mentioned by Lew, by late March 1945, on what was somewhat optimistically referred to as the 'Eastern Front', Soviet forces had reached the Oder, cutting Küstrin off. One of the last natural barriers protecting Berlin had been breached; the German capital lay just 80 km to the west. From this point on, it became imperative that the Luftwaffe do all it could to prevent the Red Army from establishing crossing points over the Oder, and if they did, to destroy the bridges. Naturally, such a task was suited to the *Mistel*. Assembly and delivery of new composites had taken place with some pace throughout February and March 1945, and they equipped II./KG 200 and I. and II./KG(J) 30.

Mistel 2 Fw 190A-8/Ju 88G-1 4D+FK of II./KG(J) 30,
Oranienburg, Germany, April 1945.

A *Mistel 2* of KG(J) 30 at its camouflaged dispersal at Oranienburg in March 1945. The Ju 88G-1 has been fitted with an SHL 3500 'short' fuse (*Sprengkopf ohne Elefantenrüssel*) and carries a 900-litre fuselage-mounted drop tank, whilst the Fw 190F-8 is fitted with a 600-litre drop tank. (EN Archive)

Typical of such missions was the one of 6 April 1945 when I./KG(J) 30 was ordered to deploy its *Mistel* during the evening against bridges across the Oder south of Stettin. The *Gruppe* was using the *Erprobungsstelle* at Rechlin as a base, the airfield being situated 150 km directly west of the targets. Oberfähnrich Georg Gutsche was an experienced bomber pilot who had undergone training on the composite. Assigned to fly this mission, he recalled:

Take-off from Rechlin was set for 1700 hrs. The aircraft were lined up one behind the other with engines running. My *Mistel* was the second one in line. From my cockpit, I could not see the horizon because the nose of my Fw 190 was pointing up too high.

The procedure was to move all three throttles forward in such a way that the *Mistel* could be kept on the runway. With sufficient airspeed, the stick could be pulled back and the undercarriage and flaps retracted. You then throttled back, adjusted the airscrews [propellers] and commenced a steady climb.

After reaching our combat altitude of 2,000 m, I could see the front, the fires and the impact of mortar explosions. The heavy haze made visual orientation very difficult, but the Oder river was easy to make out as a silvery band. The bridge that I was looking for was a dark line across this band. I was 'welcomed' by heavy anti-aircraft fire, so I put the *Mistel* into a dive, switched on the fully automatic control system, pulled down the cross hair sight and aimed at the bridge. When the target drifted out of the cross hair sight, I corrected and the automatic control system put the target squarely into the cross hair again. From about 1,000 m distance, I squeezed the trigger that automatically armed the warhead and separated the *Mistel*. My Fw 190 climbed as it released itself from the heavy weight of the Ju 88. As I pulled away, I noticed a lightning flash in the river bed below me that quickly went out. Without much difficulty I returned to my home field.

The pilots of the *Mistel* units continued to fly determined but sporadic missions against the bridges until the final days of the war, braving Soviet fighters and massed anti-aircraft defences. They succeeded, occasionally, in destroying less stable or temporary bridges or inflicting severe damage upon them, thus creating a nuisance effect on Soviet supply routes. But, ultimately, by that point nothing could hold back the enemy advance.

Fw 190/Ta 154 *MISTEL, PULKZERSTÖRER* AND *SPRENGSTOFFTRÄGER*

In early July 1944, following the appearance of the first Junkers-built *Mistel*, the RLM issued Focke-Wulf's *Entwurfsbüro* (Development Office) at Bad Eilsen with a specification for a '*Mistel* Ta 154A/Fw 190A-8' combination. The RLM hoped to develop a *Mistel* which would comprise an Fw 190A and the cheap-to-build, wooden Ta 154 fighter as its lower component. Although Focke-Wulf complied with the RLM's request to investigate such a project, it had been the company's original intention to use a Ta 154 for both upper and lower components of the planned composite. However, according to a Focke-Wulf report, 'This is not possible since there would be too much weight on the landing gear and the structure of the aircraft.'

The sleek design of the Ta 154, with its distinctive nose-wheel, was developed by Professor Kurt Tank in 1942 following a requirement from the RLM for a two-seat, twin-engined, wooden, fast attack bomber, but by October 1943, influenced by the air war waging over the Reich, this was amended to become a fast nightfighter.

On 17 July 1944, *Herr* Schöffel from Focke-Wulf visited Junkers at Dessau to observe the *Mistel* conversion process. Subsequently, he produced his feasibility study on a

This *Mistel* consisted of a war-weary Ju 88G nightfighter converted into a radio-guided flying bomb with an Fw 190 fighter attached above to guide it to the target area. The fighter roughly aimed the drone at the target, separated from the combination and then guided the drone until impact. This illustration shows one of the *Mistel* operations by KG 200 against the Oder River bridges in early April 1945.

Ta 154A/Fw 190A-8 combination. Unlike the original intention to build a twin Ta 154 combination, the Fw 190 proposal appeared more promising:

> With this [Ta 154A/Fw 190A-8] combination there are no fundamental difficulties from a structural point of view. The heavy take-off weight [up to 15 tons] is still possible for the landing gear if a concrete runway is used. A landing of the combination is not possible.

As with the Junkers designs, Schöffel proposed maintaining a normal Ta 154 cockpit for transfer flights, but replacing it with a 'large explosive charge', probably 2,500–3,500 kg, for use against 'a suitable target'. It was proposed to fit an explosive bolt to the nosewheel, which in turn would be attached to the warhead. The nosewheel would then be jettisoned after take-off. Schöffel wrote, 'For a mission as an "aerial torpedo", where total loss of the aircraft is inevitable, the inexpensive Ta 154 is especially suitable.'

When deployed operationally, it was proposed that the *Mistel* would fly to within one kilometre of the target, at which point the Fw 190 pilot, having aimed the machine, would effect separation, with the fighter's throttle set at full power, while full throttle on the

Ta 154 would be delayed for a few seconds by means of a timing switch. In his report, Schöffel also outlined the main advantages and disadvantages of the planned composite, commenting:

> Firstly, due to its lack of sophisticated instrumentation, the Ta 154 could be cheaply and relatively easily produced. Secondly, and as with the Junkers' composites, the whole configuration requires only one pilot.

During Schöffel's visit to Dessau in July 1944, *Herr* Emmert of the RLM stressed to the Focke-Wulf technician the 'urgent need' for the Ta 154/Fw 190 combination. According to Schöffel, Emmert stated, 'The RLM expects the first conversion to be ready by the end of August 1944.' Information on the required design and conversion resources were requested and completion deadlines were also required for the first, tenth and fiftieth aircraft, as well as details of where production was to take place. However, the loss of seven Ta 154s during a USAAF bombing raid on Hannover-Langenhagen airfield in early August 1944, combined with the troublesome performance of the remaining aircraft, most probably diminished further interest in the project.

In other measures of expediency, and essentially because of its cheap method of construction, it was also proposed to use the Ta 154 against Allied bomber formations either as a *Pulkzerstörer* (Formation Destroyer) or as a *Sprengstoffträger* (Explosives Carrier). In the first instance, under the pre-series variant designation Ta 151A-0/U2, a piloted aircraft laden with two tons of explosive would approach a bomber *Pulk* from the front or rear. The effect of such a quantity of explosives being detonated wirelessly by an escorting fighter in the air, in the middle of a *Pulk*, was the equivalent effect of 1,000 kg exploding on the ground, giving blast up to 140 m distance. A gyro-based three-axis control system would operate into the centre of the *Pulk* where the explosive would be detonated.

The aircraft would be unarmed and carry only the minimum of equipment, and the pilot was to bail out with a parachute through a hinged door and an opening created in the fuselage floor just aft of his usual seated position. This would take place in such a way that the pilot was effectively somersaulted in the cockpit and would therefore fall away from beneath the aircraft, head first, facing the tail. This process was designed to minimise any risk of the pilot colliding with the fuselage.

The same tactical aim was intended with the *Sprengstoffträger* version, but in this case proposals were put forward for manned and unmanned versions of the Ta 154. In the case of the former, new accommodation for the pilot was to be created in the central section of the fuselage, aft of the explosive, and the pilot was to exit the aircraft using a method similar to that proposed in the *Pulkzerstörer*. The aircraft was to be either air-towed by another Ta 154 or controlled under a *Mistel* arrangement coupled with an Fw 190 upper component. Release was to be effected when one kilometre from an enemy formation. Blast within an enemy *Pulk* would be achieved by remote control from the accompanying aircraft.

ABOVE The Jumo 211N-powered Ta 154 V7 TE+FK photographed at Langenhagen in March 1944. Proposals were put forward by Focke-Wulf to use the all-wood twin-engined fighter in a *Mistel* combination and as a *Sprengstoffträger* for remote detonation by another Ta 154 within an enemy bomber formation after the pilot had bailed out. (EN Archive)

RIGHT A Focke-Wulf sketch showing a planned Fw 190/Ta 154 *Mistel* combination to be used to steer an explosive-laden Ta 154 *Sprengstoffträger* into an enemy bomber *Pulk*. (Forsyth)

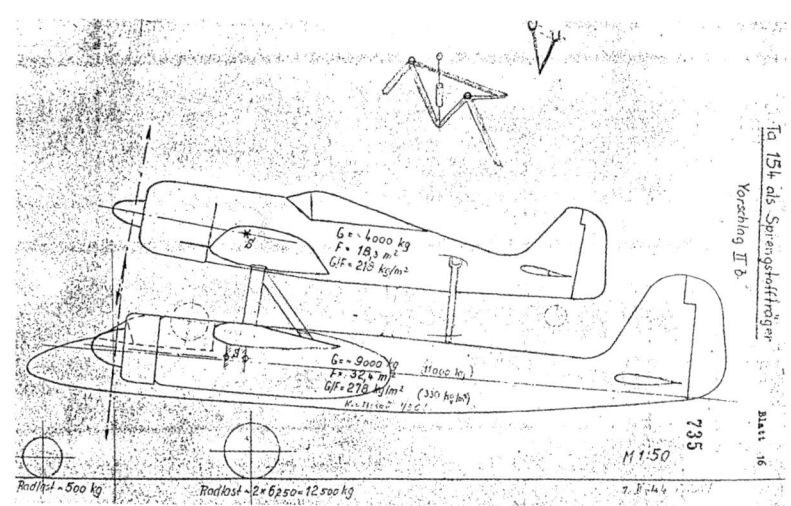

As far as is known, development did not progress beyond the drawing board for either concepts, although it is possible some basic work had commenced at Eschwege by the end of the war.

SONDERKOMMANDO 'ELBE'

In April 1945 the Luftwaffe's war against the Allied strategic bombing offensive reached desperate new heights. Following a request from Göring in early March 1945 for volunteers to take part in a radical operation 'from which there is little possibility of returning', a small group of pilots arrived in great secrecy at Stendal on the 24th to begin training as part of the so-called *Schulungslehrgang 'Elbe'*, known also as *Sonderkommando 'Elbe'*. This was the brainchild of Oberst Hajo Herrmann, the bomber ace and founder of the relatively successful *Wilde Sau* nightfighter units in 1943. His proposal had been resurrected, having been rejected on a previous occasion.

Herrmann's plan was to assemble a group of pilots who would be prepared to fly their fighters in a massed attack against a large Allied bomber formation using conventional armament, but also with the intent of ramming enemy *Viermots* to bring them down. The chances of survival would be slim, but Herrmann was encouraged by the initial call for volunteers; pilots from JG 1, II./JG 102, II./JG 103, II./JG 104, I./JG 300, II./EJG 1 and even the Me 163-equipped JG 400 put their names forward, and soon Herrmann purportedly had 2,000 names and agreement from Göring that 1,500 fighters, mainly Bf 109G/Ks with high-altitude engines and metal propellers, would be made available for the operation, which was to be codenamed *Wehrwolf*.

Herrmann's volunteers were to be trained for their mission by Major Otto Köhnke, the one-legged former *Kommandeur* of II./KG 54 who had been awarded the Knight's Cross for his actions on the Russian Front in 1942. Köhnke was known to be outspoken and critical of the Luftwaffe hierarchy, but was blessed with exemplary leadership qualities.

Köhnke arranged for the volunteers, few of whom were officers or formation leaders and many of whom had come from fighter training units, to be briefed by Oberfeldwebel Willi Maximowitz, a fearless pilot who had flown with *Sturmstaffel* 1 and IV.(*Sturm*)/JG 3 and who had shot down 15 *Viermots*. Maximowitz recommended to his anxious, if not rapt, audience that wherever possible, a 'ramming' attack should be aimed at the tail section of a bomber, as its removal would almost certainly cause it to go down. Further training, which lasted ten days, included the showing of morale-boosting films and lectures by political officers on the dangers of Jewish culture and Bolshevism. Indeed, 90 per cent of the training course consisted of political indoctrination, with the rest devoted to tactics and operations.

The *Schulungslehrgang 'Elbe'* was to use Bf 109s adapted for the mission by removing the FuG 16Z transmitter and much of the armour protection, including that around the fuel tanks. In addition, the aircraft armament was reduced to a single, fuselage-mounted MG 131 machine gun with less than 60 rounds of ammunition. Tactics were to employ aircraft in *Schwärme*, with each *Schwarm* led by an experienced pilot, the remainder being young aviators fresh from the schools. The *Schulungslehrgang* pilots were to climb to 11,000 m with the aim of out-climbing any enemy escort fighters. They were, purportedly, to receive cover from Me 262s, but there is no evidence to suggest that JG 7 or any other jet units were aware of the operation.

Once weather conditions were favourable enough for a ramming operation, the *Rammkommando* involved would be provided with high escort, make for its respective waiting area in the normal manner and operate under a Divisional VHF commentary from IX. *Fliegerkorps*. Upon receipt of the codeword *Antreten frei*, the ram-fighters were to make for the enemy formation at a height 1,500 m above the bombers. The approach was to consist of a long, shallow dive, if possible out of the sun and in line astern. Fire was to be opened at extreme range, and continued until the final steep ramming dive towards the fuselage of the bomber immediately forward of the tail unit. If possible, the pilot was then to attempt to bail out. Combat with enemy fighters was to be avoided at all costs and pilots were to climb away if attacked.

Finally, at 0930 hrs on 7 April 1945, the 120 pilots of the *Raubvögel Gruppe* of *Schulungslehrgang 'Elbe'* were placed at 30 minutes' readiness on their bases at Stendal, Gardelegen, Delitzsch and Morlitz. Fw 190s and Ta 152s of JG 301 were to provide fighter escort. With patriotic slogans broadcast into their headsets by a female voice, along with the *Horst-Wessel-Lied* and the *Deutschlandlied*, rather than a more useful navigational commentary, the fighters took off. Their targets were the 1,261 USAAF B-17s and B-24s, escorted by 830 fighters, tasked with bombing 16 airfields, ordnance depots, industrial sites and marshalling yards across a wide area of northern and central Germany.

Even those few pilots whose courage remained with them to the moment of impact with a bomber were uncertain as to whether they had been successful in their attempt to bring it down. One pilot of a Bf 109G dived into a *Pulk*, heard a loud crash and promptly bailed out at 1,000 m. He sustained severe injuries in landing and later had an arm amputated as a result. His experience was typical of many. Immediately after the raid, the Eighth Air Force reported:

> It appears that this was a desperation attempt on the part of the enemy, and although enemy aircraft fought aggressively and made determined efforts to get through to the bombers, our losses were comparatively light. Signs of desperation are evidenced by the fact that Fw 190 pilots deliberately rammed the bombers, baling out before their planes went into the bomber formations and making fanatical attacks through a murderous hail of fire. Tactics were thrown to the wind and attacks were made from all positions, mainly in ones and twos. From today's reaction, it would appear that although the enemy is fighting a losing battle, the German Air Force is preparing to fight to a finish in a fanatical and suicidal manner.

Seventeen bombers were lost during the raid, including at least five B-17s from the 3rd AD that appeared to have been rammed intentionally. In reality, and allowing for losses caused by Flak, aircraft hit by falling bombs and claims by Me 262 units also operating that day, the destruction of about 12 B-17s was attributable to *Schulungslehrgang 'Elbe'*, and recent research indicates that some 24 ram pilots were killed, eight were captured or posted missing and five returned injured. Forty-five Bf 109s were shot down or destroyed in ramming attacks – this equates to a loss rate of around 35 per cent, although the Germans lost many more aircraft during the mission. In any event, *Schulungslehrgang 'Elbe'* attempted no further operations.

Fi 103 *REICHENBERG* PILOTED FLYING BOMB

Perhaps inevitably, the perceived success of the Fieseler Fi 103/FZG 76 flying bomb against the British Isles encouraged certain individuals of influence within the Third Reich to reconsider a much previously discussed idea: namely, the concept of a manned flying bomb or, more accurately, *using* the flying bomb to create a manned weapon to be used against buildings, bridges, ships and possibly bombers.

Fi 103A-1/Re 4 Reichenberg.

An American serviceman wearing a Kapok life vest sits within the tight confines of the 'cockpit' of an Fi 103A-1/Re 4 *Reichenberg* manned flying bomb at an assembly and storage facility at Tramm, near Dannenberg, following the arrival of the US Army in the spring of 1945. This photograph illustrates how difficult it would have been to exit the *Reichenberg* in flight. (EN Archive)

According to the renowned and ideologically zealous German aviatrix Flugkapitän Hanna Reitsch, the seeds of this concept were sown at a meeting which took place at the *Deutsche Akademie der Luftfahrtforschung* in Berlin in November 1943. In attendance were a gathering of aircraft designers and engineers, scientists, technicians, military officers and torpedo experts. They concluded that the notion of such an 'aircraft-bomb' being used for *Selbstopferungs* ('SO' – self-sacrifice) operations was not beyond the realm of possibility. Despite practical objections from the *Generalluftzeugmeister*, Generalfeldmarschall Milch, Reitsch was able to secure an audience with the *Führer* at the Berghof at the time of her award of the Iron Cross First Class in February 1944. Seizing the opportunity to bring up the idea of SO aircraft and operations, she apparently won lukewarm, if not indifferent, support for such an idea. It was enough.

The path to pressing ahead with what became the Fi 103 *Reichenberg* (it is understood some documentation uses the title *Reichenbach*) has been the subject of colourful personal post-war accounts, not least of which is Reitsch's own, with personalities such as Himmler, Skorzeny, Baumbach and others all partaking in competititive intrigue within the Machiavellian atmosphere of late-war Nazi Germany. It is difficult to assess exactly what is truth and what is fiction, suffice to say that one of the few potential types for SO deployment,

German aviatrix Hanna Reitsch (left) was the main driving force behind the creation and development of a manned Fi 103 flying bomb. As a medical student, the diminutive Reitsch once aspired to be a flying medical missionary, but abandoned the idea in favour of a professional flying career. Described as 'insanely brave', the blonde, blue-eyed, fearless Flugkapitän became the darling of German aeronautics in the mid-1930s and a role model for the Nazi movement. She went on to win many air competitions and to test-fly numerous types of military aircraft, including the Me 163 rocket-powered interceptor. Reitsch also flew into a shell-cratered street in Berlin in late April 1945 to visit Hitler in his bunker days before he committed suicide, and was one of only two women to be awarded the Iron Cross First Class. She is seen here being congratulated by the *Reichsluftsportführer* Oberst Alfred Mahncke on her success in the 17th Rhön Glider Competition in the Rhön mountains on 27 August 1936. (Forsyth/Mahncke)

a glider-bomb version of the Messerschmitt Me 328 mini-fighter, proved a failure, and thus in 1944 the Henschel Flugzeugwerk at Berlin-Schönefeld was commissioned to convert a standard Fi 103 into a manned, single-seat version to be known as the Fi 103Re – the 'Re' for *Reichenberg* after a location east of Berlin. Henschel, in cooperation with Diplom-Ingenieur Robert Lusser of Fieseler, worked quickly and converted two standard flying bombs into a single-seater and a two-seat trainer, respectively, in just a few days.

By virtue of the fact that the bomb had become 'manned', it could be classified as an 'aircraft' and assigned the official designation 'Fi 103Re1'. In this form, the Fi 103, as a trainer glider, was unpowered and fitted with a long, sprung under-fuselage skid and metal ballast to simulate a warhead.

Flight-testing commenced at the *E-Stelle* Rechlin in the late summer of 1944 when Fliegerhauptingenieur Wilhelm Ziegler of *Abt.* E2 carried out some flights when released from an He 111. For these flights the *Reichenberg* remained unpowered, with no Argus 014 pulsejet engine fitted – this was the motor that powered the standard Fi 103. On one such flight, the *Reichenberg* made a hard landing and Ziegler suffered spinal injuries. The test programme was taken over by Flieger Oberingenieur Herbert Pangratz, but the Fieseler's canopy flew off on one flight and he too was injured on landing. Thereafter, in September, Hanna Reitsch herself carried out flights, along with Fliegerhauptingenieur Heinz Kensche of the RLM and Flugbaumeister Willy Fiedler of Fieseler. These flights were conducted in the unpowered Fi 103A-1/Re 2 two-seat, dual-control trainer with a short wingspan and cockpits located in the front and rear sections of the machine's body.

By November the single-seat Re 3 had been produced, which was similar to the Re 2 but was powered by a pulsejet engine and built with skid and ailerons. Kensche flew the machine for an eight-minute air test on 4 November 1944. The next day, however, he had to bail out of the 'aircraft' over the Müritzsee after 56 minutes when part of the port wing blew away as a result of vibration from the pulsejet. Further testing continued into the spring of 1945, but on 5 March test pilot Leutnant Walter Starbati was killed in an Fi 103A-1/Re 3 when, at a speed of 400–500 km/h and at 2,800 m, both wings broke away during a turn to port. With the Argus pulsejet still running, the *Reichenberg* entered into a sharp dive and, despite jettisoning the canopy, the pilot was unable to exit. Fellow test pilot Unteroffizier Schenck was also killed testing the *Reichenberg*.

By 15 March 1945, it seems common sense finally prevailed; that day, the *Technisches Amt* noted that at the suggestion of KG 200 all work on the *Re-So-Flugzeuge* (*Reichenberg* Special [for Self-Sacrifice] Aircraft) be stopped as a result of Starbati's accident.

Meanwhile, the following variant, the Fi 103A-1/Re 4, had been under production in a crude assembly facility at Tramm, near Dannenberg, south of the River Elbe. As the planned operational aircraft with an 800 kg warhead, the Re 4 was powered by an Argus 109-014 pulsejet and, in conformity with its 'one-way' tactical concept, had no landing skid. It had been planned for Henschel at Schönefeld to build 200 Re 4s, but production was transferred to another factory at Gollnow in Pommerania. The collapse of the Third Reich prevented any further work.

Fieseler Fi 103A-1/Re 4	
Wingspan	5,720 mm
Length overall	7,780 mm
Engine	Argus 109-104 pulsejet
Engine weight	153 kg
Empty weight	618 kg
Fuel weight	630 kg
Warhead weight	810 kg
Crew weight	101 kg
Equipped weight	2,190 kg
Optimum range	238 km
Service ceiling	1,800 m
Optimum cruising speed	580 km/h
Maximum speed	540 km/h

SOURCES AND BIBLIOGRAPHY

PRIMARY SOURCES

Correspondence and/or interviews between 1988–2000 with Horst Geyer, Adolf Galland, Oscar Boesch, Walter Hagenah, Franz Stigler, Willi Unger, General a.D. Walter Windisch, Fritz Buchholz, Rudi Riedl and Heinz Frommhold.

Imperial War Museum, London and Duxford

AI2(g) Report No. 1787, *R4M German Aircraft Rocket*, Sqn Ldr C. V. T. Campbell, 23 June 1945

Headquarters USAFE:
Air Staff Post Hostilities Intelligence Requirements on the German Air Force, Appendix XVI, *A History of the German Air Force Twin-Engine Fighter Arm (Zerstörerwaffe) written by Galland, Kowalewski, Nolle and Eschenauer*, 8 October 1945
Appendix XXVII – *Special Weapons for Combating 4-Engine Bombers by Day with SE and TE Fighters*, Interrogation of Galland, Dahl and Petersen, Kaufbeuern, 12–14 September 1945
Air Staff Post Hostilities Intelligence Requirements on the German Air Force, Tactical Employment Bomber Operations, Section IV, D, 10 October 1945

British Intelligence Objectives Sub-Committee (BIOS):
UNT 111.T/HEC No. 10872, *Brief History of the Development of Fighter Armament*, R. Riecker and Dr. Kokott, undated
UNT 117.T/HEC No. 10875, *Aircraft Rockets and their Installations – Development and Description of RZ.65, RZ.75, RZ.15-8, 'Panzerschreck', 'Panzerblitz' I, II, III and IV, R4M, R.100BS and R.50BS*, Schoetz, undated
UNT 118.T/HEC No. 10830, *Rocket Armament for Aircraft and Some Equipments of Modern Armament Design*, Dr. Klein, 21 June 1945
UNT 171.T/HEC No. 10859, *Air Combat Ballistics – General Discussions*, Dr. Kokott, 27 March 1946
Combined Intelligence Objectives Sub-Committee (CIOS):
Item No. 2, File No. XXX1-35, *Rheinmetall-Borsig A.F. Werk Unterlüss*, HMSO, undated
Item No. 2, File No. XXX1-63, Alexander E. Kramer, *Development of Weapons by Rheinmetall-Borsig*, HMSO, undated
Item No. 4, File No. XXVII-65, *The Rheinmetall Borsig Works and Proving Grounds Unterlüss*, HMSO, undated
Focke-Wulf Flugzeugbau, *Jagdflugzeug mit 21 cm Wurfgranate*, FW/8b/3, undated

Focke-Wulf Bericht No. 07013 *Bombenwurf auf fleigende Ziele,* Bad Eilsen, 14 July 1943

German Document Collection:

10/6128T, *Report of the Meeting on Technical Problems of Ordnance held in Erfurt on 14 and 15 September 1944: Report 182/Pt 2, Development Requirements for the Armament of Aircraft,* Buehler, Berlin

16/119, *Beschreibung der Ausrüstung einer Fw 190 für den Angriff mit gesteuerten Schuß nach oben (Gerät 116) Erstes Erprobungsergebnis 1.4.44 Parchim,* Schwetzke, *LFA Hermann Göring,* Braunschweig, 2 July 1944

Panzerbeschuß aus der Hs 129 mit SG 113A, LFA Hermann Göring, Institut für Waffenforschung, Braunschweig, 1 Feburary 1945

H.E.C. No.20/8, *Neue Entwicklungs für die Bewaffnung von Jagdflugzeugen: Auszug a us einem am 27.10.1944 vor dem Sonderausschuss Flugmechanik gehaltenen Vortrag von Fl.Ob. Stabsingenieur G. Voss*

H.E.C. No.169, *The Automtatic Triggering of Aircraft Guns by the Target. Electrostatic and Photoelectric Devices for Making Tanks or Bombers Trigger Vertically Firing Guns Mounted in Fighter Aircraft,* P. Hackemann, undated

H.E.C. No. 10774/1, UNT.9T, *Automatic Weapons and Aircraft Installations – The 50mm BK, its Ammunition and Installation in Me 410 and Ju 88,* Schoetz, Rheinmetall-Borsig, Unterlüss, 15 November 1945

H.E.C. No. 13651 (GDC 6/2887T), *Selbstätige Auslösegeräte für Flugzeugsonderbewaffnungen,* R. Schwetzke, undated

H.E.C. No. 13606/Unterlüss Report No. 181, *Harfe, Jägerfaust, Bombersäge: Recoilless and Multi-Barrel Weapons,* Dr. Grasse, Work Centre Unterlüss, 2 September 1946

H.E.C. No. 11825/6420/85, Rheinmetall-Borsig WKL, *Die Bordrakete R 100 BS zur Bekämpfung von Feindflugzeugen auf große Entfernungen,* Berlin-Marienfelde, 15 January1945 (plus miscellaneous papers)

H.E.C. No. 13661/Unterlüss Report No. 296, *Technical Report Upon the First Operational Experiences with 50 mm Automatic Weapons in Aircraft (Me 410): Discussions of the Results,*

Diplom-Ingenieur Kurt Buehler, 8 April 1947

Nav.Tech. (US Naval Technical Mission in Europe): *Automatic Weapons Division of Rheinmetall-Borsig A.G., Unterluss, Germany,* Technical Report No. 351-45, September 1945

Rheinmetall-Borsig A.G., Unterlüss, *The 50 mm B.K., Its Ammunition and Installation in Me 410 and Ju 88,* UNT.9.T., 15 September 1945

Rheinmetall-Borsig A.G., Unterlüss, *Bewaffnung Schneller Jagdflugzeuge,* Josef Schoetz, 27 July 1945 (as Halstead Exploiting Centre Developments of German Fighter Armament for the attack of Bombers, BIOS/Gr.2/ HEC/10787, UNT 5T)

Rheinmetall-Borsig A.G., Unterlüss, *SG 116 (Zellendusche) 3 cm Sondergerät für Flugzeugbau,* UNT.177, Dr. Grasse, 17 April 1946

T-2 microfilm 3576/Frame 1004, *The AA Parachute Mine SK70 and SK 106 (Verpackungsvorschrift fur Storkorper),* Rechlin, September 1940

T-2 microfilm 3576/Frame 1105, Walland, *The rocket projector as airplane armament (Angriff mit Wurfgranaten),* Arado, September 1944

T-2 microfilm 4088/Frame 245 and 246, Autobedarf Lagerlechfeld to Messerschmitt, *Vorschlag: Neues Kampfmittel,* g.Kds. 585/44, 24 December 1944

Technische Arbeitsgruppe, Der Forschungsamt des R.d.L. und Ob.d.L, *Bericht über die Flugbahnvermessung der 21 cm Wurfgranate (Schuß vom ruhenden Abschußort),* v.Düffel, Br.B.Nr.TA V40/441, undated

Unterlüss Report UNT.107.T., Diplom-Ingenieur Johannes Linke, *Halbstarres Schiessen mit und ohne Stabilisierung: Aircraft Gun Installations– Development of Larger Calibre Gun Installations for Air-to-Air Combat,* translation 12 May 1948

Unterlüss Report UNT.252.T., Diplom-Ingenieur Günther and Diplom-Ingenieur Frost, *Entwicklung einer Raketeneinrichtung für Jagdflugzeug (Development of Rocket Designs for Aircraft),* translation 6.8.1948

Unterlüss Report UNT.309, Buhler, Frost, Burgsmüller and Günther, *Die Bewaffnung der Jagdflugzeuge zur Erfolgreichen Bekämpfung von Bombenflugzeugen (Selection of the best Armament Requirements for Fighter Aircraft),* translation, 13 April 1948

Unterlüss Report UNT.376.T., Oberst Ingenieur Johannes Mix, *Entwicklung von Fliegerbordwaffen, Munition und Einbauten: Eine umfassende Übersicht des deutschen Standes bei Kriegsende (Development of Aircraft Weapons, Munitions and Installations – A Comprehensive Review of the German Position at the End of the War)*, translation, 3 August 1948 (also available as USAF Historical Research Agency Numbered USAF Study 193, *The Development of German Aircraft Armament to 1945*, by Oberst Ingenieur Mix (available at www.afhra.af.mil/Portals/16/documents/Studies/151-200/AFD-090804-091.pdf))

National Archives, Kew

AIR2/7493, HQ US 8th AF, *Defence Against the Fortress*, Special Report No. 83, 28 August 1943

AIR20/7703, *Luftflotte* 3 Situation Reports

AIR20/7708, AHB 6 Translation No. VII/124 – *Extract from report of the Goering Conference on Aircraft Production Programme*, 23 May 1944

AIR20/7709, AHB 6 Translation No. VII/140, *Extracts from Conferences on Problems of Aircraft Production*, August 1954 and AHB. 6 Translation No. VII/137, *Fighter Staff Conferences*, 1944

AIR40/117, *New German Fighter Rocket Armament and 'Oberon' Automatic Range-Finding and Firing Procedure*, AI2(g) Report No. 1774, 28 April 1945

AIR40/1460, *Der Chef des Luftwaffenfuhrungsstabes Nr. 918/45, 7 February 1945 Bekampfung von 4-mot-Verbanden durch Abwurf von Bomben (Major Stamp).* See also BA-MA RL2 II/70

AIR40/1553, *Report on Interrogation of certain German PoWs being personnel from Tarnewitz engaged on aircraft rocket weapon developments*, Oberst Ingenieur Ossenbühn and Oberst Ingenieur Dr. Franz with Gp Capt H. W. Dean, 8 November 1945

AIR40/1888, *2nd TAF Air Prisoners of War Interrogation Unit reports nos. 38-99/44*

AIR40/1890, *The Development of Aircraft Armament*, A/PWIU, 2nd TAF 81/1945, 29 June 1945

AIR40/2160, *21 cm Rocket Projectile Installation on German Fighter Aircraft*, AI2(g) Report No. 1344, undated

AIR40/2161, *The MK 108 30 mm Gun*, AI2(g) Report No. 1685, 20 November 1945, and *Penetration Trials with German MK 108 30 mm Gun*, AI2(g) Report No. 1699, 12 December 1945

AIR40/2162, *The 50mm BK 5*, AI2(g) Report No. 1780, 28 May 1945

AIR 40/3114, CSDIC.CMF. Report A., *Interrogation Report on German W.T. Mechanic*, 26 September 1944 (via Nick Beale)

HW5/400, *German section: reports of German army and air force high grade machine decrypts*, CX/MSS/T19/29, 1 December 1943 (via Nick Beale)

The X4 - German Air-Launched A.A. Rocket, AI2(g) Report No. 1773, Section E.4.D., 31 January 1945

The X.4 (8-344) German Air-Launched AA Rocket Aircraft Control Equipment, AI2(g) Report No. 1781, 23 June 1945

Miscellaneous

ADI(K) Report No. 323/1945, *The Me 262 as a Combat Aircraft* (Kitchens)

ADI(K) Reports No. 437/1944 and No. 620/1944, *GAF Experimental Units*, S. D. Felkin, 10 August and 5 December 1944

ADI(K) Report No. 620/1944, *Upward Firing Armament in German Aircraft*, S. D. Felkin, 13 November 1944

BIOS Report No. 163, BIOS Trip No.1127, *Some German Aircraft Armament Projects with particular reference to Fire Control Developments*, Flt Lt M. O. Robins, November 1945

Depts of the Army and Air Force, TM9-1985-2/TO 11A-1-26, *German Explosive Ordnance (Bombs, Fuzes, Rockets, Land Mines, Grenades and Igniters)*, March 1953

Erprobungsstelle der Luftwaffe Tarnewitz, *Schwerpunkterprobungsberichte*, July 1943– February 1945, plus associated reports including, B.Nr.: E6 5/45, *Optimale Bewaffnung und Treffwahrscheinlichkeit bei Senkrechtschuss mit automatischer Schusslösung*, Raabe, 30 December 1944

Erprobungskommando 25, Arbeitsberichte, Juni 1943–Juli 1944 (plus miscellaneous reports)

Henning, Lt E. S., *The High-Frequency War: A Survey of German Electronic Development*, Report No.

F-SU-1109-ND, Headquarters Air Materiel Command, Dayton, OH.,10 May 1946

Interview transcript Gerhard Stamp/Walter Boyne

Interview notes with Herbert Schlüter, *Kommando Stamp*, Hamburg 12 May 1979 (Manfred Griehl)

Loßberg, Oberstleutnant v., *Hs 129 mit 7.5 cm BK*, GL/C-E2 Nr.14873, 10 December 1943

Ministry of Aircraft Production, Technical Information Bureau, *German Guided Projectiles*, 1945

RAF Farnborough, *Report on Interrogation of German prisoners on Rocket Development*, Gen Armament Div., S.1 Memo No. 42, 25 October 1945

Schlüter, Herbert, *My time with the Me 262*, private recollections

War Department, Military Intelligence Division, *Handbook on Guided Missiles of Germany and Japan*, Washington, D.C., 1946

PUBLISHED ARTICLES

Buchling, Nils, 'Operational use of the 50 mm Cannon by II./ZG 26 "Horst Wessel" in the Defense of the Reich 1944', *Luftwaffe im Focus*, *Edition No. 17*, Luftfahrtverlag Start, Bad Zwischenahn, 2010

Chapman, Rick, 'Tank-Busting Midget', *Scale Models*, March 1982

'Die Bücker mit der Panzerfaust', *Jägerblatt, Nr 2 XXXI*, Februar/März 1982

Forsyth, Robert, 'Defending the Reich: Parts 1, 2 and 3', *The Aviation Historian, Issues 17–19*, Horsham, 2016–17

Hoffmann, Paul, 'Landungsflotte vor uns! Einsatz einer Besatzung der 4./KG 100 gegen der alliierte Landungsflotte vor Anzio/Nettuno, Januar 1944', *Luftwaffe im Focus No. 6*, Luftfahrtverlag-Start, Bad Zwischenahn, 2004

Kay, Antony, 'Prelude to Stand-Off – Part 1 and Part 2', *Gruppe 66, Archiv No. 4*, Winter 1966 and *No. 5*, Spring 1967

Lächler, Hans, 'Das Flugzeug als Granatwerfer: Die Werfergranate 21 im Luftwaffeneinsatz 1943-45', *Flugzeug*, Februar 1986

Norton, Mike, 'The *Panzerjagdstaffeln* equipped with the Bücker Bü 181' (unpublished, undated article)

Stapfer, Hans-Heiri, 'Panzerfaust im Gepäck', *Flugzeug Classic*, 2/2006

Urbanke, Axel, 'Die Ju 88 P-1 des *'Versuchs Kommando für Panzerbekämpfung'*, *Luftwaffe im Focus No. 6*, Luftfahrtverlag-Start, Bad Zwischenahn, 2004

PUBLISHED BOOKS

Aders, Gebhard, *History of the German Night Fighter Force 1917–1945*, Jane's, London, 1979

Balke, Ulf, *Kampfgeschwader 100 'Wiking' – Eine Geschichte aus Kriegstagbüchern, Dokumenten und Berichten 1934–1945*, Motorbuch-Verlag, Stuttgart, 1981

Beauvais, Heinrich, Kössler, Karl, Mayer, Max and Regel, Christoph, *German Secret Flight Test Centres to 1945*, Midland Publishing, Hinckley, 2002

Boiten, Dr. Theo E. W. and MacKenzie, Roderick J., *Nachtjagd War Diaries Volume Two*, Red Kite, Walton-on-Thames, 2008

Bollinger, Martin J., *Warriors and Wizards – The Development and Defeat of Radio-Controlled Glide-Bombs of the Third Reich*, Naval Institute Press, Annapolis, 2010

Boje, Walter and Krug, Helene, *Luftfahrtwissenschaft und -Technik: Wer is Wo? 1. Ausgabe: Forschung und Lehre*, Deutsche Akademie der Luftfahrtforschung, Berlin, 1939

Boog, Horst, *Die deutsche Luftwaffenführung 1935–1945: Führungsprobleme, Spitzengliederung, Generalstabsausbildung*, Deutsche Verlags-Anstalt, Stuttgart, 1982

Caldwell, Donald, *Day Fighters in Defence of the Reich – A War Diary, 1942–45*, Frontline Books, Barnsley, 2011

Caldwell, Donald and Muller, Richard, *The Luftwaffe over Germany – Defense of the Reich*, Greenhill Books, London, 2007

Chinn, Lt Col George M., *The Machine Gun: History, Evolution, and Development of Manual, Automatic, and Airborne Repeating Weapons – Volumes I and III*, US Government Printing Office, Washington, D.C., 1951 and 1953

Fleischer, Wolfgang, *German Air-Dropped Weapons to 1945*, Midland Publishing, Hinckley, 2004

Forsyth, Robert, *Jagdwaffe – Defending the Reich 1943–44*, Classic Publications, Hersham, 2004
Jagdwaffe – Defending the Reich 1944-45, Classic Publications, Hersham, 2005
with Jerry Scutts, *Battle over Bavaria – The B-26 Marauder versus the German Jets*, Classic Publications, Crowborough, 1999
Heinkel He 111 – An Illustrated History, Classic Publications, Hersham, 2014

Gellermann, Günther W., *Moskau ruft Heeresgruppe Mitte – Was nicht im Wehrmachtbericht stand – Die Einsätze des geheimen Kampfgeschwaders 200 in Zweiten Weltkrieg*, Bernard & Graefe Verlag, Koblenz, 1988

Hahn, Fritz, *Deutsche Geheimwaffen 1939–1945 – Flugzeugbewaffnungen*, Erich Hoffmann Verlag, Heidenheim, 1963

Hoffschmidt, Edward J., *German Aircraft Guns World War 1–World War 2*, WE Inc., Old Greenwich, 1969

Hogg, I. V., *German Secret Weapons of World War 2*, Arco, New York, 1970

Hölsken, Dieter, *V-Missiles of the Third Reich*, Monogram Aviation Publications, 1994

Pegg, Martin, *Hs 129 Panzerjäger*, Chandos Publications, Hull, 2019

Pocock, Rowland F., *German Guided Missiles*, Ian Allan, Shepperton, 1967

Prien, Jochen and Rodeike, Peter, *Messerschmitt Bf 109 F, G, and K Series*, Schiffer Military History, Atglen, 1995

Ransom, Stephen and Cammann, Hans-Hermann, *Me 163 Rocket Interceptor Volume Two*, Classic Publications, Hersham, 2003

Schliephake, Hanfried, *Flugzeugbewaffnung: Die Bordwaffen der Luftwaffe von den Anfangen bis zur Gegenwart*, Motorbuch Verlag, Stuttgart, 1977

Simpson, Andrew R. B., *'Ops' – Victory at All Costs*, Tattered Flag Press, Pulborough, 2012

Smith, J. Richard and Creek, Eddie J., Me 262 *Volume Three*, Classic Publications, Crowborough, 2000
Focke-Wulf Fw 190 Volume Three 1944–45, Classic Publications, Hersham, 2013

Stüwe, Botho, *Peenemünde West – Der Erprobungsstelle der Luftwaffe für geheime Fernlenkwaffen und deren Entwicklungsgeschichte*, Bechtermünz Verlag, Augsburg, 1998

Thomas, Geoffrey J., and Ketley, Barry, *KG 200 – The Luftwaffe's Most Secret Unit*, Hikoki Publications, Crowborough, 2003

Trenkle, Fritz, *Die deutschen Funk-Navigations- und Funk-Führungsverfahren bis 1945*, Motorbuch Verlag, Stuttgart, 1979

Williams, David P., *Nachtjäger Volume One - Luftwaffe Night Fighter Units 1939–1943*, Classic Publications, Hersham, 2005

Zaloga, Steven J., Osprey New Vanguard 276 – *German Guided Missiles of World War II*, Osprey Publishing, Oxford, 2019

MAIN WEBSITE SOURCES

BBC People's War at www.bbc.co.uk

Bord-Flammenwerfer "Gero" at www. deutscheluftwaffe.com/archiv/Dokumente/ ABC/s/SG%20Sondergeraete/SG%20 Bordflammenwerfer/Bordflammenwerfer.html

The Hugo Junkers Homepage at http://hugojunkers. bplaced.net

www.w2.dk - *The Luftwaffe 1933–1945* – incorporating: deZeng IV, Henry L. and Stankey, Douglas G., *Luftwaffe Officer Career Summaries* (2019 updated version) at www.ww2.dk/lwoffz. html
deZeng IV, *Luftwaffe Airfields 1935–45*, Edition: September 2015 at www.ww2.dk/lwairfields. html

INDEX

References to images are in **bold**.